ELIZABETH'S WOMEN

Elizabeth's Women

Friends, Rivals, and Foes
Who Shaped the Virgin Queen

TRACY BORMAN

BANTAM BOOKS
NEW YORK

Published in the United States by Bantam Books,
an imprint of The Random House Publishing Group,
a division of Random House, Inc., New York.

BANTAM BOOKS and the rooster colophon are registered
trademarks of Random House, Inc.

Originally published in hardcover in Great Britain by Jonathan Cape,
a division of The Random House Group Limited.

ISBN 978-0-553-80698-4
eBook ISBN 978-0-553-90786-5

Printed in the United States of America on acid-free paper

www.bantamdell.com

2 4 6 8 9 7 5 3 1

FIRST EDITION

Book design by Dana Leigh Blanchette

*Title-page illustration: Coronation portrait of Elizabeth I,
modern reproduction of a lost original, by Peter Taylor.*

To my parents, John and Joan Borman,
with love and thanks for
all their support

Contents

Illustrations

Kat Astley. By kind permission of Lord Hastings.

Princess Mary at the age of twenty-eight, by Master John. By kind permission of the National Portrait Gallery, London, UK/the Bridgeman Art Library.

Jane Seymour, 1536 (oil on panel), by Holbein, Hans the Younger. By kind permission of Kunsthistorisches Museum, Vienna, Austria/the Bridgeman Art Library.

Portrait of Anne of Cleves (1515–57), by Holbein, Hans the Younger. By kind permission of the Louvre, Paris, France/Giraudon/the Bridgeman Art Library.

Katherine Howard. By kind permission of the Royal Collection, © 2009 Her Majesty Queen Elizabeth II.

Portrait of Catherine Parr (1512–48). By kind permission of the National Portrait Gallery, London, UK/the Bridgeman Art Library.

Queen Mary I of England, by Mor, Anthonis, © Isabella Stewart Gardner Museum, Boston, MA/the Bridgeman Art Library.

SECTION 2

Coronation portrait, modern reproduction of a lost original, by Peter Taylor.

La volta. By kind permission of Viscount De L'Isle from his private collection at Penshurst Place.

Queen Elizabeth receives Dutch ambassadors, by Dutch School. Neue Galerie, Kassel, Germany/© Museumslandschaft Hessen Kassel/the Bridgeman Art Library.

Portrait miniature of Lady Katherine Seymour, née Grey, by Teerlink, Lievine, © Belvoir Castle, UK/the Bridgeman Art Library.

Lady Mary Grey (1545–78), by Eworth, Hans. By kind permission of the Trustees of the Chequers Estate/Mark Fiennes/the Bridgeman Art Library.

Mary, Queen of Scots. By kind permission of the collection of the Trustees of the 9th Duke of Buccleuch's Chattels Fund.

Unknown woman, formerly known as Margaret Douglas, Countess of

Lennox, by unknown artist. By kind permission of the National Portrait Gallery, London.

Elizabeth Talbot, Countess of Shrewsbury, by unknown artist. By kind permission of the National Portrait Gallery, London.

Lady Arabella Stuart, aged thirteen, by Rowland Lockey after English School, 1589, Hardwick Hall. By kind permission of the Devonshire Collection, © NTPL/John Hammond.

Portrait of Elizabeth Fitzgerald, after Stephen van der Meulen (1632–90). By kind permission of Private Collection/the Stapleton Collection/the Bridgeman Art Library.

Lady Mary Dudley, Lady Sidney, circle of Hans Eworth. By kind permission of Petworth House, the Egremont Collection, © NTPL/Derrick E. Witty.

A young lady aged twenty-one, possibly Helena Snakenborg, later Marchioness of Northampton, © Tate, London 2009.

Elizabeth Knollys, attributed to George Gower, Montacute, the Sir Malcolm Stewart bequest (the National Trust), © NTPL/Derrick E. Witty.

Elizabeth Vernon. By kind permission of the collection of the Trustees of the 9th Duke of Buccleuch's Chattels Fund.

Bess Throckmorton. By kind permission of the National Gallery of Ireland. Photo © National Gallery of Ireland.

Queen Elizabeth in old age, English School/Corsham Court, Wiltshire/the Bridgeman Art Library.

Funeral procession of Elizabeth I, 1603. By kind permission of William Camden British Library, London, UK, © British Library Board. All rights reserved/the Bridgeman Art Library.

Introduction

Elizabeth once famously declared: "I know I have the body of a weak and feeble woman." This line, and the apparent regret with which she uttered it, has been taken to represent her conformity to the view that her sex was naturally subject to "womanly weakness." Although she is often hailed as a shining beacon for womanhood, the embodiment of feminism before that term was even invented, Elizabeth was deeply conventional in her views of the female sex. When a foreign visitor to court complimented her upon her ability to speak many languages, she retorted "that it was no marvel to teach a woman to talk; it were far harder to teach her to hold her tongue."[1]

It is partly for this reason that Elizabeth is universally accepted as being a man's woman. As well as taking every opportunity to deride her sex, she loved to flirt with the many ambitious young men who frequented her court. Her liaison with Robert Dudley is well documented, as is her infatuation in old age with his stepson, the Earl of

Essex, and her more sober relationships with trusted advisers such as Lord Burghley. Yet this tells only part of the story. Elizabeth deliberately showcased these relationships in order to carve out a place in what was essentially a man's world. In her own private world, the story was very different. Here it was the women, more than the men, who held sway.

Elizabeth was born into a world of women. No man had been admitted to the presence of her mother, Anne Boleyn, during her confinement at Greenwich Palace, childbirth being a strictly female mystery in the sixteenth century. As a child, she was served by a predominantly female household of attendants and governesses, interspersed with occasional visits from her mother and the wives who later took her place. As queen, Elizabeth was constantly attended by ladies of the bedchamber, maids of honor, and other members of her household. They clothed her, bathed her, and watched her while she ate. Among her family, it was her female relations who had the greatest influence: from her half sister, Mary, who distrusted and later imprisoned her, to her cousin Mary, Queen of Scots, who posed a constant and dangerous threat to her crown for almost thirty years.

Elizabeth met, corresponded with, and was influenced by hundreds if not thousands of women during the course of her long life. I have focused the story upon those women who help to reveal Elizabeth the woman, as well as Elizabeth the Queen. From her bewitching mother, Anne Boleyn, to her dangerously obsessive sister, Mary Tudor, and from the rivals to her throne such as the Grey sisters and Mary, Queen of Scots, to the "flouting wenches" like Lettice Knollys, who stole her closest male favorite, these were the women who shaped the Virgin Queen, and it is through their eyes that the real Elizabeth, stripped of her carefully cultivated image, is revealed.

Researching the life of Elizabeth and the women who surrounded her has taken me to some fascinating places, including magnificent palaces such as Hampton Court and Hatfield House, and the treasure trove of national and local archives containing her correspondence, most notably the British Library. I have consulted both original material and the wealth of published correspondence that exists for the period. When an original manuscript is cited, I have retained the contemporary spelling. This may be idiosyncratic in places, but it also reveals something of the writer. Elizabeth often referred to Mary Tudor as "sistar" in her letters, which provides a clue to the way that she might have spoken. Where possible, I have therefore preserved such details because they give a wonderful sense of the period.

There are no doubt many other women whose stories could have been told in this book—women such as the tragic Amy Robsart, whose death effectively put paid to any hopes that Elizabeth might have had of marrying Robert Dudley; Sybil Penn, the woman who helped nurse her through an attack of smallpox that almost killed her; or Lady Mary Herbert, a brilliant writer and literary patron whose intellectual talents were on a par with Elizabeth's own. But I have focused upon those women who had the greatest influence on Elizabeth: those who forged her opinions in childhood, trained her for queenship, and helped her to achieve legendary status as Gloriana, the Virgin Queen. I hope that the women whose lives are explored in this book will delight and intrigue the reader in their own right, as well as for the light that they shed upon one of the most iconic women in history.

ELIZABETH'S WOMEN

CHAPTER 1

Mother

Giovanni Michiel, the Venetian ambassador to England during "Bloody" Mary Tudor's reign, noted with barely concealed distaste that the Queen's younger sister, Elizabeth, "is proud and haughty . . . although she knows that she was born of such a mother."[1] Clearly he, and many others like him at the Marian court, believed that the Lady Elizabeth ought to be ashamed of being the offspring of Henry VIII's disgraced second wife, the infamous Anne Boleyn—variously referred to as "the concubine" and "the whore." After all, Mistress Boleyn had usurped the place of the rightful queen, Mary's mother, Catherine of Aragon. Her subsequent alleged infidelities had led to her downfall and execution, and to her only child, Elizabeth, being declared a bastard. Little wonder, then, that the Venetian ambassador marvelled that this child should grow up apparently either oblivious to or, worse, not caring about the scandal of her mother's past. Surely she ought rather to hide herself away in perpetual shame at being the daughter of an infamous adulteress? Yet here she was, displaying all the traits with which Anne had so beguiled her male courtiers—not to mention King Henry himself. And her coal-black eyes were an uncomfortable reminder that for all her Tudor traits (most notably her abundant red hair), she was very much her mother's daughter.

Yet the common view of Elizabeth that has developed over the centuries since her death is that she had little regard for Anne Boleyn, preferring to gloss over that shady side of her history and instead boast about the fact that she was the daughter of England's "Good King Hal." "She prides herself on her father and glories in him," remarked one observer at court.[2] The many references that she made to Henry VIII, and the way in which she tried to emulate his style of monarchy when she became queen, all support this view. By contrast, she is commonly believed to have referred directly to her mother only twice throughout the whole of her life, and neither of these references is particularly significant or revealing. Unlike her sister, Mary, she made no attempt to restore her mother's reputation when she became queen, either by passing an act to declare Anne's marriage to Henry lawful or by having her remains removed from the Tower and reburied in more fitting surrounds. One might therefore be forgiven for concluding that Elizabeth was at best indifferent toward, and at worst ashamed of, her mother. Far from it. It would be her actions rather than her words (or lack thereof) that would betray her true feelings.

Anne was the second of three surviving children born to the ambitious courtier Thomas Boleyn and his wife Elizabeth, daughter of Thomas Howard, second Duke of Norfolk. A combination of shrewd political acumen and advantageous marriages had transformed the Boleyn family from relatively obscure tenant farmers into titled gentry with a presence at court. Thomas's marriage to the Duke of Norfolk's daughter had served him well, both politically and dynastically. "She brought me every year a child," he noted, and even though only three of these survived into adulthood, there was the vital son, George, to carry on the family line. The two daughters, Mary and Anne, might prove useful in the marriage market.

The date of Anne's birth was not recorded, but it is estimated at being around 1500 or 1501.[3] From the outset, she and her sister, Mary, were groomed to make marriages that would boost their family's aristocratic credentials and enable Thomas to move further up the political ladder. Anne soon emerged as the more intelligent of the two girls, and her father noted that she was exceptionally "toward" (an adjective that would later be applied to her daughter, Elizabeth), and resolved to take "all possible care for her good education." As was customary for girls at that time, Anne received a good deal of "virtuous instruction," but it was in the more courtly accomplishments of singing and dancing that she really excelled. She played the lute and virginals with a skill beyond her years, and also became adept at poetry and verse. The more academic subjects of literature and languages completed her education, and by the age of eleven, she could speak French extremely well.

All of this was quite typical of the education received by other girls of her class, but in 1512 an opportunity arose to set herself apart from her peers. It was in this year that her father was appointed ambassador to the Regent of the Netherlands, Margaret of Austria. Margaret's court was renowned for being the most sophisticated and prestigious in Europe, an ideal training ground for young aristocratic men or women who wished to enhance their social standing. Thomas used his skills in diplomacy and charm to persuade the archduchess to take Anne under her wing. And so, at the tender age of twelve, Anne set sail for the Netherlands. She was quick to absorb the full range of skills expected of a court lady. By all accounts, Margaret was delighted with her and wrote how "bright and pleasant" she was for her young age.

But it was in France that Anne's education in court life reached its zenith, and her experiences there would have a profound effect upon her character and demeanor. This time, Thomas Boleyn used his political contacts to secure places for both Anne and her elder sister, Mary, in the household of Mary Tudor, sister of Henry VIII, who had recently been betrothed to the aged King Louis XII. The Regent Margaret was sad to lose this lively and engaging addition to her court, but Anne shared her father's ambition and was delighted at the prospect of serving Henry VIII's sister, a renowned beauty. She travelled straight to

France from the Netherlands, arriving there in August 1513. It was to be a brief service, however, for Louis died just three months after the wedding (some said the exertion of satisfying his young bride had led to his demise), and Mary caused a scandal by marrying her brother's best friend, Charles Brandon, in secret, before hastily returning to England. Anne had acquired a taste for life in France, however, and so remained there after Mary's departure, transferring her service to Queen Claude, wife of the new king, Francis I.

Her sister, Mary, preferred the diversions on offer at the king's court, which was one of the most licentious in Europe. Francis was even more notorious a philanderer than his great rival across the channel, Henry VIII, and it was not long before the alluring Mary Boleyn caught his eye. She proved so easy a conquest that he nicknamed her his "English mare" and "hackney," whom he had the pleasure of riding on many occasions. By the time she returned to England, her reputation had preceded her, and, never one to be outdone by his French rival, Henry VIII also took her as his mistress. Like Francis, he quickly tired of a bait so easily caught.

Meanwhile, in stark contrast to her sister, Anne was earning a reputation as one of the most graceful and accomplished ladies of the queen's household. She thrived in the lively and intellectually stimulating French court and developed a love of learning that continued throughout her life. Among her closest companions was Margaret of Navarre, sister of Francis I, who was regarded as something of a radical for her views on women, and she encouraged Anne's interest in literature and poetry. It was here that Anne also developed a love of lively conversation, a skill that would set her apart from the quieter, more placid ladies at the English court when she made her entrée there.

So entirely did Anne embrace the French manners, language, and customs that the court poet, Lancelot de Carles, observed: "She became so graceful that you would never have taken her for an Englishwoman, but for a French woman born." Another contemporary remarked: "Besides singing like a syren, accompanying herself on the lute, she harped better than King David and handled cleverly both flute and rebec."[4] Anne was particularly admired for her exquisite taste and the elegance of her dress, earning her the praise of Pierre de Brantôme, a seasoned courtier, who noted that all the fashionable ladies at court tried to emulate her style, but that she possessed a "gracefulness that rivalled Venus." She was, he concluded, "the fairest and most bewitching of all the lovely dames of the French court."[5]

Anne had certainly blossomed during her years in France. Her slim, petite stature gave her an appealing fragility, and she had luscious dark brown hair, which she grew very long. Her most striking feature, though, was her eyes, which were exceptionally dark—almost black—and seductive, "inviting conversation." But for all that, she was not a great beauty. Her skin was olive colored and marked by small moles at a time when flawless, pale complexions were admired. The Venetian ambassador was clearly bemused by Henry VIII's later fascination with her. "Madam Anne is not one of the handsomest women in the world," he wrote, "she is of middling stature, swarthy complexion, long neck, wide mouth, bosom not much raised, and in fact has nothing but the English King's great appetite, and her eyes, which are black and beautiful, and take great effect on those who served the Queen when she was on the throne."[6] Even George Wyatt, who was to write an adulatory account of Anne during Elizabeth's reign, admitted: "She was taken at that time to have a beauty not so whitely as clear and fresh above all we may esteem."[7] She also had small breasts, a large Adam's apple "like a man's," and, most famously, the appearance of a sixth finger on one of her hands.[8] But it was undoubtedly her personal charisma and grace, rather than her physical appearance, that gave her the indefinable sex appeal that was to drive kings and courtiers alike wild with frustrated lust. Wyatt observed that her looks "appeared much more excellent by her favour passing sweet and cheerful; and . . . also increased by her noble presence of shape and fashion, representing both mildness and majesty more than can be expressed."[9]

Anne Boleyn's allure, honed to perfection at one of the most sophisticated courts in the world, set her apart when she made her entrée into Henry VIII's court in 1522. Her father had secured her a position in Catherine of Aragon's household, and she swiftly established herself as one of the leading ladies of the court. While the women admired and copied her fashions, the men were drawn to her self-confidence and ready wit, but more particularly to her provocative manner, which made her at once playfully flirtatious and mysteriously aloof. George Wyatt later said of her: "For behaviour, manners, attire and tongue she excelled them all."[10] She had first come to notice in a court pageant organized by Cardinal Wolsey for the king on Shrove Tuesday 1522, in which she played the part of Perseverance—particularly fitting given the events that later unfolded.

Among Anne's suitors was the poet Thomas Wyatt, whose ardent expressions of love were hardly restrained by the fact that he was already married. Rather more eligible was Henry Percy, later sixth Earl of Northumberland, who grew so besotted with her that he tried to break a prior engagement in order to marry her. It was apparently some time, though, before Anne attracted the attention of the king himself.

The early relationship between Henry VIII and Anne Boleyn showed little sign of the intensity that it would later develop. The very fact that Anne had been at court for some four years before there was any sign of an attachment suggests that it was hardly a case of love at first sight. Rather, the affair appears to have developed gradually out of a charade of courtly love. By late 1526, all the court knew that Lady Anne was the king's latest inamorata. But this was very different from Henry's previous infidelities, for Anne proved to be the most unyielding of mistresses. She persistently held out against his increasingly fervent advances, insisting that while she might love the king in spirit, she could not love him in body unless they were married. It was a masterstroke. Perhaps having learned from the example of her sister, who had given way all too easily and had been discarded just as easily, Anne sensed that Henry would lose interest as soon as she had succumbed to his desires, so she kept her eyes focused on the main prize: the crown of England. It was an extraordinary goal even for one born of such an ambitious family, for Henry already had a queen—and a popular one at that. But Anne knew that he was tiring of his wife, Catherine of Aragon, who, at forty, was some five years older than himself and now unlikely to bear him the son he so desperately needed. Anne, meanwhile, was in her midtwenties, with every prospect of fertility.

At first she rebuffed the king's advances altogether, refusing to accept either his spiritual or physical love. Henry complained that he had "been more than a year wounded by the dart of love, and not yet sure whether I shall fail or find a place in your affection," and begged Anne to "give yourself, body and heart, to me."[11] He even promised that if she assented, he would make her his "sole mistress," a privilege he had afforded to no other woman. But Anne was determined to hold out for more, and told him: "I would rather lose my life than my honesty . . . Your mistress I will not be." She proceeded to play the king with all the skill and guile that she had learned at the French court, giving him just enough encouragement to keep him interested but rebuffing him if he tried to overstep the mark. Thus, one moment Henry was writing with gleeful anticipation of the prospect of kissing Anne's "pretty duggs [breasts]," and the next he was lamenting how far he was from the "sun," adding mischievously, "yet the heat is all the greater."[12]

The longer their liaison went on, the greater Anne's influence at court became. She was constantly in the king's presence, eating, praying, hunting, and dancing with him. The only thing she did not do was sleep with him. As her status grew, so did her pride and haughtiness. She became insolent toward her mistress, Catherine of Aragon, and was once heard to loudly proclaim that she wished all Spaniards at the bottom of the sea. A foreign visitor to the court noted with some astonishment: "there is now living with him [the king] a young woman of noble birth, though many say of bad character, whose will is law to him."[13] But Henry cared little for the resentment toward Anne that was building at court, and as his love for her drove him increasingly to distraction, he began to think the unthinkable: if marriage was the only way he could claim her, then he would seek an annulment from the Queen. This was precisely what Anne had been angling for, and she encouraged the king in his new resolve. It would take him almost six years to achieve it, and nobody could have predicted the turmoil that would ensue. Inspired by his pursuit of marriage to Anne, Henry would overturn the entire religious establishment in England, wresting the country from papal authority and appointing himself head of the Church. The religious, political, and social ramifications would be enormous, reverberating for decades and laying the foundations for discord in all of his children's reigns.

Anne actively supported the king in his religious reforms, realizing that they held the key to her future. She introduced Henry to William Tyndale's writings and kept a copy of his English translation of the New Testament in her suite for anyone who wished to read it. She also befriended a number of leading reformers at court, and it was through her influence that they were later appointed to powerful bishoprics. It was said that men such as Hugh Latymer, Nicholas Shaxton, Thomas Goodrich, and even Thomas Cranmer, who was appointed Archbishop of Canterbury in 1533, owed their positions to her. A posthumous account of Anne, written by the reformist cleric William Latymer, described her as "well read in the scriptures" and "a patron of Protestants."[14]

In espousing the reformist religion, Anne made some dangerous enemies at court. The Catholics were in no doubt that the king's alarmingly radical religious reforms were down to her. Eustace Chapuys reported to his master, Charles V, that "the concubine" had told the king "he is more bound to her than man can be to woman, for she extricated him from a state of sin . . . and that without her he would not have reformed the Church to his own great profit and that of all the people."[15] Anne also alienated large swathes of the population who were already sympathetic to Queen Catherine.

Catherine's daughter, Mary, was herself the subject of pity. She had returned to court in 1527, aged eleven, after a two-year sojourn in Wales, as was traditional for the heir to the throne. Until then, she had been the king's cherished only child, "much beloved by her father," according to the Venetian ambassador.[16] She had been feted at court and proudly shown off to foreign ambassadors, who all praised her appearance and intelligence. Her long red hair was "as beautiful as ever seen on human head," and another observer complimented her delicate, "well proportioned" figure, as well as her "pretty face . . . with a very beautiful complexion." Gasparo Spinelli, a Venetian dignitary, told of how the young princess had danced with the French ambassador, "who considered her very handsome, and admirable by reason of her great and uncommon mental endowments."[17]

From the tender age of two, Mary had been a highly prized pawn in the international marriage market, betrothed first to the Dauphin of France, and three years later to the Holy Roman Emperor, Charles V. Young as she was when all of these negotiations were being conducted, she had learned to cherish high expectations of her future life. Her education had reinforced this view. During her early years, she had learned the typical courtly accomplishments of playing the lute and virginals, singing, dancing, and riding. She had later been tutored by the celebrated humanist Juan Luis Vives.

Upon her return to court in 1527, Mary learned that her father had become enamored of Anne Boleyn, but it did not cause her any immediate concern—there had been mistresses before, and no doubt there would be more to follow. Mary's ally Chapuys warned that Anne "is the person who governs everything, and whom the King is unable to control."[18] Still Mary clung doggedly to the belief that her mother's position was unassailable.

Anne agreed to become Henry's wife later that year, but she continued to refuse his advances throughout the long years during which he and his ministers tried to secure the annulment of his first marriage through protracted negotiations with the Pope. Henry's sexual frustration mingled with Anne's increasing bouts of temper to often explosive effect. Keenly aware that time was passing her by and that she could have been married with children by now, Anne threatened to leave the king. Her behavior became increasingly erratic, and she lashed out at the slightest provocation, such as when she discovered that Catherine of Aragon was still mending her husband's shirts. Even though she had triumphed over the beleaguered queen, she had no sympathy for her and told one of Catherine's ladies that "she did not care anything for the Queen, and would rather see her hanged than acknowledge her as mistress."[19]

By 1529, with the prospect of success still frustratingly out of reach, Anne fixed her wrath upon the king's chief minister, Cardinal Wolsey, whom Henry had appointed to secure the annulment, but whom she suspected of deliberately impeding matters. She threw her weight behind the faction at court (led by her father and uncle, the Duke of Norfolk) that was plotting to get rid of Henry's chief minister. In the event, she helped secure his downfall, but not the divorce.

In 1531 Catherine of Aragon was banished from the court, and
Anne was established as queen in all but name. Princess Mary was now
forced to choose between her duty to Henry as her father and her king,
and the love and loyalty that she felt for her mother. For her, it was an
easy choice. She instantly sided with the beleaguered queen and
avoided any accusations of disobedience to the king by placing all of
the blame on that "concubine," Anne Boleyn. But while she professed
her continuing devotion to her father, this once cherished daughter
was gradually slipping from his favor. Anne exacerbated the situation
by doing everything she could to keep them apart, determined to focus
her royal lover's mind on his annulment from Catherine and marriage
to her—and with it, the promise of his longed-for son. She treated
Mary with barely concealed disdain, emphasizing the power that she
now had over her. "The said Anne has boasted that she will have the
said Princess for her lady's maid . . . or to marry her to some varlet,"
reported Chapuys, "but that is only to make her eat humble pie."[20]

Although Mary steadfastly defended her mother and suffered no
weakening of resolve, the psychological toll of watching her parents'
marriage unravel and the ever more cruel indignities inflicted upon her
mother had a devastating effect upon the young girl's health. She suffered increasing bouts of nausea and on one occasion was unable to
keep any food down for three weeks, causing panic among her attendants. In the spring of 1531, when she was recovering from one of her
frequent stomach upsets, she wrote to her father, saying that nothing
would speed her recovery more than to visit him at Greenwich. Her request was peremptorily refused, as Chapuys believed, "to gratify the
lady [Anne], who hates her as much as the Queen, or more so, chiefly
because she sees the King has some affection for her."[21] It seemed that
Henry, too, had become cruelly indifferent to his daughter's suffering.
Knowing how much comfort she derived from spending time with her
mother, later that year he banished her from Catherine's presence. He
even forbade her from writing to her mother.[22] Thenceforth, the two
women were forced to be strangers.

Mary wondered how she had suddenly come to this after being
cherished and lauded throughout her childhood. Coinciding with her

most formative teenage years, this first great crisis of her life had a profound effect. The formerly confident, lively young girl was now beset with melancholy and depression, worn down by fear about what the future would hold. But the crisis also strengthened certain aspects of her character and beliefs. As a show of support for her sainted mother, she identified herself strongly with the Spanish cause, throwing in her lot with Chapuys and his Imperial master, Charles V. She also fervently embraced her mother's Roman Catholic faith. Both of these moves set her in direct conflict with Anne Boleyn.

Meanwhile, the subject of Mary's hatred had made a decision that would turn the course of history. Late in 1532, Anne Boleyn resorted to what was for her the most desperate of all measures: to relinquish her former strategy and sleep with the king. In so doing, she was gambling on the by no means certain prospect that if she became pregnant, he would overcome all of the remaining obstacles and marry her. After all, even though he had pursued her for years, the fact that she had remained just beyond his grasp was a large part of her allure. If she gave that up, then she might well lose his interest for good.

But the gamble seemed to have paid off. Henry was, at least initially, even more besotted with Anne now that she had become his mistress in body as well as in name. The Imperial ambassador, Eustace Chapuys, was aghast when he discovered that "the King cannot leave her for an hour." By December, she was pregnant. Her royal lover now had to act fast if the baby was to be born legitimate. He therefore set aside the ongoing wranglings with the Church and married Anne in secret on January 25, 1533, in his private chapel at Whitehall. His marriage to Catherine was annulled shortly afterward.

Anne was formally recognized as queen on April 12, 1533, and her coronation followed six weeks later. This was a lavish affair, full of iconography and symbolism designed to emphasize the legitimacy of her position and her suitability as queen. The theme was the Assumption of the blessed Virgin Mary. Throughout the procession, the city of London was displayed as a kind of "celestial Jerusalem," with Anne as the Virgin, dressed in white and with her long dark hair worn loose around her shoulders.

Along the route, a tableau was built with a castle in the foreground against the backdrop of a hill. As Anne's procession passed by, a stump on the hill poured out a mass of red and white roses (symbolizing the Tudor dynasty), and then a painted cloud opened up to release a white falcon, which swooped down onto the flowers. As a final touch, an angel descended from the same cloud and placed an imperial crown upon the head of a white falcon. Anne had adopted this bird as her emblem in 1532, in preparation for her marriage to Henry, when she had been granted a crest of her own: a white falcon alighting upon roses. The message was clear: With the accession of Anne, already pregnant, new life would burst forth from the Tudor stock.

The coronation ceremonies lasted for four days and were clearly intended to enhance Anne's status: For all that she had recently been created Marquess of Pembroke, she was still just the daughter of an English aristocrat—only the second such queen since 1066. The coronation was also a test of loyalty for the court and the people. Although the only notable not to attend was Sir Thomas More, the lord chancellor (who thereby helped seal his own fate—he was executed for high treason two years later), most of the others were there under duress and bitterly resented the woman whom they viewed as a usurper. The citizens of London who turned out to watch the procession evidently felt the same. Chapuys described the coronation as "a cold, meagre and uncomfortable thing, to the great dissatisfaction, not only of the common people, but also of the rest."[23] Dissatisfaction soon turned to open mockery. Everywhere along the processional route were Henry's and Anne's initials intertwined. But this cipher was turned to parody, and as the new queen passed, cries of "ha-ha" could be heard among the disdainful crowds.

Another reason for their scorn was that this new queen was very obviously pregnant, and further advanced than one might expect for a lady who had been married for barely four months: a bastard child growing within a usurper's belly. Yet this child was Anne's chief hope of securing her position as queen and of retaining the king's notoriously fickle affections. The pregnancy was announced in May, by which time it was already widely known. The addition of an extra panel to Anne's skirts to accommodate her increasing girth removed any lingering doubt. The following month, Archbishop Cranmer told an acquaintance that the Queen "is now somewhat big with child."[24]

Although she triumphed in the expectation of giving Henry a son and heir, Anne was distracted by more immediate concerns and complained about the loss of her famously slender figure. This may have been due to more than her accustomed vanity, for she no doubt feared that as her attractiveness waned, the king would seek diversion elsewhere. Her fears were well grounded. In August 1533, as Anne entered the eighth month of her pregnancy, rumors of a secret liaison between the king and a "very beautiful" woman began to spread throughout the court. By the time they reached Anne's ears, the tale had been embellished: Henry had slept with at least one other woman, probably more. Lying in her chamber, her body heavy and ungainly, Anne must have been tortured by the thought that her husband had strayed so soon after the marriage. Chapuys noted with barely concealed satisfaction that the new queen was "very jealous of the King, and not without legitimate cause."[25] Furious at such humiliation, Anne confronted Henry with what she had heard. Rather than comforting his heavily pregnant wife, he spat back that she must "shut her eyes and endure," as her betters had done. Just a few short months before, Anne had been the sole focus of the king's attention, the woman whom he had worshipped for years and moved heaven and earth to attain. Now it seemed that she was just like any other woman to him. As Chapuys observed: "She ought to know that it was in his power to humble her again in a moment, more than he had exalted her before."[26] For Henry, it seemed that the thrill had been entirely in the chase.

The quarrel between King Henry and his new wife lasted for several days and was the talk of the court. Having made many enemies on her path to the throne, there was little sympathy now for Anne, who was left to seethe and fret in the confines of her chamber. Perhaps she reasoned that the only way to regain her husband's affection and avoid sliding into obscurity would be to give him the son for which he had so long craved.

In the middle of August, Henry and Anne moved from Hampton

Court to Windsor, and from there to Greenwich, the king's favorite palace, which had been appointed for Anne's confinement, or "lying in." His daughter, Mary, was ordered to join the ladies who had assembled there to attend Queen Anne. Mary's feelings at being so cloistered with the woman whom she saw as the architect of all the evils that had befallen her and her mother can only be imagined. Thanks to Anne, she was no longer a princess but simply "Lady Mary," the king's bastard daughter. And now she was forced to stand by and witness firsthand this whore's ultimate triumph as she gave birth to a prince.

Meanwhile, there was frenzied activity at Greenwich Palace as preparations were made for Anne to "take to her chamber." A queen's confinement was subject to an elaborate set of conventions—part religious, part medical—that stretched back hundreds of years. They had been refined in the fifteenth century by Lady Margaret Beaufort, who had drawn up strict ordinances for "the deliverance of a queen." These dictated that a queen would effectively go into seclusion some four to six weeks before the birth was expected. As one foreign observer noted with some bemusement: "This is an ancient custom in England whenever a princess is about to be confined: to remain in retirement forty days before and forty after."[27] She would be confined to her chamber, which was actually a suite of rooms based upon the privy chamber apartments usually found at court (to which only the most privileged persons would gain access), but with certain modifications. For example, an oratory would be installed so that prayers could add necessary succor in an age when knowledge of obstetrics was limited, together with a font to provide a quick baptism for a sickly baby.

The expectant queen would herself select the room in which she wished to give birth. This received the greatest attention, being hung with heavy tapestries—"sides, roof, windows and all"—depicting scenes from romances or other pleasant subjects, so as not to upset mother or child. The theme for the tapestries in Anne's chamber was the story of Saint Ursula and her eleven thousand virgins. It would prove a peculiarly fitting one. Once the tapestries had been hung, the floor would be "laid all over with thick carpets," and even the keyholes would be blocked up to keep out any glimmer of light from the world beyond. Finally, a specially constructed bed of state upon which the precious infant would be born was installed. This would comprise a mattress stuffed with wool and covered with sheets of the finest linen, and two large pillows filled with down. The bed prepared for Anne's confinement at Greenwich was bedecked with an elaborate counterpane, "richly embossed upon crimson velvet," lined with ermine and edged with gold. It was rumored to have formed part of the ransom of the Duke of Alençon, who had been captured at Verneuil in 1424. If this was true, then perhaps Queen Anne wished to be reminded of the country in which she had spent so much of her youth.

A crimson satin tester and curtains embroidered with gold crowns completed the effect, with the Queen's arms being added as another reminder of her lineage—and, therefore, her right to the throne. The final touch was the installation of two cradles: one a "great cradle of estate," richly upholstered with crimson cloth of gold and an ermine-lined counterpane to match that of the Queen's bed; the other a more modest carved wooden cradle painted with gold.

The whole effect of this richly arrayed birthing chamber was designed to impress. But it would also have been stifling and oppressive for those within, with its heavy tapestries that shut out all light, and the thick velvet fabrics that smothered the bed, especially given that it was the middle of August. This was made worse still by the braziers, which were lit some days before the Queen entered her chamber, and also by the rich perfumes that filled the air from the unstoppered bottles that were scattered around the room.

While these preparations were under way, Anne made a request of her own regarding the birth of her child. She asked her husband to procure from his former wife the "rich triumphal cloth" that Catherine of Aragon had brought with her from Spain for the baptism of her future children. This cloth, a painful reminder of all her children who had been stillborn or died within days of birth, was one of the few possessions that Catherine had left, and she was outraged when she heard of Anne's request. Although it was undoubtedly a callous, cold-hearted act on Anne's part, she was perhaps driven by more than sheer vindictiveness. As the hour of her lying-in grew closer, she was determined to prove the legitimacy of her child, which she knew was the subject of increasingly vociferous whispers that it was a bastard, conceived out of lawful wedlock. In her jaundiced view, the baptismal cloth of her predecessor, who was still revered by the people as their true queen, was a symbol of legitimate royal blood, and she was desperate to secure it for her unborn child. But Catherine held firm, and Anne was eventually forced to relent, perhaps aware—for once—of the widespread resentment that would follow if she got her own way.

On August 26, Anne formally took to her chamber. As custom dictated, she heard Mass in the palace chapel before hosting a banquet for all the lords and ladies of the court in her great chamber, which had been richly decorated for the occasion. There "spices and wine" were served to Anne and her guests, and soon afterward she was escorted to the door of her bedchamber by two high-ranking ladies. Here she took formal leave not just of the king but of all the male courtiers, officials, and servants, and entered an exclusively female world, in which women were to take over all the positions in her household usually occupied by men. As Lady Margaret Beaufort's ordinances dictated: "women were to be made all manner of officers, as butlers, panters, sewers."[28] Any provisions or other necessary items would be brought to the door of the great chamber and passed to one of the female attendants within. Even the king was refused entry.

All of this was intended to emphasize that childbirth was a purely female mystery. In a male-dominated society, this was the only sphere in which women held precedence. But there was a price to pay for this temporary superiority: at the end of the elaborate, exclusively female ritual, the Queen must produce a male heir. Anne herself seemed confident enough of this. She had ordered a letter announcing the birth to be written in advance. Clearly not overly concerned about tempting fate, she thanked God for sending her "good speed, in the deliverance and bringing forth of a prince."[29] The king shared his wife's optimism and had already decided that the boy would be christened Henry or Edward. He also spent what should have been anxious days awaiting news in planning a splendid joust to mark the safe delivery of his son. One courtier remarked that he had never seen His Majesty so "merry."

If the astrologers and soothsayers were to be believed, then he had good reason, for all bar one had predicted the birth of a prince. The exception was the renowned "seer" William Glover, who had dared to tell Queen Anne that he had had a vision of her bearing "a woman child." This had not been well received.

Quite apart from the sex of her child, there must have been some concern about its chances of survival. Childbirth was fraught with danger in Tudor times and often resulted in the death of the mother, child, or both. Around a quarter of children died at birth, and the same number died in infancy. Worse still, Anne's closest female relations had suffered an unfortunate history in this respect. Her mother had lost several babies in infancy, and her sister, Mary, had borne a son with mental disabilities whom Anne would not suffer to be at court. But in her favor was the fact that her health was generally good, and as one observer remarked, she seemed "likely enough to bear children." What was more, she had become pregnant almost immediately after becoming Henry's lover, which surely augured well—for both this and all future conceptions.

On September 7, the eve of the Feast of the Virgin, just twelve days after entering her confinement, Anne went into labor. This was much earlier than anticipated, so it was assumed that either the baby was premature or the midwives had miscalculated. Or perhaps Anne had bent the truth a little when telling them the date of conception. She and Henry had started sleeping together at least a month before their marriage, but of course it would not do to reveal this fact when questions were already being raised about the child's legitimacy.

The king and his courtiers waited eagerly for news as the labor progressed throughout the morning and early afternoon. Meanwhile, inside the Queen's bedchamber, women rushed to and fro in the cloistered darkness, bringing the necessary provisions and equipment for the midwives and keeping a tense vigil. The past seven years had been building up to this moment. The waiting, frustrations, turmoil, and hostility that Anne had endured would all be swept away in one glorious moment.

Shortly after three o'clock in the afternoon, the baby was born. Just as Anne had hoped, this child would one day bring England to such glory and power that its name would echo down the centuries as one of the greatest monarchs who ever lived. But in the stifling confines of the birthing chamber on that hot September day, none of this could have been predicted, for the child that Anne had borne was not the hoped-for prince. It was a girl.

After all the upheaval that the king and his country had endured to attain an heir, this was surely a disaster. No woman had sat upon the throne of England for centuries, and then it had been a catastrophe, plunging the country into civil war.[30] Besides, the king already had a female heir (albeit an illegitimate one, thanks to the annulment of his marriage to Catherine), and he would not welcome another.

Amidst their quiet consternation, Anne is alleged to have declared: "Henceforth they may with reason call this room the Chamber of Virgins, for a virgin is now born in it on the vigil of that auspicious day when the church commemorates the nativity of our blessed lady the Virgin Mary."[31] Even if this quote is erroneous, it would have been entirely in character for Anne to have brazened it out. After all, had she not been delivered of a perfect, healthy child, who, with her flame-red hair, bore all the marks of the Tudor dynasty? Moreover, the labor had been straightforward (albeit "particularly painful," according to her earliest biographer, William Latymer), and there were no signs to suggest that she might not bear the king many more children.

In the meantime, a herald had announced the news to the waiting courtiers that "the queen was delivered of a fair lady," and the letter that had been prepared to announce the arrival of a prince had to be hastily amended with an additional s.[32] The king, on the surface at least, showed little of the fury that historians have since assigned to him upon hearing that his long quest for a male heir was still not over. Upon visiting his newborn daughter for the first time, he remarked to the Queen with a sanguinity similar to her own that as they were both still young, they might confidently expect to have sons in due course. He then announced that the girl would be named Elizabeth, after both his mother and Anne's.

According to Chapuys, such optimism on the part of the royal cou-

ple was little more than a front. On the day of the christening, he
wrote to his Imperial master, Charles V: "the King's mistress was delivered of a daughter, to the great regret both of him and the lady, and
to the great reproach of physicians, astrologers, sorcerers, and sorcer-
esses, who affirmed that it would be a male child. But the people are
doubly glad that it is a daughter rather than a son, and delight to mock
those who put faith in such divinations, and to see them so full of
shame."[33] He later added that the new queen had shown "great disappointment and anger" at the birth of her daughter.

If, as Chapuys claimed, the king was furious when he learned of the
baby's sex, then it would have been understandable: He had, after all,
moved heaven and earth in his frustrated attempts to secure an annulment from Catherine of Aragon so that he could marry Anne, and all
to achieve his desperate desire for a male heir. But the only direct evidence for his and the Queen's dismayed reaction to Elizabeth's birth
comes from Catholic or pro-Spanish sources, both of which may well
have been layering their own prejudices onto the accounts they gave.
The historical narratives written in the centuries after the event have
often exaggerated how disastrous it was because they had the benefit
of knowing that Elizabeth would turn out to be the only living child
that Anne was able to provide her husband. In fact, George Wyatt's account, written in Elizabeth's reign, may have carried equal merit. According to him, the king was delighted at the birth of a healthy
daughter and "expressed his joy for that fruit sprung of himself, and his
yet more confirmed love towards her [Anne]."[34]

There is very little contemporary evidence to suggest that giving
birth to a girl irrevocably damaged the relationship between Henry
and Anne. It is therefore tempting to conclude that Anne's failure to
produce the hoped-for male child at the first attempt would have been
seen as a temporary setback—albeit a bitterly disappointing one, after
all the anticipation—rather than an unmitigated disaster.

But there was more to it than that. Although Elizabeth's birth had
not destroyed Anne's marriage, it had significantly weakened her position in the eyes of her people—and, indeed, of the world. Throughout
the arduous negotiations for an annulment of Henry's marriage to

Catherine, whose childbearing years seemed to be over, Anne had represented youth and fertility, and the whole prospect of her marrying the king had rested on the premise that she would give him a son. Without it, she was just the daughter of a family whose prominence was based upon trade. Even the elaborate symbolism of her coronation had merely papered over the cracks. She was still a usurper in the eyes of most people—including those at the center of political power. Giving birth to a son would make her virtually invincible, certainly in the eyes of the king, who could hardly forsake the mother of his legitimate heir. It would even help Anne to face down the might of Catherine's Habsburg supporters across Europe and of her daughter, Mary, whose claim to the throne would have withered away against that of a boy.

The gamble had failed—at least on the first throw of the dice. With a mere daughter, Anne was no better than Henry's rejected first wife; indeed, in the eyes of Catholic Europe and most of her English subjects, she was a good deal worse. The child who should have been her security threatened to be her undoing, and Anne was plunged back into a world of uncertainty and hostility. Her enemies at court and abroad now had still more ammunition against this pretender to the throne. Even her husband was hedging his bets, and within weeks of Elizabeth's birth, he had summoned his illegitimate son, Henry Fitzroy, home from France. Either he wanted to make a statement and emphasize the fact that he could father sons, or, more worryingly for Anne, he planned to keep the boy in reserve in case she failed in her duty.

The king's new daughter was christened on September 10 in the Chapel of the Observant Friars at Greenwich, with notables from across the kingdom in attendance, including the Dukes of Norfolk and Suffolk, who escorted the baby Elizabeth to the chapel, and the Dowager Duchess of Norfolk, who carried her in her arms under the canopy of estate. Elizabeth was wrapped in a purple mantle with a long train edged with ermine, which was borne by the Countess of Kent. The baby's half sister, the Lady Mary, who had been at Greenwich during Anne's confinement, was also in attendance. As custom dictated, neither the king nor the Queen was present, as this was an occasion primarily for the godparents, but after the ceremony, their child was brought to them in a procession that made its way through corridors lit by five hundred torches. Waiting in the Queen's apartments was Anne, robed and lying on the magnificent bed on which she had given birth three days before, with Henry by her side. The couple showed every sign of rejoicing when they saw their little daughter, and celebrated heartily with their guests.

As a public relations exercise, it was faultless. But cracks had already begun to appear. Although the christening was observed with all due ceremony, it somehow lacked conviction as a celebration of the king's new heir, and Chapuys described it as "cold and disagreeable."[35] Furthermore, it had not been followed by the jousts, fireworks, and bonfires that would have been staged for a prince. There were also rumblings among the people, who were still disapproving of Queen Anne and thought still less of her for producing a useless daughter after everything they had gone through on her behalf. Elizabeth herself became the subject of hatred. Two friars were arrested for saying that the princess had been christened in hot water, "but it was not hot enough."[36] Meanwhile, the Spanish referred to the "concubine's daughter" as the "little whore" or "little bastard." They had greeted her arrival with barely concealed amusement, delighted at what they perceived to be God's punishment for the English king's expulsion of papal authority.

Although the Lady Mary shared their sentiments, any satisfaction that she felt at witnessing the birth of a mere daughter to Anne Boleyn was short lived. Her new half sister had been immediately proclaimed as the king's first legitimate child, glossing over the fact that he already had a daughter. Elizabeth, not Mary, was now the sole heir.

CHAPTER 2

"The Little Whore"

When the furor of Elizabeth's birth had begun to subside, Anne was left to contemplate her new role as the mother not of a prince but of an unwanted girl. After everything she had been through, it would be logical to assume that she felt more than a degree of bitterness and resentment toward Elizabeth. But often motherhood defies logic, and any disappointment that she might have felt was quickly overshadowed by much deeper maternal instincts. Anne was not an obviously maternal woman: Her guile, seductiveness, and self-interest seemed somehow at odds with such emotions. And yet these feelings were apparently more than just the result of giving birth, because some time before she fell pregnant, she had told Henry that she longed for children, as they were "the greatest comfort in the world."

Anne's reaction to her newborn daughter seemed to bear this out. She lavished affection upon Elizabeth and could hardly bear to be apart from her. When she returned to court after her confinement, she took her daughter with her. Courtiers looked on in astonishment as Anne carefully set the baby down on a velvet cushion next to her throne under the canopy of estate. It was highly unusual for a queen to keep her child with her: surely it ought to be bundled off to the royal nursery, as was customary? But Anne had never been one to abide by convention, and she went one step further by expressing her intention to breast-feed Elizabeth herself. This was going too far, even for Henry, who might have been inclined to indulge his wife during these first few weeks after the birth. It was unheard of for a queen to breast-feed her offspring; even noblewomen would enlist the services of a local wet nurse. The king insisted that Anne do the same for propriety's sake, and she reluctantly assented. A Mrs. Pendred was duly assigned as Elizabeth's wet nurse.

A far greater sacrifice was on the horizon, for Anne knew full well that she would not be able to keep Elizabeth with her forever. Sooner or later the child would have to be set up in her own establishment away from court, as tradition dictated for royal offspring. Henry was already making plans for this and had appointed the palace of Hatfield, some twenty miles from London, as the most suitable place. As well as being within convenient reach of the court, this pleasant retreat in Hertfordshire, with its gently rolling countryside and plentiful woodland, was also well away from the unhealthy, plague-ridden air of London. This latter consideration prompted the move to take place. On December 2, 1533, the Privy Council—the king's chief advisory body—met to consider, among other pressing items, "a full conclusion and determination for my Lady Princess's house." With Christmas approaching, it was agreed that the risk of infection at Greenwich was too high, because people from across the city and country would come to court.

A few days later, when Elizabeth was barely three months old, she was removed to Hatfield in all due state, along with an army of nursemaids, governesses, stewards, and other household staff who would become her surrogate family. As was appropriate for a princess, her household was largely female. There was a wet nurse to suckle her and four "rockers" to attend her in her cradle, as well as numerous other ladies to nurse, bathe, amuse, and protect her. It was a premise of Tudor childhood that infants would be marked for life with the character of the women who nursed them. All of these women were therefore subject to the utmost scrutiny, and this was particularly so for those who cared for royal children.

The impact that the loss of her daughter had upon Anne was great indeed. During those three short months at Greenwich, she had forged an extremely close bond with Elizabeth, doting on her in public and showering her with gifts. This went beyond maternal affection. Young as she was, Elizabeth was an ally against Anne's enemies at court, for she symbolized her mother's fertility and thus the hope of future children. She was also the king's only legitimate heir. Little wonder that Anne was heartbroken to see her go.

It is hard to imagine that the three-month-old Elizabeth would have had any lasting memory of her mother from those earliest days of her infancy. Any bond that did exist would most likely have been instinctive rather than based upon remembered affection, but even then it would have had to have been stoked regularly in order to avoid losing it altogether. This is precisely what Anne intended to do. For a start, she used her influence to insure that her daughter would be surrounded by members of the Boleyn family. The Queen's aunts, Lady Shelton and Anne Clere, took general charge of the household, and Margaret Bourchier, Lady Bryan, was appointed Elizabeth's "Lady Mistress." Through these ties of kinship, Anne no doubt hoped to maintain some hold over her daughter, albeit from a distance.

Anne seemed to share a particularly close affinity with Lady Margaret Bryan, who was the half sister of her mother. Lady Bryan had been chosen by Henry VIII because of her competence in caring for his first daughter, Mary. She was a woman of excellent credentials. The widow of Sir Thomas Bryan and the sister and heiress of Lord Bourchier, she had been a member of Catherine of Aragon's household. Her son, Sir Francis Bryan, was one of the king's closest companions at court and wielded some considerable influence there. Sir Francis was part of the Seymour faction, which at once set him at odds with Anne Boleyn. But his mother apparently did not share his sympathies, for she was on good terms with the new queen.

At the venerable age of sixty-five, Lady Bryan had had many years' experience in child care and was ideally suited for the role of Lady Mistress to the king's new daughter. She had proved so adept in caring for Elizabeth's half sister, Mary, for six years that Henry had rewarded her with the title of baroness. Having been accustomed to treat Mary as heiress to the throne, it must have been with some embarrassment and sympathy that she had witnessed the girl suffering the humiliation of being declared a bastard and ordered to yield precedence to the baby Elizabeth. But Lady Bryan was as much a pragmatist as her young charge would prove to be, and she was no doubt consoled by the trust that Henry had placed in her with this new appointment. It is to her credit that she subsequently encouraged Mary to look with affection upon her younger sister, despite the myriad reasons the elder daughter had to despise this usurper to her father's favor, not to mention her own title of princess.

Lady Margaret helped to ease the wrench that Anne and Elizabeth felt at first being parted in December 1533. Of a naturally warm and caring disposition, she looked after the infant princess with maternal affection. She referred to herself as Elizabeth's "mother," and the tone of her letters to the Queen attests to the fondness that she felt toward this pretty red-headed child. Margaret was effectively an extension of Anne and carried out her orders with such care and assiduity that Elizabeth herself came to view her as a second mother.

This could easily have sparked a fit of jealousy in the new queen, who was only rarely able to see her daughter, but instead it brought the two women closer together, united by their affection for the child. By the end of Lady Margaret's first year in charge of the royal nursery, she was believed to have such influence with Anne that courtiers sought her advice when trying to ingratiate themselves with the Queen. Lady Lisle, for example, agonized over what to buy the latter as a New Year's gift in 1534, and after consulting with Lady Bryan, she chose a little dog to add to her collection of pets. Anne liked it so much that she immediately snatched the dog from the messenger's arms without waiting for him to utter the customary request to accept it.[1]

As well as relying upon Lady Bryan, Anne strengthened her ties with her infant daughter by sending tangible reminders of herself to

Hatfield. From the moment of her child's birth, she had lavished expensive presents upon her. Anne's love of clothes was passed on to her daughter, as she created a miniature version of herself, dressing Elizabeth in the finest silks and richest velvets.

The best account of these gifts is provided by a memorandum of "Materials Furnished for the use of Queen Anne Boleyn and the Princess Elizabeth" between January and February 1535, when Elizabeth had been at Hatfield for just over a year.[2] Every detail of the child's costume was considered by her mother: from the "velvet blak" collar of a dress made from "Russet velvet," to some purple sarsenet "for lyning of a sleve of purpull satten ymbrotheryd ffor my Lady prynses." Anne had always had an impeccable sense of style, and she set off her own dark coloring with rich fabrics in tawny, damask, and leaf green. The records show that she was spending some £40 per month on clothes and accessories for herself and her young daughter (equivalent to around £13,000 today), which was a considerable sum compared to her other expenses.

Having been supplanted in the succession by the loathsome child of the king's "Great Whore," the Lady Mary had refused to suffer the further humiliation of yielding precedence to her and continued to refer to herself as princess. When a message arrived to say that the king ordered that she should "lay aside the name and dignity of Princess; and commanded her servants no longer to acknowledge her such," Mary refused to accept it because it was not delivered by a "person of honor." She then "boldly" told the messenger that "she was the King's true and lawful daughter and heir" and that "her servants would not take notice of this order upon the same reason."[3] Her father was beside himself with rage, admonishing her for "forgetting her filial duty" with such "pernicious" behavior. Undeterred, she insisted that she was his "lawful daughter, born in true matrimony," but assured him that "in all other things you shall find me an obedient daughter."[4]

Courageous though it may have been, Mary's behavior merely served to bring an even greater punishment upon her. A month after Elizabeth's birth, Henry announced that he intended to disband his elder daughter's establishment and make her serve the new princess when she moved to Hatfield that December. Mary's allies at court were aghast. "The King, not satisfied with having taken away the name and title of Princess, has just given out that, in order to subdue the spirit of the Princess, he will deprive her of all her people . . . and that she should come and live as lady's maid with this new bastard," reported Chapuys, adding that Mary was "mightily dismayed" by this turn of events. Like Mary, he was convinced that Anne was behind it all. "I do not understand why the King is in such haste to treat the Princess in this way, if it were not for the importunity and malignity of the Lady."[5]

If Mary already despised her new stepmother, she now had greater cause when she reluctantly obeyed her father's command and made her way to Hatfield. It was clear from the start that she would be treated with all of the indignity and disgrace of a bastard. Elizabeth had been conveyed to her new home in a velvet litter, escorted by the Dukes of Norfolk and Suffolk, together with a large retinue of ladies and gentlemen. In order to emphasize her status as England's new heir, she had been paraded in front of the people of London as her magnificent entourage had made an unnecessarily circuitous route through the capital. "There was a shorter and better road," complained Chapuys, "yet for greater solemnity, and to insinuate to the people that she is the true Princess, she was taken through this town."[6] By contrast, Mary had been forced to travel in a humble litter of leather, not the royal velvet. Just a short time before, she had been used to being followed by a long train of servants clad in gold-embroidered coats; now she was accompanied by "a very small suite."[7] She had been forced to leave behind almost all of the ladies who had been her constant companions since childhood, including the Countess of Salisbury, who had first been appointed to her household fourteen years before. Equally fond of her charge, the countess had asked to be allowed to accompany her. Chapuys observed that it was "out of the question that this would be accepted; for in that case they would have no power over the Princess."[8] Now Mary had only the company of her stepmother's relations, the Sheltons, to look forward to. Anne incited her aunt, Lady Shelton, to treat the girl as harshly as she herself would have done in

her place. On one occasion, upon hearing that Mary had stubbornly refused to pay her half sister due reverence as princess, she ordered Lady
Shelton to box her ears "as the cursed bastard she was."⁹

From the moment that she arrived at Hatfield, Mary set out to be as
intransigent as possible. Clearly there under extreme duress, she refused to be bowed by the petty indignities and outright cruelty that she
suffered on a daily basis. Upon her arrival, she was ordered to go pay
her respects to the princess. She retorted that she "knew no other
Princess in England except herself, and that the daughter of Madame
de Penebroke [Pembroke] had no such title." The most that she would
concede was to call the child "sister," considering that her father had
acknowledged her to be his, just as she called the Duke of Richmond
"brother."¹⁰ Anne was furious when she heard this, and instructed her
aunt to take every opportunity to reinforce the girl's inferiority. Thus
Elizabeth was given a place of honor in the dining hall, while Mary
was forced to sit at a lower table. But the more they tried to subdue
her, the more she rebelled. Rather than suffering this indignity, she
took to eating her meals in her room. This was reported to Anne, who
duly ordered Mary back to the dining room. And so the war of attri-
tion continued.

Increasingly troubled by the news of his elder daughter's willful behavior, the king decided to go in person and force her to see the error
of her ways. In January 1534, barely a month after her removal there,
he prepared to make his way to Hatfield. But when his new wife heard
of this, she urged him not to bestow such an honor upon the ungrateful girl and suggested that he should send his chief minister, Thomas
Cromwell, instead. Chapuys surmised that her real motive for doing so
was that she was deeply insecure about the king's relationship with his
elder daughter and feared that "the beauty, virtue and prudence of the
Princess might assuage his wrath and cause him to treat her better." To
Anne, Mary embodied everything that threatened her own position.
She was a fervent Catholic, with the might of Spain and the Holy
Roman Empire behind her. Worse still, as Henry's firstborn and the
daughter of the people's beloved Catherine of Aragon, many looked
to her, and not Elizabeth, as the rightful heir to the throne. Anne knew

that many still persisted in referring to Mary by her former title. Chapuys made little secret of it, and later reported to his master, Charles V: "The King went lately to see his bastard daughter, who is twenty miles away, and the Princess with her."[11]

If Anne already despised Mary as a symbol of her husband's first marriage and of her own questionable legitimacy as queen, then how much more intense her hatred became when fuelled by the fierce maternal protectiveness toward her own daughter. From the moment of Elizabeth's birth, questions had been raised about her legitimacy, and she had been compared unfavorably to her elder half sister, whose blood was entirely royal, not half so. Anne therefore immediately resolved to do everything in her power to undermine Mary's position—if not destroy her altogether—by endeavoring to take everything that was hers and give it to Elizabeth. This even included her name: Anne had argued fiercely that her newborn daughter should be christened Mary, claiming that it was entirely appropriate considering that she had been born in the "chamber of the virgins" and on the eve of the Virgin Mary's nativity. Her true motive was clear, however: she wanted to give the people a new Mary so that the old one would be forgotten.[12] The king had refused, more sensitive than Anne to the hostility this would spark among his subjects.

Henry was more amenable to Anne's request regarding his impending visit to Hatfield and sent orders that Mary was to be kept from him. Cromwell accompanied his royal master so that he might speak to the girl on the king's behalf. But the Lady Mary would not be bowed by the most powerful man in her father's council. She stubbornly refused to acknowledge the new queen and her daughter, and simply pointed out that she had "already given a decided answer and it was labor wasted to press her."[13]

Mary's determined behavior succeeded in winning over the king, despite all the best efforts of his wife. As he was preparing to mount his horse prior to departing, she suddenly appeared on a terrace at the top of the house and knelt down in reverence to him. Disarmed by such a touching display of filial affection, the king bowed to her and put his hand to his hat. All of those present duly followed suit and "saluted her reverently with signs of good will and compassion." It was a small but significant victory for Mary, for it had proved that her father still loved her, despite all the persuasions of the new queen. However, by the time he returned to court, he seemed to have resumed his former stance and complained to Chapuys of Mary's obstinacy, "which came from her Spanish blood." But when he was reminded by another ambassador present that the girl had been well brought up, "tears came into his eyes, and he could not refrain from praising her."[14]

When she heard of this, Anne flew into a rage and was more determined than ever to remove this troublesome girl from the king's affections. It was even rumored that she planned to do away with her for good. "A gentleman told me yesterday that the earl of Northumberland told him that he knew for certain that she had determined to poison the Princess," reported Chapuys with some alarm. He was later told by another informant that Anne had boasted that she would "use her authority and put the said Princess to death, either by hunger or otherwise," adding that she "did not care even if she were burned alive for it after." Mary herself seemed to fear that this might happen, for she told Cromwell that her keepers "were deceived if they thought that bad treatment or rudeness, or even the chance of death, would make her change her determination."[15]

For all her defiance, those first few months at Hatfield were the most miserable of Mary's life. Separated from her mother, ostracized by her father, and forced to pay court to the baby sister who had supplanted her, she was beaten, scorned, and humiliated relentlessly by those around her. As well as depriving her of most of her servants, the king also drastically reduced her expenses, and within weeks she was described as being "nearly destitute of clothes and other necessaries."[16]

In the spring of 1534, Mary had to suffer a further torment when her despised stepmother paid a visit to Hatfield herself. Sensing that the girl's stubbornness would only increase the more cruelty she endured, Anne changed tack by trying to coax her into acknowledging Elizabeth's legitimacy and her own position as Henry's lawful wife and consort. Upon arriving at Hatfield, she sent a message to Mary inviting her to come and honor her as queen. By means of persuasion, she added that if Mary agreed to do so, she would make sure that she was "as well received [at court] as she could wish," and would regain the king's "good pleasure and favour" for her. Mary's reply was curt and defiant. She retorted that she "knew not of any other queen in England, than madam, her mother," but that if "Madam Boleyn" wished to intercede for her with the king, she would be most grateful.[17] Anne was outraged by such an insult and returned to court vowing to do Mary as much harm as she could.

But it seemed that the more Anne schemed, bullied, and cajoled, the firmer Mary's resolve was to defend her own legitimacy as princess and that of her mother as queen. In March 1534, when she refused to accompany Elizabeth on her removal to Eltham Palace in southeast London, "she was put by force by certain gentlemen into a litter with the aunt of the King's mistress, and thus compelled to make court to the said Bastard." As a further punishment, she had all of her royal jewels confiscated. Afraid of being secretly put to death at the orders of her stepmother, Mary was also taunted by Lady Shelton, who told her that the king "would make her lose her head for violating the laws of his realm."[18] Little wonder that Mary and her Spanish allies hatched plans for her to escape to the Continent—all of which came to nothing.

Meanwhile, Anne was overjoyed to see that her baby daughter was thriving. Elizabeth was, according to one contemporary, "as goodly a child as hath been seen."[19] As well as spending time with her at Hatfield, Anne and the king visited their daughter at Eltham, and Anne derived intense satisfaction from the obvious delight that her husband took in their daughter. "Her grace is much in the King's favour," observed one courtier who was present.[20] The royal couple could not be at Eltham for the whole of Elizabeth's visit, but Anne insured that the child would have a constant reminder of her mother by ordering her emblem to be installed in the stained-glass windows of the gallery where she played, at a cost of a shilling each.[21]

Elizabeth spent much of her early childhood moving from palace to palace. As well as Eltham, she and her household stayed at Hunsdon, Langley, the More, and Richmond. For the most part, her mother was obliged to keep track of her well-being by letter rather than visits, for she was greatly preoccupied with court affairs—not to mention the pressure to produce a male heir. Elizabeth was occasionally brought to see her at court, such as in the spring of 1535, but these visits were all too rare. Anne therefore corresponded regularly with Lady Bryan and ensured that her daughter had everything necessary for her proper upbringing.

In the autumn of that year, Anne and Margaret conferred over the weaning of Princess Elizabeth. Margaret had reported that the child was now old enough to drink from a cup and therefore no longer needed a wet nurse. This was exactly in accordance with the accepted wisdom of the time, which stated that children should be fully weaned after two years. They would then continue to be fed largely on milk, and only gradually would poultry and other white meats be introduced. Rich food was considered unsuitable, even for royal offspring, and the records show that Elizabeth enjoyed no exception to this rule. The matter was then referred to the king and the council, who agreed that "my lady princess" should be weaned "with all diligence."[22] Lady Bryan was put in charge of the task, but Anne sent her a private letter, possibly with her own maternal instructions about how it should be done.

Anne was no less assiduous in her instructions regarding the Lady Mary, but these were to insure that she was as miserable and uncomfortable as possible. To Mary's credit, her behavior was remarkably restrained. Rather than lashing out at her tormentors, she simply accepted their taunts and humiliations with the patience borne of a natural martyr. Like her mother, Mary was not accustomed to taking the easy path. If both women had acceded to Henry's demands, they would have enjoyed a far more comfortable life and might even have been accorded the honor that Anne of Cleves later enjoyed for being so pliable. But unlike hers, the consciences of Catherine and her daughter were utterly inflexible. They had an unshakeable belief both in the justice of their cause and in the Roman Catholic faith. It was simply a matter of weathering the storm until the king came to his senses.

After a time, Mary's dignified behavior won her the respect not just of the people but of her keepers at Hatfield. Even Lady Shelton softened toward her when she saw that she would not be bullied by cruelty. This earned her a severe reprimand from the Duke of Norfolk and Queen Anne's brother, George Boleyn, who admonished her for "behaving to the princess with too much respect and kindness, saying that she ought only to be treated as a bastard." Likewise, it was reported that "one of the principal officers of the Bastard [Elizabeth] has been removed because he showed some affection to the Princess and did her some service." When Mary was seen walking along a gallery at Hatfield, the "countrypeople . . . saluted her as their princess." She was kept strictly out of view as a result, her windows being "nailed up through which she might have been seen."[23]

It was clear that her keepers were weakening, however. In August 1534, the household prepared to move again, this time to Greenwich. As before, Mary was required to concede precedence to Elizabeth. She duly waited as the infant was carried into her litter and it had moved off. But when she had mounted her horse, the comptroller of the household whispered to her that she might "go before or after, as she pleased." She seized her chance to assert what she saw as her rightful position and "suddenly pushed forward," overtaking Elizabeth's litter and arriving at Greenwich about an hour before her. Later, when the party prepared to enter the barge to the palace, Mary "took care to secure the most honorable place."[24]

But this was at best a minor victory in a prolonged war that her stepmother and half sister looked set to win. Anne had evidently heard of Mary's small defiance on the road to Greenwich, and when her daughter's household removed the following month, she made sure that her stepdaughter would take the inferior place in the procession once more. Since Mary was already indisposed, this "increased her illness."[25]

Nevertheless, the ranks of Mary's supporters seemed to grow daily. Toward the end of October 1534, Anne paid a visit to Elizabeth and Mary at Richmond Palace, accompanied by the Dukes of Suffolk and Norfolk. She was dismayed to find that no sooner had she entered her daughter's apartments than the dukes excused themselves in order to pay court to Mary. Such an obvious show of allegiance to one of her greatest rivals by two of the most powerful courtiers (one of them her own uncle) was humiliating in the extreme. Little wonder that Anne came to believe that as long as Mary lived, she and her daughter would never be recognized as the true queen and heir. Mary must be dealt with or Anne must face her own downfall. "She is my death, or I am hers," she once lamented.[26]

To Anne's horror, Henry's attitude toward his elder daughter also began to change. There were even rumors that he would restore her to the succession and oust the "little bastard" Elizabeth. In October 1534, Chapuys excitedly related to Charles V that in an interview with Cromwell, the king had said that he "loved the Princess [Mary] more than the last born, and that he would not be long in giving clear evidence of it to the world." This can be given little credence, however, for just a few months later, another court dignitary reported that Henry had denounced Mary "for the bastard she is, and he will have no other heir but the Princess."[27] While it is difficult to determine the king's true feelings toward his two daughters, it is certain that he became more lenient toward the elder. During the first few months of 1535, he sent her gifts of money amounting to some "sixty or eighty ducats."[28]

If Henry's former love for Mary was being revived, then this was in direct proportion to the decline of his passion for Anne Boleyn. Anne sensed this and despised her stepdaughter all the more for it. The bitterness of her hatred toward Mary would have abated considerably if she had been able to produce an undisputed heir to obliterate the claim of this "cursed bastard" for good. The signs had initially been promising. Within months of Elizabeth's birth, Anne had fallen pregnant again. But in July 1534, all the renewed hope of an heir was crushed when the child was stillborn. Over the next twelve months, Anne was under increasing pressure. Her hold over the king seemed to be slipping each day as his disappointment with her grew and he sought consolation with other ladies at court. Worse still, he harbored a growing conviction that Anne's failure to produce a son was a sign from God, just as it had been with Catherine of Aragon—and look how that had ended.

In these dark months, Anne became ever more isolated, her sole comforts the little daughter whom she saw but rarely and the intimate circle of friends and admirers who surrounded her at court. In her desperation to regain Henry's affection (and thereby her former power), she took to following him about "like a dog its master," as one courtier observed mockingly.[29] How different this was from the *belle dame sans merci* that she had played to perfection for seven years in order to ensnare her royal lover. Neither did she have the beguiling looks that had once so bewitched the king. The considerable stress under which she had labored as she tried in vain to claw back her power at court after Elizabeth's birth had started to show on her face. A portrait of around 1535 forms a startling contrast to that painted just two years before, when Anne was at the height of her powers. Her famously seductive eyes have become sunken and tired; her high cheekbones have disappeared beneath skin that is beginning to sag; and her pretty, smiling lips that the king once longed to kiss have grown thin and pinched with disappointment. That same year, in a dispatch to the Doge and Senate, the Venetian ambassador described the Queen as "that thin old woman." She would then have been thirty-five years old at most.

Painfully aware that she could no longer rely upon her feminine charms to maintain Henry's affection, Anne switched her attention to her daughter, Elizabeth. The king had already acknowledged her as his heir and seemed delighted with this pretty, precocious little girl. But Anne knew that this was not enough: his favor was notoriously fickle, and Elizabeth's place in it was as fragile as her own. She therefore had to strengthen her daughter's position. If it was against convention for female heirs to rule, then they could at least prove useful in the political power games of Europe by being married off to foreign princes. England and France had long been on hostile terms, fuelled by frequent bouts of war. But there was currently a truce between them, and in July 1535, Henry's great rival, Francis I, finally agreed to enter negotiations for a marriage between Elizabeth and his third son,

Charles. Anne was no doubt instrumental in bringing this about, for she still felt a great deal of affinity with the country in which she had spent so much of her youth. What was more, allying her daughter with a scion of the ancient Valois lineage would inject some much-needed royal blood into Elizabeth's own children.

With her daughter's prospects apparently much improved, Anne accompanied her husband on their customary summer progress— a tour of selected parts of the kingdom—in 1535. In September they honored Sir John Seymour with a visit to Wolf Hall in Wiltshire. The Seymours were of an ancient lineage that stretched back to the time of William the Conqueror, whom they were said to have accompanied to England in 1066. Sir John's wife, Margery, was of equally distinguished birth, being descended from Edward III. Their sons, Edward and Thomas, were already carving out careers for themselves at court. But it was their eldest daughter, Jane, whom the king was most particularly eager to see.

Jane Seymour was one of the Queen's ladies-in-waiting and had first appeared at court around 1529. It is likely that her introduction had been thanks to Sir Francis Bryan, son of Princess Elizabeth's Lady Mistress, who was connected to the Seymours by marriage. She had been appointed a lady-in-waiting to Queen Catherine of Aragon, whom she greatly admired, and had remained in her service until the latter had been exiled from court two years later. She had then been transferred to the service of Catherine's archenemy, Anne Boleyn. Henry had become acquainted with her immediately, and she had been included in the list of Anne's ladies who received a gift from him at Christmas 1533. However, it had apparently not been until late in 1534—by which time Henry was beginning to tire of his tempestuous second wife—that Jane had caught his eye. She was then about twenty-seven years old, some seven or eight years younger than her royal mistress, but still a late age to remain single at a time when most girls were married off at fifteen or sixteen.

That Mistress Seymour had attracted the king's attention was a source of some astonishment to contemporaries at court. True, she had good breeding to recommend her, but she seemed to have little else. She was plain and sallow faced, "so fair that one would call her rather pale than otherwise."[30] A portrait painted of her around 1536 shows her to have had a large, plump face with a double chin. Her eyes are small and beady, her lips thin and closely compressed, and she wears a cold, detached expression. One onlooker at court dismissed her as being "of middle stature and no great beauty."[31] Neither did Jane have the sparkling wit and intelligence of her predecessor; in fact, she was barely literate. Even Chapuys, who was predisposed to favor this rival to the hated "Concubine," was at a loss to explain what the king saw in her. He could conclude only that she must have a fine "enigme," meaning "riddle" or "secret," which in Tudor times referred to the female genitalia.[32]

But this archetypal plain Jane was exactly what the king needed. Her looks were mirrored by her demeanor. While Anne was tempestuous and flirtatious, Jane appeared meek, docile, and placid. She was so calm and quiet that scarcely any of her words are recorded in contemporary accounts—in stark contrast to the outspoken Queen Anne, who provided copious fodder for ambassadors' scandalized letters home. While Anne created dissent and faction at court, Jane was renowned as a peacemaker. "As gentle a lady as ever I knew," wrote one courtier, and Henry himself claimed that she was "gentle and inclined to peace."[33] She also set great store by her virtue and was unquestionably chaste. Thomas Cromwell described her as "the most virtuous lady and veriest gentlewoman that liveth," and many others echoed his views.[34] The quality that may have appealed most to Henry, however, was her submissiveness. Adopting the motto "Bound to obey and to serve," Jane carried it out to the letter. She had none of Anne's feistiness and independence; for her, Henry's will was all that mattered. Only in safeguarding her chastity did she defy him, but in doing so she earned even more of his respect. Little wonder that one courtier observed: "the King hath come out of hell into heaven for the gentleness in this, and the cursedness and unhappiness in the other."[35]

For all her apparent mildness and passivity, Jane was every bit as ambitious as her brothers and had a streak of cold ruthlessness that gave her little sympathy for Anne. She had long been a supporter of Catherine of Aragon, admiring her queenly decorum and sharing her religious faith. Her loyalty to the fallen queen had not ceased when she had been transferred to Anne Boleyn's service. Indeed, she had evidently resolved to do what she could to bring her new mistress down. Coached by her brothers, Jane played her part in court intrigues to perfection, quietly drumming up support for Anne's enemies while maintaining a veneer of quiet detachment. Almost from the moment that the king had started paying Jane attention, she had sent messages to Catherine's daughter, Mary, urging her to have courage because her troubles would soon be at an end.

The courtship between Henry VIII and Jane Seymour had been conducted discreetly to begin with. However, it had not escaped the notice of the ever-vigilant Imperial ambassador, Chapuys, who in October 1534 noted that the king had become "attached" to "a young lady" of the court whose credit was increasing as that of Queen Anne declined.[36]

The king's growing infatuation with Jane could not have come at a worse time for her royal mistress. Rumors had begun to circulate about the nature of Anne's relationship with certain young men at court, including her own brother. Henry himself had begun to question her purity soon after their marriage, complaining that she had seemed more experienced than a virgin ought to be. He was apparently at a loss to explain why he had ever been so attracted to her and even whispered to one confidant that he thought it might have been witchcraft.[37]

But just as Anne's situation appeared desperate, something happened that looked set to secure her future with the king forever. During that summer progress of 1535, she fell pregnant once more. Two years had passed since the birth of Elizabeth and a year since that of her stillborn child. Surely now it would be a case of third time lucky? Her very survival depended upon it. Although delighted at his wife's condition and outwardly solicitous of her every need, Henry could not disguise the distaste that he had come to feel for her, and courtiers noticed that in private he "shrank from her."[38]

All of this would be put aside if Anne gave Henry a son. At a stroke, it would secure the king's lasting favor and would finally legitimize her in the eyes of the world. It is an indication of how much Anne's confidence had been damaged by the uncertainties and betrayals of the past two years that rather than triumphing in her condition, she plunged into a depression, plagued with an intense fear of what might happen if she failed. She was also tormented by jealousy, knowing full well that her husband was pursuing Jane Seymour even as she herself was suffering the sickness and fatigue of early pregnancy.

But as the new year arrived, things turned more decisively in Anne's favor. On January 8, 1536, Catherine of Aragon, the woman whom most of England still regarded as the rightful queen, died at Kimbolton Castle in Cambridgeshire. Her daughter was devastated. The king, perhaps encouraged by his wife, had refused Mary's heartfelt pleas to be allowed to go to her mother as she lay dying. Now she would never see her again. Mary's hatred of Anne was more implacable than ever.

By contrast, Henry and Anne were overjoyed, both relieved that this enduring challenge to the legitimacy of their marriage had finally disappeared for good. For one who set so much store by these things, it was surely a sign that God had not forsaken the king after all. He immediately ordered great festivities at court. A delighted young Elizabeth was summoned from Hatfield and arrived to behold her father, "clad all over in yellow from top to toe," in great high spirits. The princess was immediately conducted to Mass, accompanied by "trumpets and other great triumphs."[39] After giving thanks to God, a sumptuous banquet was staged, and Elizabeth, who had so recently been weaned, might have enjoyed her first taste of the rich foods of court: spit-roasted boar, peacock, and swan, venison pies, sweetmeats, marchpane, and spiced fruitcake. The bland milk and white meat of her diet at Hatfield would never seem quite the same again.

When the feasting was over, the king processed into an adjoining chamber, where dancing was already under way, and "there did several things like one transported with joy."[40] Anne looked on in triumph as he lifted their daughter in his arms and proudly paraded her in front of the whole court. The pride she felt in Elizabeth was matched only by the hope she felt for the child that was now growing inside her.

But if Anne had learned anything from her years at Henry's court, it was how quickly things could change. Barely three weeks after the celebrations that seemed to crown her triumph as Henry's queen and the mother of his heir, disaster struck. On January 29, Catherine of Aragon was laid to rest at Peterborough Cathedral. What should have been a joyful day for Anne was marred when she discovered her husband cavorting with Jane Seymour. Her fury was immediate and uncontrollable. She raged and lashed out in an increasing frenzy as her shocked attendants looked on, fearful for her unborn child. They were right to be afraid. That evening, overcome with fevered exhaustion, Anne miscarried. This time, God had surely shown his hand, for the fifteen-week-old fetus had all the appearance of being a boy. Chapuys was quick to convey the news to Charles V. With barely suppressed satisfaction, he told his master: "the Concubine had an abortion which seemed to be a male child which she had not borne three and a half months, and on which the King has shown great distress." His conclusion was brutal but accurate: "She has miscarried of her saviour."[41]

According to Chapuys, Anne immediately put the blame on her uncle, the Duke of Norfolk, for deliberately shocking her with news that the king had suffered a bad fall while jousting. But this accident had not proved too serious and had in any case happened six days before the miscarriage. Furthermore, news of it had been broken to Anne "in a way that she should not be alarmed or attach much importance to it."[42] Chapuys preferred to put the blame on Anne's "incapacity to bear children," a view that was shared by many at court. It is possible that she had gynecological problems, given her mother's many miscarriages and stillbirths, and her sister's history. But it seems at least equally likely that the miscarriage had happened as a result of the acute stress under which she had labored throughout this difficult pregnancy, together with the constant dread that Henry would find a means to get rid of her. Chapuys noted that many at court attributed it to "a fear that the King would treat her like the late Queen."[43]

Whatever the cause, things now began to unravel rapidly for Anne. This second miscarriage convinced Henry that their marriage had offended God, and that for as long as it continued, He would deny him the male heir that he so craved. Chapuys was perhaps exaggerating when he claimed that "for more than three months this King has not spoken ten times to the Concubine . . . when formerly he could not leave her for an hour."[44] But there was no denying that he harbored a growing resentment toward his second wife and treated her with barely concealed distaste.

The king now switched his attentions firmly to his new mistress, showering her with "great presents." In plotting to ensnare the king's affections for good, Jane Seymour employed some of the same tactics that she had seen Anne Boleyn put to such powerful effect. She knew that a mistress could become a queen, and was determined to follow suit. For a start, she refused to yield her virginity and met all of Henry's advances with a show of maidenly modesty. When in April 1536 he sent her a purse of money with an accompanying declaration of love, Jane reverently kissed the letter before sending it back un-opened, begging the king to consider that there was "no treasure in the world that she valued as much as her honor, and on no account would she lose it, even if she were to die a thousand deaths." She added cun-ningly that if the king wished to send her such a present in future, then he should wait "for such a time as God would be pleased to send her some advantageous marriage."[45]

If Henry experienced an uneasy feeling of déjà vu, he did not show it. Jane's ploy worked just as successfully as Anne's had done. Before long, the king's passion for her was known throughout the court. In a striking repetition of history, courtiers now flocked to Jane in the hope of advancement, just as they had to Anne. Henry appointed rooms for Jane next to his own in Greenwich Palace, and also installed her brother Edward and sister-in-law Anne there so that they could act as chaperones when the couple met.

Meanwhile, Queen Anne was forced to endure the humiliation of seeing gifts and love messages arriving for her lady-in-waiting. Occa-sionally it all became too much, and she lashed out at the placid crea-ture with slaps and curses. Years later, her daughter, Elizabeth, would use similar treatment toward her ladies when they provoked her.

It was upon Elizabeth that Anne now lavished all her affection, perhaps seeing her little daughter as the only friend she had left in the world. During those bleak early months of 1536, while her enemies at court were plotting her downfall and the king was seeking solace with his new mistress, Anne turned her back on all of it and busied herself with ordering pretty new clothes for her infant daughter. In April she was overjoyed when the king agreed that the child could visit her at Greenwich, where the court was then residing. Anne sought Elizabeth's company a great deal during this time, playing with her and dressing her in new velvet frocks and embroidered satin caps.

But all the while, the king's chief minister, Thomas Cromwell, was quietly gathering evidence that would rid his master of Anne's irksome presence for good. The Queen had delighted in surrounding herself with lively, flirtatious, and attractive courtiers—men such as Henry Norris, who was Cromwell's main rival; Mark Smeaton, a court musician; and her brother, George Boleyn, with whom she had always enjoyed a close relationship. Her flirtations with these men were almost certainly harmless: Anne had far too much to lose to risk adultery. Besides, her ability to keep Henry at bay for the seven years of their courtship had proved that she was not lacking in self-control. But Cromwell had the means he sought to bring her down, and he wasted no time in collecting innocent tales that could be twisted into damning evidence.

Anne, preoccupied with her daughter at Greenwich, knew nothing of the horror that was about to unfold, and even the king was kept in ignorance until Cromwell judged that he had a suitably compelling case to take to him. Finally, on the first day of May, the minister confronted Henry with the evidence. Outraged, dismayed, but—sadly for Anne—not disbelieving, he ordered that his wife's alleged lovers be thrown into the Tower. The news spread like wildfire, and all too soon it had reached the ears of the Queen herself.

Gathering her daughter in her arms, she ran to the king, desperate to convince him of her innocence. The scene was witnessed by Alexander Ales, a Scottish theologian and protégé of Cromwell, who was then visiting court. He later recounted what he had seen in a letter to Elizabeth, written soon after she had ascended the throne. "Never shall I forget the sorrow which I felt when I saw the most serene Queen, your most religious mother, carrying you, still a little baby, in her arms and entreating the most serene King, your father, in Greenwich Palace, from the open window of which he was looking into the courtyard, when she brought you to him." Ales had not been close enough to hear what had passed between them, but he judged that from "the faces and gestures of the speakers," it was clear that an argument had ensued and that the king had been very angry.[46]

Perhaps Anne had seized upon Elizabeth as being the best means of persuading her husband that she was innocent. The charges had hinted that the child might not be his, but with her fiery red hair and long, straight nose, she was the very image of Henry. If Anne had used her daughter in a last, desperate attempt to save her own life, it was in vain. The king remained steadfast.

Following Henry's discovery of his wife's alleged adultery, events at court moved with bewildering speed. On May 2, Anne was arrested and taken to the Tower. Her trial took place a little over two weeks later, and she faced a string of lurid and scandalous charges. Her crime, they said, was not just adultery, but incest and perversion. Driven by her "frail and carnal lust," she had kissed her brother by "inserting her tongue in his mouth, and he in hers," and had incited others in her entourage to yield to her "vile provocations."[47] She had taken Henry Norris to her bed just six weeks after giving birth to Elizabeth. In vain, Archbishop Cranmer defended her to the king, telling him that he could not believe her guilty of the charges against her because "I had never better opinion of woman."[48] His was virtually a lone voice amidst the growing tide of accusations.

As the details of her supposed crimes grew ever more explicit, Anne remained impassive. When the time came for her to speak, however, she presented a spirited and articulate defense, giving "so wise and discreet aunswers to all thinges layde against her, excusinge herselfe with her wordes so clearlie as thoughe she had never bene faultie to the same."[49] Her daughter would inherit this talent for oration and use it to much greater effect than Anne was able to on this occasion. In the event, it did nothing to move the hearts either of her accusers or of the king himself, who, upon hearing of her bravery, remarked: "She hath a stout heart, but she shall pay for it!" According to Chapuys, even if his wife had been found innocent, he had already resolved to abandon her.[50]

The jury returned the verdict that was expected of them: Anne was convicted of high treason and sentenced to death. When she was escorted back to her rooms in the Tower, she gave way to hysteria, telling the lieutenant, Sir William Kingston: "I heard say the excutor was very gud, and I have a lytel neck," before putting her hands around it and "lawynge [laughing] hartelye." Aghast, Sir William exclaimed that "this lady hasse mech joy and plesure in dethe."[51] The night before her execution, Anne chattered and joked endlessly, telling her astonished companions that it would not be hard for her enemies to think of a nickname for her when she was dead, for they could call her "la Royne Anne sans teste [tête]." She then "laughed heartily, though she knew she must die the next day."[52]

On May 19, 1536, as Anne stood on the scaffold, stripped of her title as queen and all the honors that had accompanied it, she gave a last, dignified speech to the hushed crowds that had gathered at the Tower. Rather than bemoaning her fate and rejecting the charges against her, she was full of praise for the king, lauding him as "one of the best princes on the face of the earth." Such a calm acceptance of her impending death could hardly have been expected of a woman whose frequent bursts of temper had become notorious at court, and who had often complained bitterly to her husband about much more trivial matters than those of which she now stood accused. Surely now, with the sword about to strike, she had nothing to lose in railing against the man who had so easily accepted the trumped-up charges against her in order to rid himself of her for good? That she chose rather to praise him could have been to protect those whom she left behind—none more so than Elizabeth. She knew that things already looked bleak for her daughter, who had been rendered illegitimate by the dissolving of Henry's marriage to Anne. She might therefore have resolved to do anything she could to soften the king's heart toward herself, and thereby their child.

Anne made no recorded mention of Elizabeth during her impris-onment in the Tower, but there is evidence to suggest that she took great care to protect her daughter's future, even as she saw her own crumbling into the dust. In 1535, when her favor with the king was de-clining rapidly, she had written a conciliatory letter to her stepdaugh-ter, Mary. Perhaps sensing which way the succession was turning, she wished to ensure that her cruel treatment of the young woman would not prejudice her against her half sister, Elizabeth.[53] Then in late April 1536, just a few days before her arrest, she had had an earnest discus-sion with her chaplain, Matthew Parker. According to Parker, she com-mended her daughter to his spiritual care and shared her hopes for Elizabeth's education. It may be that he subsequently exaggerated the importance of this conversation when it turned out to be their last. But the fact that Parker was one of the most fervent reformers at court and Anne herself had shown sympathies in that direction suggests that she wished her daughter to follow the same path. In the event, Elizabeth's intellectual and religious upbringing would be assigned to the care of others, but when queen, she would appoint Parker as her first Arch-bishop of Canterbury. Her religious leanings would prove her to be very much her mother's daughter.

Anne would have a far greater influence upon her daughter than has long been supposed. Even at her young age, Elizabeth already re-sembled her, and she would grow to do so more strikingly as the years passed. She would also inherit some of her mother's personal traits, notably tenacity, self-discipline, and charisma. Equally, there would be flashes of Anne's cruelty and vindictiveness. But above all, it would be the example provided by Anne's life—and in particular its end—that would prove her greatest legacy to Elizabeth. From this, her daughter learned not to trust expressions of love and devotion; she learned to guard her reputation fiercely; and she learned to be a self-reliant, polit-ical pragmatist. Anne had had qualities that would have made her a great queen, but she had also had a number of fatal flaws. It was in learning from both that Elizabeth was able to become the queen that her mother was never able to be.

Her final speech over, Anne knelt on the scaffold with great composure and commended her soul to God. A highly skilled executioner had been brought over from Calais and used a sword in the French fashion, rather than the traditional axe—the only mercy that Henry showed toward his estranged wife. With a clean strike, Anne's head was severed from her body. The sombre crowd looked on aghast as her eyes and lips continued to move, as if in silent prayer, when the head was held aloft. She was apparently as bewitching in death as she had been in life. When the spectators had finally dispersed, Anne's weeping ladies sought in vain for a coffin in which to lay their mistress's body. In one final indignity, they were compelled to use an old arrow chest, and it was in this that Henry's second queen was laid to rest in the Tower chapel of St. Peter ad Vincula.

For her daughter, Elizabeth, life would never be the same again.

CHAPTER 3

The Royal Nursery

The blood of her mother, soon put to death by the King,
sprinkled even to her cradle with the blot of bastardy.[1]

When the sword struck off her mother's head, Elizabeth was just two
years and eight months old. Blissfully unaware of the ghastly event
that had just taken place at the Tower, she was in the company of Lady
Bryan at Greenwich Palace, playing or learning her letters like any normal royal offspring. But it soon became obvious, even to a child of Elizabeth's tender years, that something was badly wrong.

Her father had ordered that she be kept to her rooms in the days
immediately following Anne's execution. Elizabeth's Master of the
Horse, Sir John Shelton, wrote to Cromwell upon receiving his instructions: "I perceive by your letter that my lady Elizabeth shall keep
her chamber and not come abroad, and that I shall provide for her as I
did for my Lady Mary when she kept her chamber."[2] Although the

king may have wished to shield his youngest daughter from the scandal surrounding her mother's trial and execution, it seems more likely that it was because she was an uncomfortable reminder of a painful episode that Henry would rather forget.

Lady Bryan, who was no longer a second mother to her but the only mother the child had, stayed with her at all times. Did she tell Elizabeth the truth during this time, or did she judge it best to wait until the girl was of an age to better understand what had happened? The surviving evidence provides little clue. Any conjecture must rest upon what is known of Lady Bryan's character and her approach to her duties. She was certainly competent: Henry VIII would not have tolerated anything less than excellence in the woman to whom he had entrusted the upbringing of his heirs. Her letters to Anne Boleyn also suggest that she was a warm and caring person, with a genuine affection for the precocious young girl in her charge.

Lady Bryan's correspondence with the court, in particular the letters that she exchanged with the king's chief minister, Thomas Cromwell, infer a highly organized, no-nonsense approach to her duties, and a formidable will that would brook no challenge to her authority from the rest of the household. If, therefore, Lady Bryan had decided that it was in Elizabeth's best interests to shield her from the truth about her mother during the immediate aftermath of the execution, then it can be reasonably assumed that the other household members would have been sworn to secrecy. The fact that the child had to ask why her status had changed suggests that this was the case. "Why Governor," she demanded of the hapless Sir John Shelton, "how happs it yesterday Lady Princess and today but Lady Elsabeth?"[3] Quite how long Lady Margaret chose, or was able, to maintain the pretense cannot be known. She at least seems to have done so during those turbulent days at Greenwich.

Henry, meanwhile, had moved to Hampton Court, and there is an account—almost certainly apocryphal—that Lady Bryan had initially taken Elizabeth there so that she might be comforted by her father. The story goes that when she approached the king with her young charge in her arms—just as Anne Boleyn had done shortly before her death—and asked him if he wished to see his daughter, he bellowed: "My daughter? My daughter? You old devil, you witch, don't dare to speak to me!" Terrified by this outburst, Lady Bryan apparently fled with Elizabeth and went in search of Cromwell, who counselled her to take her charge to Hatfield until the king's anger had abated.[4] Although this is a touching tale, it is not substantiated by any of the contemporary accounts, which all attest that Elizabeth was moved directly to Hunsdon a few days after her mother's death.

In the weeks and months after her household's removal to the country, it seems likely that the inquisitive and precocious young Elizabeth picked up scraps of whispered conversations from members of her household, gradually piecing them together until the full, horrific picture of her mother's death emerged before her eyes. Not all of her household would have shared Lady Bryan's sensitivity. Indeed, they may have felt duty bound to prevent the girl from falling into the same disgrace as her mother by revealing the latter's fate in all its grisly horror. The Tudors were not squeamish about death. Alexander Ales, the Protestant refugee who had described Anne's final plea to her husband, also told Elizabeth of his premonition of her mother's death in what today seems an astonishingly insensitive manner. "On the day upon which the Queen was beheaded," he wrote, "at sunrise, between 2 and 3 o'clock, there was revealed to me (whether I was asleep or awake I know not), the Queen's neck after her head had been cut off, and this so plainly that I could count the nerves, the veins and the arteries."[5] Elizabeth was just shy of her twenty-sixth birthday when she read this account, by which time she had experienced enough violence and bloodshed to prevent any undue delicacy.

With no contemporary evidence to tell us of Elizabeth's reaction when she eventually learned the truth, it is tempting to layer modern perceptions onto the mind of a girl who was at an impressionable age. Historians and psychologists alike have speculated that such a traumatic realization must have had a deep and enduring impact upon her outlook and behavior. "The harm done to Elizabeth as a small child resulted in an irremediable condition of nervous shock . . . In the fatally vulnerable years she had learned to connect the idea of sexual inter-

course with terror and death."[6] Another argues that by depriving her
of a female role model in her formative years, Anne's death inhibited
Elizabeth's feminine attributes, leading to "a lively dread of pregnancy
and childbirth."[7] In short, the execution of her mother caused Elizabeth to cling to that most famous trait: her virginity.

Certainly the loss of her mother must have had a significant impact
upon the young Elizabeth's outlook and development. But this would
have been lessened considerably by the fact that for almost all of her
early childhood, Anne had been a distant figure, making only occasional visits to her daughter. What's more, Elizabeth had followed the
traditional path of royal children and had been set up in her own
household, thus fostering a sense of separation, even independence,
from her parents at court. While she would have missed the steady
supply of new clothes and other gifts, it is unlikely that she felt bereft
at her mother's now permanent absence. Indeed, at her tender age, she
would probably have felt a greater wrench if one of her nursemaids or
governesses had been suddenly taken away from her.

The practical impact of her mother's death upon Elizabeth was
more immediate. Her father had taken Jane Seymour as his third wife
within days of Anne's execution, and in her apparently placid good nature and sweet submissiveness, he hoped to lose himself—and gain an
heir. His marriage to Anne had been annulled shortly before her execution on the grounds that his earlier relations with her sister, Mary,
had rendered it invalid. This meant that Elizabeth was now illegitimate. Worse still, rumors about her parentage were now reaching
fever pitch, and even the council declared that it believed she was the
offspring of an incestuous affair between Anne and her brother,
George Boleyn. Others, such as the outspoken Imperial ambassador,
Chapuys, thought that another of Anne's alleged lovers, Henry Norris,
was the more likely father.

In the first week of July, Parliament repealed the statute declaring
Elizabeth the king's lawful heir, and formally pronounced her illegitimate. "In what ill case the young Lady Elizabeth now was, any one
may guess: she being degraded into a meaner condition upon the
Queen her mother's divorce and death." So observed the seventeenth-

century chronicler John Strype.[8] As her question to Sir John Shelton demonstrates, Elizabeth was quick to grasp the change in her situation. Being but a lady when before she had been a princess was galling, even to one of her tender years. If she had understood the meaning and implications of her illegitimacy at the time, she would have been even more miserable. She evidently soon came to appreciate them, however, and it was an issue about which she would be forever sensitive.

In the immediate aftermath of Anne Boleyn's execution, things had looked much more promising for Elizabeth's half sister, Mary. While Elizabeth had been kept to her rooms, Mary had been summoned to court by her father, who had "made much of her" and given her "many jewels belonging to the unjust Queen."[9] But if she thought that she would now be automatically restored to her place in the succession and given the title of princess, she was mistaken. Henry had no intention of revoking the annulment of his marriage to Catherine of Aragon after all the religious and political upheaval that it had caused, and he persisted in trying to force Mary to accept her illegitimate status.

At first Mary refused to give in, determined to honor her mother's memory. A frustrated Cromwell admonished her for such unforgivable defiance: "To be plain with you, I think you the most obstinate woman that ever was." He warned that if she did not conform to her father's will, "I will never think you other than the most ungrateful, unnatural, and most obstinate person living."[10] The Dukes of Norfolk and Suffolk were sent to demand her submission, and when she continued to resist, they brutally told her that "if she was their daughter, they would beat her and knock her head so violently against the wall that they would make it as soft as baked apples."[11]

Resolute and principled though she was, Mary was not entirely devoid of political awareness, and, submitting to the persuasions of Chapuys—who urged her to push home her advantage and be restored to the king's favor—she reluctantly agreed to acknowledge that her parents' marriage had been invalid and her father was the supreme head of the Church. She was said to be deeply grieved at what she saw as a betrayal of her late mother, but the letter that she subsequently wrote to her father was full of humble deference and begged his forgiveness for her intractability. She even apologized for her refusal to acknowledge her half sister's status as heir. "Concerning the Princess (so, I think, I must call her yet, for I would be loth to offend), I offered at her entry to that name and honour to call her sister, but it was refused unless I would also add the other title unto it; which I denied not then more obstinately than I am now sorry for it, for that I did therein offend my most gracious father and his just laws; and now that you think it meet, I shall never call her by other name than sister."[12]

It seems that these were more than mere words. Mary's attitude toward her half sister had softened considerably upon the death of Anne Boleyn. Having been restored to the king's favor, she was given back her household. This was still a joint one with Elizabeth, but it was now reorganized to reflect Mary's new status. Although both girls were illegitimate, Mary naturally took precedence as the elder. The younger sister's retinue was reduced accordingly, and Mary could now "detach" her chamber staff if she went to court or visited another royal residence on her own. All of this considered, Mary could afford to be magnanimous to the child whose world, like her own some three years before, had suddenly fallen apart. In a startling about-face, whereas previously she had not had a good word to say about the "little bastard," now she commended her precociousness to their father. "My sister Elizabeth is well," she wrote to him in July 1536, "and such a child toward, as I doubt not but your Highness shall have cause to rejoice of in time coming."[13]

It appeared that Mary would be rewarded for her newfound obedience. A few days after sending this letter, it was reported that she was "every day better treated, and was never at greater liberty or more honourably served than now . . . she has plenty of company, even of the followers of the little Bastard, who will henceforth pay her Court."[14] But if her efforts toward Elizabeth were merely an attempt to persuade the king to restore her to the succession, it failed. Later that month, he formally declared both of his daughters illegitimate in favor of the children from his new marriage to Jane Seymour.[15]

Although this was a bitter blow to Mary, it paved the way for a genuine rapprochement with Elizabeth. United in their illegitimacy, the two daughters were now on an equal footing for the first time, and this improved their relationship considerably. Furthermore, Mary had always been fond of children and pitied the little girl's motherless state. She therefore resolved to show Elizabeth the love and affection of a protective older sister.

By contrast, the usually unflappable Lady Bryan appeared to have been thrown into some confusion during the weeks after Anne Boleyn's death. Whereas before, she had received regular instructions from the Queen and had been able to seek her advice or approval on all aspects of Elizabeth's upbringing, now she was left, as she put it, "succourless . . . as a redeless creature."[16] Her sense of isolation was increased by the loss of her second husband, David Soche, who died about a month after the disgraced queen. There were signs of dissension within Elizabeth's household as her Lady Mistress's authority was challenged by Sir John Shelton. Like Margaret, Sir John was both related to and had been appointed by the late queen Anne, and therefore no doubt saw himself as at least equal to the Lady Mistress, if not superior by virtue of his sex, even though his position was in theory subordinate to hers. Worse still, nobody at court seemed to have given any consideration to even the basic needs of the king's younger daughter.

Their negligence, coupled with her own distress, prompted Lady Bryan to show her mettle. At the beginning of July 1536, just a few weeks after the household's departure from court, she wrote a furious letter to Thomas Cromwell, complaining bitterly at not being informed of Elizabeth's new status and being left to get on with the important business of managing the young girl's affairs without any guidance from the council. Lady Bryan was a stickler for matters of propriety, and this was something she was not prepared to tolerate. "Now as my Lady Elizabeth is put from that degree she was in, and what degree she is at now, I know not but by hearsay," she began. "I know not how to order her or myself, or her women or grooms." With more than eight years' experience in charge of the king's children, as well as considerably more from raising her own, Lady Margaret cannot have been so at a loss as to how to proceed. Rather, one suspects that this was prompted by resentment at being kept in the dark about Elizabeth's situation, when she was used to being consulted on all matters relating to the child's upbringing.

Evidently intending to shock or shame Cromwell into action, Lady Bryan then turned her attention to the matter that she thought deserved most urgent redress: the fact that her young mistress had barely any clothes to wear. At almost three years old, Elizabeth would have been growing fast, and now more than ever she needed a regular supply of new dresses, shoes, and other accessories. In the past, this need had been more than adequately supplied by the regular parcels sent by her mother. But now this was apparently yet another matter that the council had overlooked. "I beg you to be good to her and hers, and that she may have raiment," Lady Bryan implored, "for she has neither gown nor kirtle nor petticoat, nor linen for smocks, nor kerchiefs, sleeves, rails [nightdresses], body stitchets [corsets], handkerchiefs, mufflers, nor begins [nightcaps]. All thys her Graces mostake I have dreven [driven] of as long as I can, that, be my trothe, I cannot drive it any lenger."[17]

The contemporary sources tell us that Anne had ordered a whole suite of new clothes for her daughter in the months leading up to her death. Lady Bryan's letter was written just six weeks later. Had Elizabeth really grown so fast that she had exhausted all of her considerable supply of clothes? It seems more likely that her Lady Mistress was exaggerating in order to make a point. She was desperate to insure that Elizabeth was not permanently neglected by the king and his council. At heart a kind woman, she was no doubt motivated at least partly by a genuine concern for the child's welfare, conscious of her role as surrogate mother. But she was also ambitious and had no desire to become sidelined in a court where she was used to enjoying some status as chief custodian of the royal heirs. Her letter to Cromwell should therefore be taken at a great deal more than face value; it was a carefully crafted attempt to restore prestige and attention to Elizabeth's household, and thereby to her own position as its head.

In order to push her point home still further, Lady Margaret went on to describe the shameful disorder into which other household affairs had fallen, particularly those relating to mealtimes. It was in this matter that her archenemy, Sir John Shelton, had been causing trouble. He had insisted that Elizabeth should dine in state, rather than privately in her rooms. Perhaps he wished to uphold this Boleyn descendant's regal status for as long as possible, and no doubt also to ingratiate himself with the infant who, in the fickle world of Tudor politics, might still be restored to the succession. "Mr Shelton would have my Lady Elizabeth to dine and sup every day at the board of estate," she exclaimed. "It is not meet for a child of her age to keep such rule. If she do, I dare not take it upon me to keep her Grace in health; for she will see divers meats, fruits, and wine, that it will be hard for me to refrain her from." Lady Margaret thought it would be much more appropriate—and economical—if Elizabeth took her meals in her rooms, away from the temptations of grown-up food, and she urged Cromwell to see that this was ordered.

Lady Bryan ended the letter on a softer note, betraying her fondness for the motherless child in her care. "My Lady has great pain with her teeth," she wrote, "which come very slowly," and admitted: "This makes me give her her own way more than I would." She went on to praise Elizabeth's character and precociousness, saying that she was "as toward a child and as gentle of conditions as ever I knew in my life." She added a hope that the child might be "set abroad" (in other words, go out in public) on special occasions, and predicted "that she shall so do as shall be to the King's honour and hers." Lady Bryan evidently hoped that Cromwell would relay this to the king and thereby soften his heart a little toward his younger daughter. She signed off with an apology for her "boldness in writing thus," which was a little insincere, given everything that had gone before.[18]

Lady Bryan's letter had the effect that she desired—at least in some respects. The records contain no further pleas for clothing, which suggests that Cromwell had put this to rights. He also supported Lady Margaret's request for Elizabeth to eat in her rooms and wrote to Sir John Shelton accordingly. Cromwell no doubt found it distasteful to involve himself in such petty household squabbles, but he knew that Lady Bryan was not a woman to be gainsaid.

As the upheaval of Anne Boleyn's execution began to subside and the petty disputes in Elizabeth's household were resolved, the child's upbringing settled into a more ordered and comfortable pattern. Lady Bryan supervised the household with assiduity and efficiency, quick to spot anything that detracted from the accustomed routine. Her letter to Cromwell had proved her to be a stickler for etiquette, which was understandable when one considers that it was an important part of her job to ensure that Henry VIII's children were brought up with beautiful manners. The prolific nineteenth-century historian Agnes Strickland credits Lady Bryan with forging Elizabeth's character during these critical years. "Much of the future greatness of Elizabeth may reasonably be attributed to the judicious training of her sensible and conscientious governess."[19] Certainly Lady Margaret's influence must have been considerable: she had provided a steady, constant maternal figure, and her continued presence during the turmoil of Anne's downfall and execution provided some much-needed stability in the life of the young Elizabeth.

If she had succeeded in reestablishing some order in Elizabeth's household and insuring that the child was not forgotten at court, Lady Margaret had not managed to restore her to the king's affections. This was perhaps too unrealistic a hope, given that the very name of Anne Boleyn, and anything connected with her, was still anathema to Henry. By contrast, he doted on his new wife. The Imperial ambassador, Eustace Chapuys, surmised that Henry's relief at finding a more suitable queen was akin to "the joy and pleasure a man feels in getting rid of a thin, old and vicious hack in the hope of getting soon a fine horse to ride."[20] The people of England were similarly relieved, and the wedding, which took place on May 30, was greeted with widespread rejoicing. With both of her predecessors dead and no legal bars to their union, Jane's legitimacy was beyond question. Now all she had to do was produce the male heir that the king—and his country—so longed for.

The person who had least cause to rejoice at the king's marriage to Jane Seymour was his younger daughter, Elizabeth. Her mother's disgrace and execution had already rendered her a bastard, but now a

woman had come to the throne who was openly opposed to everything Anne had stood for. If Elizabeth had been old enough to understand how Jane had usurped her mother's place and hastened her
downfall, she would have resented her bitterly. As it was, at just under
four years old, it is unlikely that she realized the full implications of
Jane's rise to power. Any resentment she may have felt would therefore
have been caused by the favoritism that the new queen showed toward
her half sister, Mary.

Jane made no secret of her admiration for the late Catherine of
Aragon, and shrewdly praised her as widely as possible in order to enhance her own popularity. Although she was in theory stepmother to
both Elizabeth and Mary, it was upon the latter that all of her attention
was bestowed. Even before she became queen, Jane had shown "great
love and reverence" toward Mary.[21] She had also started to lay the foundations for a reconciliation between her future husband and his elder
daughter. "I hear that, even before the arrest of the Concubine, the
King, speaking with Mistress Jane Semel [Seymour] of their future
marriage, the latter suggested that the Princess should be replaced in
her former position," reported Chapuys. Her attempts had been met
with a rebuke from the king, who told her "she was a fool, and ought
to solicit the advancement of the children they would have between
them, and not any others."[22]

Thenceforth, Jane played a much more cautious game, anxious not
to overstep the mark with Henry like her predecessor had done.
Rather than repeating her suggestion of restoring Mary as princess and
heir to the throne, she praised her virtues to the king whenever the opportunity arose. Less than a week after her wedding, she confided to
Chapuys that she had "spoken to the King as warmly as possible in
favour of the Princess, putting before him the greatness and goodness
of all her kindred."[23] In the meantime, she wrote often to Mary, assuring her that she would soon be restored to favor. Mary responded with
effusive gratitude and affection, thanking Jane for her "motherly joy"
and "most prudent counsel." She soon began to address her as her
"good mother"—a mark of respect that she never accorded to Anne
Boleyn.[24] In July 1536, both women had their wish. It was at least partly

thanks to Jane's encouragement that Mary had at last made a complete submission to her father by accepting her illegitimacy. As a result, she had been welcomed back to court for the first time in almost three years. Jane was delighted at this first triumph of her reign and was also pleased to have a companion of her own rank at court. She presented Mary with a "beautiful diamond" to mark the occasion.[25]

By contrast, the new queen showed scant interest in her husband's younger daughter, the "little whore," Elizabeth. Although her campaign to restore Mary to favor had been supported by the powerful Imperialist faction at court, there were no such supporters for Elizabeth, and Jane was not willing to put her neck on the line for a friendless bastard—particularly the daughter of a woman whom she had despised. Despite her passive demeanor, Jane had a keen sense of her status and was cool and distant with her inferiors, demanding that they show her all due reverence as queen. Even Chapuys, who was inclined to favor the woman who had ousted the detested "concubine," admitted that she was "proud and haughty."[26]

Most accounts have painted a harmonious picture of Jane befriending first Mary, and then Elizabeth, but there is little evidence to support this, beyond the traditional exchange of gifts between the three women at New Year's 1537. The fact that Elizabeth had been invited to court that Christmas was due not to the new queen's kindly intervention but to that of the Lady Mary, who was Elizabeth's most valuable (if not her only) advocate at court. Mary had little control over what happened once they were there, however. During the festivities, while she took her place opposite Henry and Jane at the top table, her younger sister was seated out of sight.[27]

Apart from her visit to court at Christmas 1536, Elizabeth remained at her household as it made its regular removes to different country palaces. Then, in February 1537, news arrived from the court that threatened to disrupt the newfound stability of Elizabeth's regime. Her stepmother, Queen Jane, was pregnant. The king was transported with joy, convinced that this time God would grant him a son. The news was formally announced at court and celebrated at a Mass in late May. Jane's pregnancy progressed without incident, and in the middle of September, she began her confinement at Hampton Court. Everything depended upon the sex of the child. The king and his courtiers waited anxiously for news as Jane's labor dragged on for two days and three nights. Finally, at about two o'clock in the morning of October 12, the child was born. It was a boy. Henry's long struggle for a male heir was over at last.

There was great rejoicing throughout the court and beyond. England had a male heir; her troubles would surely now be at an end. But for Elizabeth, the news was a good deal less welcome. Her status, which had long been subject to doubt, had been struck a blow from which it seemed unlikely ever to recover. She was now not only a bastard, she was an insignificant girl. The king had a son with whom to carry on the Tudor dynasty. At a little over four years old, Elizabeth looked set to be relegated to the sidelines of history.

The birth of Edward did have a more positive impact on Elizabeth by bringing her closer to Mary. Any lingering rivalry between the sisters seemed to have been removed now that the king had an undisputed male heir, born in lawful wedlock. It was even less likely that either Mary or Elizabeth would inherit the throne, and this took the pressure off their relationship.

Both girls were invited to court for their brother's christening on October 15. While Mary was accorded the honor of acting as Edward's godmother, Elizabeth was too young to play any formal role and instead was carried to the christening by the Queen's brother, Edward Seymour. Mary led her young sister by the hand when the ceremony was over.

Scarcely had this event taken place than Mary and Elizabeth learned the shocking news of their stepmother's death. Jane had never recovered from the birth, and died, possibly of puerperal fever, some twelve days later. "Divine Providence has mingled my joy with the bitterness of the death of her who brought me to this happiness," lamented the king.[28] Having had barely a year to come to terms with her new stepmother, Elizabeth's young life was once more thrown into uncertainty.

Worse was to come. Elizabeth learned that her Lady Mistress, the woman who had cared for her throughout almost all of her infant life, was to be taken from her and transferred to the household of the new prince. All the attention she had received from her surrogate mother was passed on to her baby brother. The lesson was simple to one of Elizabeth's precociousness: being male was what counted.

In the event, Lady Bryan did not leave Elizabeth's household altogether. The death of Jane Seymour just days after giving birth to Edward made it expedient for the boy to leave court and join his half sisters' establishment. With Mary still there, this meant that Lady Margaret had charge of all of the king's children, but it was clear that her main focus would be Edward. The alacrity with which she transferred her affections to the new prince meant that Elizabeth had now lost a second mother in just over a year.

Lady Bryan no longer worried about Elizabeth's wardrobe and instead made sure that Edward had every luxury he could wish for. "His grace . . . was as full of pretty toys as ever I saw child in my life," she wrote proudly to Cromwell, but even then she was not satisfied and begged the minister to procure "a good jewel to set on his cap."[29] She doted on the boy, her affection no doubt fuelled by the fact that he was England's undisputed heir, the boy upon whom were pinned the country's hopes. As he grew into a demanding toddler, she indulged his every wish. "The minstrels played, and his grace danced and played so wantonly that he could not stand still," she proudly told Cromwell in 1539, when the boy was two years old.[30]

By now Lady Bryan had learned not to become too attached to the royal offspring in her charge. The Lady Mary had been abruptly removed from her care after six years, and the apparent neglect that had been shown toward Elizabeth after Anne Boleyn's death must have made Margaret doubt how much longer she would be responsible for the girl's upbringing. Elizabeth had no such experience to draw upon. Customary it may have been, but to be suddenly deprived of her Lady Mistress in order to make way for her infant brother must have seemed a gross injustice to her young mind. But the brutal world of Tudor politics had taught her a valuable lesson, one that she would put into practice when queen: it was not wise to become too attached to those around you. "Affection," as she would later say herself, "is false."[31] As a result, she came to prize one quality above all others, and only after it had been proved beyond doubt would she return it in like sort. That quality was loyalty.

The loss of Lady Bryan accelerated Elizabeth's development, sharpening her wits and giving her a wariness beyond her years. But then, children were generally made old before their time in the Tudor world. Many encountered its cruelty when mothers and siblings suffered premature deaths from childbirth or disease, while a few experienced the worse fate of losing family members to the executioner's axe. They were lectured in the classics, when today children would still be reading fairy tales, and dressed as miniature versions of their parents. Elizabeth was no exception. From infancy, she had worn uncomfortable corsets and petticoats beneath the pretty but restrictive dresses that her mother provided. Her intellectual talents were already outstripping those of her governess, and members of her household remarked upon her "towardness" with a mixture of astonishment and admiration.

After the initial shock and disruption caused by Lady Bryan's transfer to Prince Edward's care, Elizabeth adapted to life in her changed household. Mary remained with her, and they enjoyed a close and harmonious relationship. Over the next few years, Mary paid Elizabeth a great deal of attention, supervising her lessons and teaching her card games and other entertainments. She also delighted the little girl with trinkets and gifts, such as a box embroidered with silver, a kirtle of yellow satin, a gold pomander with a clock inside it, and regular supplies of pocket money.[32] This was the first time since Anne Boleyn's death that Elizabeth had been so spoilt, and she responded with eager affection.

Both girls were delighted with their new brother—Mary for maternal reasons, and Elizabeth because it gave her a playmate nearer to her own age. By March 1537, the Sheltons had no longer been in charge of the girls' household, which must have been a relief to Mary.[33] The domestic arrangements of the household were flexible as long as their father remained unmarried, for there was no expectation that either Mary or her sister would be required to attend a new stepmother. They continued to move regularly from palace to palace, Elizabeth sometimes sharing a residence with Mary, and sometimes with Edward.

Apart from her siblings and attendants, Elizabeth had other company in the form of well-born young children who joined her household from time to time. Among them was Lady Elizabeth Fitzgerald. Known as "Fair Geraldine," Lady Fitzgerald was herself of royal blood. Her mother, Lady Elizabeth Grey, was the granddaughter of Edward IV's queen, Elizabeth Woodville, and first cousin to Henry VIII. Her father was Gerald Fitzgerald, ninth Earl of Kildare, a former favorite of the king.

The Kildares had fallen on hard times after leaving their estates in Ireland to visit Henry's court in 1533. In their absence, Elizabeth Kildare's half brother, Lord Thomas Offaly, had seized the earldom for himself, but he was later arrested for treason and beheaded at Tyburn. His rebellion had ruined the house of Kildare, for both title and estates were confiscated and the family had been forced to seek permanent refuge in England.

Elizabeth Kildare's childhood was as turbulent as that of the king's younger daughter. Just a few months after her arrival in England, her father was imprisoned in the Tower on corruption charges and died there on September 2, 1534. Her mother was forced to seek help from her brother, Lord Leonard Grey, who agreed that she might take refuge at his estate in Beaumanor, Leicestershire. After about two years there, King Henry took pity on his cousin by inviting her youngest daughter, Elizabeth, to join the household of his two daughters at Hunsdon. Although she was to share in their education, the gulf in their ages meant that she was probably afforded separate tuition. She was some eleven years younger than Mary, and six years older than Elizabeth. Although evidence of the time that she spent with the king's daughters is scarce, it suggests that she enjoyed all the comforts of a royal household, and one source noted that "she tasteth costly food."[34] She stayed with them until they moved to a different residence and evidently created a favorable impression, for she was invited back to Hunsdon when they returned there in 1538. It has been suggested that she acted as an intermediary between the two sisters whenever they quarrelled—which, as Elizabeth grew older, was ever more frequently.[35] Henry VIII's younger daughter in particular had formed an attachment to her pretty cousin, and she would become one of her closest favorites as queen.

Meanwhile, the king had gradually begun to take more notice of his youngest daughter. The fact that he now had a male heir no doubt removed some of the resentment he had felt toward her after Anne's death. Elizabeth's moral development was a cause of particular concern to Henry, given the sin that he believed her mother had fallen into. He made it known that he preferred "ancient and sad persons" for her household, and when a young girl of good gentry stock applied to enter her service, he refused in favor of a "gentlewoman of elder years," claiming that his daughter was already surrounded by too many young women.[36]

Among these was a woman whose youth, by Tudor standards, was passing her by, and whose quiet assiduity and impeccable morals made her an ideal guardian for his daughter. Her apparently genuine devotion and loyalty to her young charge had quickly secured her a place in Elizabeth's affections. Her name was Blanche Parry, and she would prove one of the most influential women in Elizabeth's life.

Blanche Parry, or "Apparrie," as her name often appears in contemporary records, was born around 1508 in the Herefordshire village of Bacton. Her father was Henry Parry of Newcourt, one of the oldest estates in the country, and his first language was Welsh, although he had learned English in order to carry out his official duties as sheriff and steward of Dore Abbey. He had married an English lady, Alice Milborne, of nearby Tillington, who was of the same social status as Henry, being the daughter of an esquire. Blanche was raised in a bilingual household, and she and her seven siblings enjoyed an idyllic childhood in the midst of a loving, devoted family. They were imbued with a keen sense of their ancestral pedigree, for the Parrys had a long history of service to the Crown. In the previous century, they had played an important role in the Wars of the Roses, fighting on behalf of King Edward IV. Using the lessons from their family's past, Blanche and her siblings were taught to serve those in power with loyalty and diligence.

The royal court must have seemed a distant cry from this idyllic backwater in the Welsh marches. But Blanche would be introduced into this privileged world when, aged fourteen, she joined her aunt and namesake, Blanche Milborne, Lady Troy, in the service of Elizabeth Somerset, Countess of Worcester. The countess rose to prominence at court, becoming a close associate of Anne Boleyn, and Lady Troy basked in her reflected glory, winning a reputation as a cultured and trustworthy servant. She evidently came to the notice of King Henry himself, for in 1531 she was appointed to the Princess Mary's household when the latter was separated from her mother, Catherine of Aragon, and set up in her own establishment at Ludlow. Lady Troy performed her tasks so well that she was subsequently transferred to Elizabeth's household and was involved in the important task of choosing a wet nurse for her. The evidence suggests that she went on to play a role in both Elizabeth's and Edward's education, helping to teach the children their letters. When Lady Margaret Bryan took charge of the prince's household in 1537, Lady Troy was appointed Lady Mistress to Elizabeth, overseeing the domestic side of her household.[37]

It was through Lady Troy's influence that her niece was appointed to the princess's household. Blanche may have been among the original ladies to serve the infant Elizabeth. She was certainly there when the latter was still a baby, for she was put in charge of the four "rockers" of her cradle. This was an important task, for it would keep the child quiet and amenable, and thus insure that favorable reports of her could be sent back to court. Lady Troy gave her niece all the training necessary to carry out her duties successfully. Indeed, the evidence suggests that she was grooming Blanche to succeed her in the princess's household, as she herself was nearing retirement.

Blanche quickly struck up a close relationship with Elizabeth and doted upon the child. It has been suggested that she sang her to sleep with Welsh lullabies and taught her the rudiments of that language as she grew older. Blanche's love of Wales was well known, and one contemporary praised her as a "singular well willer and furtherer of the weale publike" of that country.[38] She accompanied Elizabeth on all her frequent changes of residence, and provided much-needed stability in a fragile and turbulent world. The young girl came to trust in her steady kindness and was flattered by her humble reverence. Blanche also acted as something of a playmate, sharing Elizabeth's passion for riding. This was encouraged as Elizabeth grew older. The household accounts show that Blanche kept her own horse and was paid an allowance for "horse-meat," which was over and above her already generous wage. She was the only one of Elizabeth's women to be given this additional sum, which suggests that riding with her royal mistress had become part of her official duties.

However much the young Elizabeth might have valued the steady, uncomplaining Blanche, there is little evidence to suggest that Mistress Parry had a profound influence on her emotional or intellectual development. In fact, her influence would be far greater when Elizabeth was queen. She was an important presence during her royal mistress's childhood, but perhaps because she never attained preeminence as her Lady Mistress, she remained in the background. Unswervingly loyal and diligent, she had always put the princess first, but, at least while there were others in this most senior of posts, Elizabeth did not return the compliment. Things would be very different with the woman who now rose to prominence in Elizabeth's household.

CHAPTER 4

Stepmothers

> We are more bounde to them that bringeth us up wel than
> to our parents, for our parents do that wiche is natural for
> them, that is bringeth us into this worlde but our brinkers
> up ar a cause to make us live wel in it.

Elizabeth quoted these words of Saint Gregory in 1549 when she was facing the greatest crisis of her young life since the death of her mother. The "brinker up" to whom she was referring had been in charge of her upbringing for almost thirteen years, and had had a profound effect on the girl's intellectual, spiritual, and, above all, emotional development. Those who saw them together were astonished by the close bond that existed between them. Many said that she was the mother that Elizabeth had never had.

Katherine (or "Kat," as Elizabeth affectionately called her) Champernowne had been appointed to her royal charge's household in the autumn of 1536. Like all those with responsibility for the king's children, she had excellent credentials. She was the daughter of Sir Philip Champernowne, a landowner in Devon, and Katherine, daughter of Sir Edmund Carew of the same county. Both were ancient families with a pedigree stretching back hundreds of years, and their union meant that Kat was related to all the leading gentry of the West Country. Her father was immensely proud of this fact, and engaged in learned discussions about his family's history with the antiquarian John Leland.

As well as her pedigree, Kat had the advantage of being raised in an enlightened household where the benefits of a good education were fully believed in and promoted. It was common for aristocratic households to employ private tutors for their children, and their education began from about the age of five. Unusually for the time, Sir Philip was as committed to his daughters' education as he was to his sons'. Kat developed an interest in humanism and classical scholarship, which had become popular in the Tudor court from the early sixteenth century. Her sister Joan, meanwhile, became a passionate advocate of the rights of women in society—a deeply shocking concept in the early sixteenth century.

All of this was in stark contrast to the education received by most aristocratic women, which was woefully inadequate when compared to that received by men. The king's late wife, Jane Seymour, had been barely literate. The primary aim of a woman's education was to produce wives schooled in godly and moral precepts, as well as in household management, sewing, embroidery, dancing, music, and riding, rather than to promote independent thinking. So long as a woman was able to act as a charming hostess and manage the household efficiently, then her husband would have no cause for complaint. Such social skills were easily passed on from one generation to the next, and most girls were educated by their mothers or other female guardians. Katherine Pole, Countess of Huntingdon, considered that her four daughters had left her care "literate, although not overstuffed with learning." All of this was a natural and satisfactory state as far as men were concerned. Indeed, some compared a woman with a good education to a madman with a sword: She would handle it without reason, and only as violent fits of illness compelled her.

Kat's own experience aside, there were other exceptions to this general rule. Sir Thomas More saw to it that all his daughters received an education on a par with that given to young men, and Lady Jane Grey was renowned as an excellent scholar. Mary Herbert, Countess of Pembroke, became, during Elizabeth's reign, a poet and literary figure in her own right. But even those women complied with the most important lesson taught to all their sex: that their position in society was one of subservience to men. Sixteenth-century society was shaped by the Church, which taught the misogynistic lessons of Saint Paul. Women were the authors of original sin; instruments of the devil. Their only hope for salvation was to accept their natural inferiority to men; as the Calvinist preacher John Knoxe declared: "Woman in her greatest perfection was made to serve and obey man." The young Elizabeth would have cause to rejoice that Kat had not conformed to this view.

Kat's introduction to court was facilitated by the career of her eldest brother, John, who was said to have been a favorite of Henry VIII because of his "quaint conceits." It was also helped by the advantageous marriage that her sister Joan had made to Sir Anthony Denny, a gentleman of the king's privy chamber, which gave her a further entrée into court circles. Kat soon became known for her humanist principles, which automatically won her favor with the most powerful minister at court, Thomas Cromwell. He was sufficiently impressed to recommend her to the king as a suitable companion for his daughter.

Kat had joined the Lady Elizabeth's household by October 1536, although it is not clear in what capacity. She was probably then in her late twenties or early thirties, and certainly unmarried. She wrote to thank Cromwell for his favor in "reporting well of her to the King" and securing a position for her. However, it was with some embarrassment that she also had to beg him to procure her a salary, for she had no other means to sustain herself and was loath to burden her father with her upkeep, "who has as much to do with the little living he has as any man."[1] It seems that she was successful, for her name appears in the household accounts some time later as a salaried employee.[2] When Lady Margaret Bryan was transferred to Prince Edward's household shortly afterward, Kat was presented with an ideal opportunity to advance her career. As an unmarried woman, she could not assume the position of Lady Mistress, and this title went to Lady Troy. But she was more than consoled by being appointed governess to the three-year-old Elizabeth—an honor for one of her youth and slender experience.

Kat was an instant hit with the young Elizabeth. She was of an altogether different character than the formidable Lady Bryan. Her enlightened education had given her a much stronger sense of independence than most women of her age, and her keen intellect was appealing to the precocious child for whose education she was now responsible. She was also of a warm, kind-hearted disposition and had a lively sense of fun, which occasionally bordered on irreverence. The only known portrait of her shows an attractive but homely woman with large dark eyes and an expression that suggests some secret amusement. She is dressed in fine but sombre clothes, as befits her station. Beside her is a painted skull, which was a symbol both of humility and of the need to meditate upon one's salvation. Further proof of her moral uprightness is given in the inscription, which affirms the sitter's ability to resist temptation. The overall image is exactly appropriate for the governess of a king's daughter. But then, appearances can be deceptive.

No doubt it was less Kat's sound moral purpose and piety that impressed the young Elizabeth than her lively humor and challenging intellect. For the first time since the death of her mother, she had found a woman who just might be a fitting replacement. One of the strongest attractions was the fact that Kat was deeply unconventional in her approach to child care. Various contemporary tracts provided strict guidelines for governesses in raising young girls. For a start, they should be encouraged to play with girls their own age. The surviving evidence suggests that this was something Elizabeth rarely did; Kat was the closest she had to a playmate, as well as being a role model during her most formative years. Governesses were also discouraged from displaying affection toward their charges. Kat almost certainly failed in this respect: the closeness of the bond she shared with Elizabeth cannot have rested

upon intellectual compatibility alone. Finally—and most important—
girls were to be kept away from all men (with the exception of tutors
and clerics), on the basis that "our love naturally continues toward those
with whom we have passed our youth."[3] As the years to come would
prove, this last lesson was one that Kat would have done well to heed.

Not everyone in the household at Hunsdon was as impressed with
Mistress Champernowne as her young charge. The bond that had developed so quickly between them excited a degree of jealousy among
the other women of the household. Kat did not help matters by appearing to shun all other rivals to Elizabeth's affections. It was customary for the Lady Mistress and one other female servant (usually
Blanche Parry) to sleep in Elizabeth's bedchamber. However, a number of Elizabeth's other ladies later attested that Kat had ousted Lady
Troy and her niece from this duty, saying that she "could abide nobody
there but herself."[4] It was further alleged that Lady Troy eventually
tired of Kat's domineering influence and resigned her post. Blanche
Parry endured it more stoically, being of a calmer temperament, and
established a relatively harmonious—if not overly friendly—working
relationship with the governess. Nevertheless, Kat's presence was said
to have caused an uncomfortable atmosphere in the household.

Meanwhile, at court there was hardly any mention of Elizabeth and
her sister, Mary, during the months immediately following the death
of Jane Seymour. The king was content to leave the direction of their
households to others. By contrast, he took a keen interest in the health
and upbringing of his new heir, Prince Edward. He and his ministers
were also preoccupied by the progress of the Reformation and the
country's relations with its Continental neighbors. It was the latter
consideration that drove Thomas Cromwell to seek a new bride for his
royal master in 1539. After the turmoil of his marital history so far,
Henry was prepared to let his chief minister arrange his fourth marriage as a matter of state, free from the complications of love and desire. Besides, he now had a male heir, which considerably reduced the
pressure on any such union.

In January 1539, Pope Paul III had reissued the bull of excommunication against Henry, which had first been presented upon the latter's divorce from Catherine of Aragon. This, coupled with a dangerous alliance between the two greatest powers in Europe, France and Spain, prompted Cromwell to seek an alliance with a Protestant state in order to neutralize the increased Catholic threat. He alighted upon the Duchy of Cleves, a strategically important principality of the Holy Roman Empire that connected the Habsburgs' dominions in the Netherlands with their Italian territories. Although nominally part of the Empire, the Duchy was virtually independent from Charles V's authority and was renowned for its reformist Catholic tendencies, which were exactly in line with those held by Henry in England. Cromwell therefore pursued the alliance as a means of consolidating the English Reformation.

The Duchy offered a potential bride for Henry in the form of Anne, daughter of the late duke, Johann III, and sister of his successor, Wilhelm. Anne was then twenty-three years of age and had already been used as a pawn in the international marriage market when she had been betrothed to François, heir to the Duchy of Lorraine, in 1527. This had come to nothing, leaving her free to marry elsewhere. Cromwell duly began negotiations in March 1539, and by late summer an agreement had been reached.

Although the English emissaries praised Anne for her beauty, Hans Holbein was dispatched to paint a portrait of her so that Henry could see what he was letting himself in for. He received this a short while later and was said to have been delighted as he surveyed the pretty, doll-like face that looked back at him, with its fair hair, delicate eyes, mouth, and chin, and demure, maidenly expression. The match was confirmed, and a treaty was signed in October. A few weeks later, Anne embarked upon her journey to England.

On New Year's Eve, she arrived at a stormy, windswept Rochester Castle in Kent. The next day, following the custom favored by Renaissance monarchs who were betrothed to foreign brides whom they had never met, Henry hastened to greet her in disguise. He was horrified with what he saw. "I like her not! I like her not!" he shouted at

Cromwell when the meeting was over. It seemed that Anne had been rather flattered by her portrait. In contrast to the petite stature of his first three wives, she was tall, big boned, and strong featured. Her face was dominated by a large nose that had been cleverly disguised by the angle of Holbein's portrait, and her skin was pitted with the marks of smallpox. She also suffered from a body odor so strong that it was remarked on by several members of the court even at a time when personal hygiene was by no means fastidious.

However abhorrent his new bride might be to Henry, there was no going back. It would have caused a major diplomatic incident if he had reneged on the treaty, and England could ill afford to lose allies. The wedding duly took place on January 6, and the king now had to do his duty by consummating it. To be fair, although much has been written of the physical shortcomings of the "Flanders Mare," Henry himself cannot have presented a very alluring prospect. He was more than twice his young bride's age, his girth had increased considerably in recent years, and he suffered from stinking sores on his legs, caused by injuries inflicted while jousting in his younger days. But Anne was an apparently willing bride and gave every appearance of joy in her new husband. She did not have much to compare him to, for as well as being chaste, she was entirely innocent of the ways of the world and apparently had no idea what was involved in consummation.

Thanks to the events that happened afterward, a detailed account of the wedding night exists among the records of Henry's reign. The king had run his hands all over his new wife's body, which had so repelled him that he had found himself incapable of doing any more. The following morning, he told Cromwell that he found Anne even more abhorrent than when he had first beheld her, bemoaning: "she is nothing fair, and have very evil smells about her." He went on to claim that there had been certain "tokens" to suggest that she was no maid, not least "the looseness of her breasts," which he had apparently examined closely. As a result, the lady had been "indisposed to excite and provoke any lust" in him, and he concluded that he had "left her as good a maid as I found her." For her part, Anne's innocence had been proved by her confiding to her maids that she believed she might be pregnant because the king had "kissed her good night."[5]

To the outside world, everything was as it should be. Anne wrote to her family assuring them that she was very happy with her husband. She was probably telling the truth: her innocence prevented her knowing that anything was amiss. Meanwhile, Henry made sure that he appeared in public with his new queen as often as could be expected. A month after the wedding, they rode by barge to Westminster and were feted by the Londoners who lined the route. Anne received gifts from her new subjects and entered into the court festivities with all due alacrity. The royal couple attended the May Day tournaments, and the king dined in his new wife's apartments—a sign of intimacy.

But behind the scenes, Henry was working to secure an annulment. Not only did Anne repel him physically, but her character and accomplishments were not what was expected of a Queen of England. The education of noble ladies in Cleves was very different from that in her new country. The English ambassador there noted that "they take it here in Germany for a rebuke and an occasion of lightness that great ladies should be learned or have any knowledge of music."[6] As a result, Anne could neither dance nor play a musical instrument, and her ignorance and shyness rendered her an embarrassment in the sophisticated world of the Tudor court. The sooner Henry could be rid of her, the better.

None of this was known to the king's younger daughter, Elizabeth, who was at her residence of Hertford Castle, eagerly awaiting an invitation to come meet her new stepmother. At six and a half years of age, Elizabeth had clearly matured a great deal since the death of her first stepmother in 1537. Having been in the political wilderness for three years, she seemed to appreciate the necessity of ingratiating herself with her father's new wife in order to enhance her own position. She therefore wrote to Anne shortly after the wedding, entreating her: "Permit me to show, by this billet, the zeal with which I devote my respect to you as queen, and my entire obedience to you as my mother." In words of the utmost deference and courtesy, she went on to say: "I am too young and feeble to have power to do more than felicitate you with all my heart in this commencement of your marriage. I hope your Majesty will have as much goodwill for me as I have zeal for your service."[7] It was a masterly composition, worthy of one much older in years than Elizabeth. She had evidently learned a great deal from the vicissitudes of her father's favor and the ever-changing pattern of his marital relations.

Anne was charmed when she received this letter. In her guileless state, she immediately showed it to the king and asked if Elizabeth might come to court. Expecting him to share in her delight at his daughter's precocity, she was taken aback when her husband angrily denied her request. Irritated by this reminder of another marriage that he would rather forget, he gave the letter to Cromwell, ordering him to write a suitable reply. "Tell her that she had a mother so different from this woman that she ought not to wish to see her," he added spitefully.[8]

Whether Anne succeeded in persuading Henry to reconsider is not certain, but she subsequently became acquainted with all of his children. Although she was shy and uncultured, she was also sensible and kind, and rapidly established a good rapport with them. At twenty-four, she was close to Mary's age, and the two struck up an apparently warm friendship. Although she sent gifts to Prince Edward, Anne evidently reasoned that he had nurses and governesses enough to fuss over him, for the child that she focused most attention upon was Elizabeth. Having been predisposed to like the young girl by that charming note she had received soon after her marriage, she perhaps also came to pity her for her motherless state and the king's cruel neglect.

But events at court looked set to deprive Elizabeth of yet another benign female influence. By June 1540, Henry had formed an attachment to Katherine Howard, a pretty young lady-in-waiting in his wife's household. It was not long before Anne herself found out. On June 20, she confided to the Cleves ambassador, one of her few friends at court, that she knew all about her husband's affair. Just four days later, she received orders from the council to remove herself from court and go to Richmond Palace. Henry had apparently found a way to rid himself of his unwanted new wife.

A short while later, Anne learned that her marriage to the English king had been called into question because Henry was concerned about her prior betrothal to the Duke of Lorraine, and had therefore refrained from consummating the union. An ecclesiastical inquiry was duly commissioned, and a delegation of councillors arrived at Richmond in early July to seek Anne's cooperation. She was so shocked by this sudden turn of events that she fainted. When she had sufficiently recovered herself, she steadfastly refused to give her consent to the inquiry. But eventually, perhaps fearing a fate similar to Catherine of Aragon or, worse still, Anne Boleyn, she resolved to take a pragmatic approach. The marriage was duly declared illegal on July 9, and the annulment was confirmed by Parliament three days later. Anne wrote a letter of submission to the king, saying that "though this case must needs be most hard and sorrowful unto me, for the great love which I bear your most noble person, yet, having more regard to God and his truth than to any worldly affection . . . I knowledge myself hereby to accept and approve the same." She agreed that the marriage had not been consummated, referring to "your Majesty's clean and pure living with me," and offered herself up as his "most humble servant."[9]

Anne was to be richly rewarded for her compliance. She was given possession of Richmond Palace and Bletchingly Manor for life, together with a considerable annual income. This was further boosted by her right to keep all of her royal jewels, plate, and goods in order to furnish her new properties. Moreover, she was to be accorded an exalted status as the king's "sister," taking precedence over all his subjects, with the exception of his children and any future wife that he might take. Henry later granted her some additional manors, including Hever Castle, the former home of Anne Boleyn. This was to become her principal residence, and she lived a very comfortable life there on the fringes of public life. She was also permitted to visit court from time to time, and her former husband favored her with several visits, which by all accounts were very convivial. It says much for Anne's strength of character that she managed to accept and adapt to her new life with dignity.

The demise of Anne of Cleves's marriage to Henry VIII by no means ended the friendship that she had begun to form with his younger daughter; indeed, if anything, it strengthened it. Now it was Elizabeth's turn to feel pity for her estranged stepmother, and the two established a bond that would continue for the rest of Anne's life. As soon as the annulment was settled, Anne requested the king's permission to invite Elizabeth to visit her from time to time, assuring him that "to have had [her] for her daughter would have been greater happiness to her than being queen."[10] It is a measure of the affection Anne had for Elizabeth that this was the only favor she asked of her estranged husband after the annulment. Henry agreed with a good deal more alacrity than he had greeted her first request to meet his younger daughter.

Elizabeth subsequently made frequent visits to Anne and was a source of much comfort to her father's rejected wife. Her former stepmother, in turn, would prove to be a positive influence during Elizabeth's formative years, and the two became very close. The same could not be said of Elizabeth's next stepmother, who replaced the last with bewildering speed.

Katherine Howard was the opposite of Anne of Cleves in almost every respect: immature, foolhardy, and reckless, she was also seductive, vibrant, and—to the king—utterly irresistible. The effect she had upon his younger daughter also ran counter to that of her predecessor. It was almost entirely negative: the ultimate lesson in what not to do. It had a profound effect upon the young Elizabeth's emotional development and helped to imbue her with a characteristic that would become her most famous.

Katherine shared a close kinship with Elizabeth. Her father was the younger son of Thomas Howard, second Duke of Norfolk. She was therefore a cousin of Anne Boleyn, although it is unlikely that the two women ever met. The date of Katherine's birth is uncertain. Contemporary sources cite it as being anywhere between 1518 and 1524, but it was probably nearer to the latter. Perhaps the reason for the uncertainty is that Katherine was the youngest of ten children. Her mother died when she was young, and she was raised by her father's stepmother, the Dowager Duchess of Norfolk. The duchess set little store by the new fashion of giving highborn young ladies a wide-ranging education, and instead insured that her protégée would receive only the rudimentaries of reading, writing, and music. Katherine showed little inclination for academic pursuits in any case; she was barely literate, and preferred the frivolous pastimes of dancing and gossiping with the other young girls in the duchess's household. She was sexually aware from a very young age, thanks in part to the inappropriate attentions of her music teacher, who was found to have taken advantage of his position. Although the duchess reprimanded the pair when she found them embracing, she had no time to keep a close eye on Katherine— a lapse that would have fatal consequences.

When the household moved to the duchess's palatial London house at Lambeth in 1538, Katherine became acquainted with Francis Dereham, a kinsman who had recently joined her grandmother's service. She was then fourteen years old at most and had little sense of morality, for she regularly welcomed Dereham into her bedchamber, and before long they became sexually involved. Katherine was far from the corrupted innocent she is often portrayed as. The evidence suggests that she was as much a sexual predator as Dereham and knew exactly what she was doing. Young as she was, she had already gained enough sexual knowledge to avoid getting pregnant, and once boasted: "A woman might meddle with a man and yet conceive no child unless she would herself."[11] The duchess soon discovered her wayward charge's indiscretions and beat her severely. As her moral guardian, it was her duty to ensure that Katherine remained chaste, for a "spoilt" bride was of little value in the aristocratic marriage market. But she managed to hush up the whole affair and the following year secured Katherine a place as maid of honor to Anne of Cleves, thanks to the intervention of Katherine's uncle, the third Duke of Norfolk.

There was little that was maidenly about Katherine when she arrived at court. She was greatly excited at the prospect of a dazzling life filled with potential lovers, and was blissfully unaware of the dangers and intrigues that simmered beneath the surface. One of the most powerful intriguers was her own uncle, who from the very beginning used Katherine as a pawn in his political games. Alluring and vivacious, she provided a perfect foil to the ugly new queen, whom Henry VIII was already trying to get rid of. Norfolk therefore set his sights on securing the greatest prize of all for his niece: marriage to the king himself.

Henry apparently first met Katherine shortly before his marriage to Anne of Cleves. In December 1539, he travelled to Greenwich to await his new bride's arrival. Katherine was among the ladies who were assembled there, ready to serve this royal mistress from across the seas. It was said that Henry was struck by her from the moment he first saw her, and was determined to have her. The fact that he was already betrothed to another was an annoying inconvenience. Reluctantly going through with the marriage to a woman whom he found repugnant, Henry nevertheless maintained his interest in the seductive young maid of honor. This interest became increasingly marked, and by spring 1540 it was known throughout the court. The king "crept too near another lady," wrote one observer.[12] In the tradition of his other courtships, expensive gifts soon followed. Tutored by her uncle, Katherine was careful to play the maid with her royal lover and never allowed him into her bed. With the prospect of an annulment from Anne of Cleves seeming ever more likely, she held out for the main prize.

She did not have long to wait. With the same bewildering speed that characterized so much of Henry's recent marital history, he had his marriage to Anne annulled and took Katherine as his new bride within the space of just over two weeks. Henry and Katherine were married in secret at Oatlands Palace in Surrey on July 28, 1540, the very same day that his minister, Thomas Cromwell, whose downfall was caused by the part that he played in arranging the king's marriage to Anne of Cleves, was executed.

The contrast to his previous marriage cannot have been greater. While Anne's unattractiveness had rendered Henry impotent, the nubile charms of his new teenage bride drove him to distraction. "The King is so amorous of her that he cannot treat her well enough and caresses her more than he did the others," exclaimed the French ambassador.[13] It is easy to understand what Henry saw in Katherine. She was small, with a curvaceous figure, auburn hair, and sparkling eyes that resembled those of her late cousin, Anne Boleyn. She had a playful exuberance and vitality that must have appealed to the aging king, who was desperate to recapture his own youthful vigor. If she was repelled by her royal suitor, who was more than thirty years her senior and had a physique that bordered on the grotesque, she was wise enough not to show it. She gave him a sexual excitement that he had not felt since courting her cousin. Infatuated with his alluring new plaything, Henry apparently never thought to question how she had gained such skill.

Katherine revelled in her newfound importance as queen and threw herself into all the diversions the court had to offer. She was particularly fond of dancing, and showed off her figure in beautiful new gowns, together with an array of priceless jewels that the king lavished upon her. These frivolous pursuits did little to improve her knowledge of court politics, however, and she seemed to think that she could fritter away her days as queen with not a care in the world.

The king's elder daughter, Mary, bitterly resented her empty-headed new stepmother, who was some eight or nine years younger than herself. Katherine's high spirits and decadence were at odds with Mary's sober piety, and the latter failed to pay her the respect that was due to her as queen. Katherine took her revenge by trying to reduce her stepdaughter's establishment. She apparently had some success: Chapuys reported that one of Mary's ladies had died of grief after being dismissed by the king. This may have been an exaggeration, but relations between the two women remained frosty.

By contrast, Elizabeth showed a great deal more respect toward the new queen. Katherine was the closest to her in age of all her stepmothers, and they were also bound by ties of kinship. But Elizabeth's good grace was no doubt due to the pragmatic approach that she had learned from Anne of Cleves. Her objective was to be received back into the court—and her father's favor—and she knew that flattering the queen was an effective way of achieving this.

For her part, Katherine seemed delighted with her younger stepdaughter. On the day that she was publicly acknowledged as queen at Hampton Court, she asked that Elizabeth be placed opposite her at the celebratory banquet because she was "of her own blood and lineage."[14] This favor was repeated on many other occasions, for Katherine always insisted that Elizabeth should take the place of honor nearest her own. As well as being a great privilege for Elizabeth, it was also a deliberate slight to Mary, and many believed that Katherine would try to get Elizabeth reinstated as heir to the throne, ahead of her dour half sister.

During the months following her marriage to Henry, Katherine continued to show great favor toward his younger daughter, who enjoyed more visits to court than ever before. In early May 1541, Katherine invited Elizabeth to stay with her at the king's riverside mansion in Chelsea, and arranged for the young girl to be transported in great state in the royal barge.[15] A few days later, she insured that Elizabeth was included in the royal family's visit to Prince Edward at Waltham Holy Cross in Essex. She also expressed her affection by giving Elizabeth various little gifts of jewelry, such as beads with "crosses, pillars and tassels attached." Although the royal accounts noted that these were "little thing worth," it was the thought that counted.[16]

Katherine was clearly enjoying the glittering new world that had unfolded before her when she became Henry's wife. Her days were filled with dancing, feasting, and visits to her husband's many beautiful palaces. There was more than enough to occupy the time of this pleasure-seeking young queen. But it was not long before she fell prey to other temptations. The gout-ridden king was unlikely to have satisfied her voracious sexual appetite, and by the spring of 1541, she had started an affair with a young lover: Thomas Culpepper, a gentleman of her husband's privy chamber. When she accompanied the king on his progress to York that summer, Culpepper went with them. Every evening, he would creep up the back stairs to Katherine's apartments and spend the night cavorting with her. To make matters worse, she also met her former lover, Francis Dereham, who had returned from exile in Ireland, and took a staggering risk by appointing him as her personal secretary.

Nothing could remain a secret at court for long, and by the time the entourage returned to London, gossip about the Queen's indiscretions was rife. The ambitious Howards had made many enemies at court, who now seized upon this information to plot their downfall. By October, the king's closest confidant, Thomas Cranmer, had been informed of Katherine's adulterous liaisons, and he was given the unenviable task of breaking the news to Henry. The king's first reaction was disbelief. Katherine had embodied everything that he had looked for in a queen: She was charming, beautiful, and vivacious, and he worshipped her. To be told that she was an arch-deceiver with the morals of a whore was too much for him to bear. He ordered an immediate inquiry, praying that the allegations would prove false. But they were corroborated by several witnesses, and on November 8, the Queen herself confessed. Dereham, she said, had used her "in such sort as a man doth use his wife many and sundry times."[17] She also admitted to having sexual encounters as a girl with her music master, and later confessed to sleeping with Culpepper.

Confronted with the shocking and sordid burden of truth, the king fell into a deep distress. When he had discovered the allegations against his second wife, Anne Boleyn, he had experienced some relief because he had long since tired of her. But he was still infatuated with Katherine when the scandal broke, and had no notion of her infidelity. The certainty of it broke his heart. Katherine was "found an harlot" before she married, and "an adulteress" after it.[18] There could be only one fate. She was condemned to death for treason, and taken to the Tower on February 10, 1542. Three days later, she faced her execution. So weak that she had to be supported as she climbed the steps of the scaffold, she gathered enough strength to confess that her sentence was justified. Moments later, the axe fell. Katherine's remains were buried in the chapel of St. Peter ad Vincula, along with those of her disgraced cousin, Anne Boleyn. She may have been as young as eighteen at the time of her death.

Although Elizabeth had witnessed a succession of stepmothers come and go over the past few years, none of them affected her so deeply as Katherine Howard. It was not that she had felt any great affection for the young queen. Much as she had revelled in her favor, she had not known her long enough to form a lasting attachment. Rather, Katherine's sudden and brutal demise was so similar to that of her own mother that she was profoundly shocked by the experience. A precocious eight-year-old, she would have understood a great deal more of what it meant to be condemned for treason and beheaded than when Anne Boleyn had fallen prey to the same fate. What's more, the two queens had met their deaths at exactly the same spot, in that dread fortress by the Thames. The only difference was that Anne had been afforded the privilege of a sword, whereas Katherine suffered by the traditional axe.

At her impressionable age, Elizabeth was horrified by the events that had unfolded so rapidly and so catastrophically. One moment she had been basking in the Queen's affection and watching her carefree frivolity at court. The next she had learned that this same queen had been beheaded. It was apparently enough to convince Elizabeth, young as she was, never to marry. Some twenty years later, her favorite, Robert Dudley, would confide to the French ambassador that he had known Elizabeth since she was eight years old, and that from that time she had always declared: "I will never marry." Her views were echoed by her closest adviser, William Cecil, who remarked that "marriage with the blood royal was too full of risk to be lightly entered into."[19] The notion that marriage was inextricably bound up with death had first formed in the young Elizabeth's mind as she gradually learned the truth about her mother. Katherine Howard's fate confirmed it. By taking a husband, royal women, exalted though their status might be, were placing themselves at the mercy of men. As they uttered their marriage vows, they were relegating themselves to the status of pawns, to be used in the dangerous game of high politics.

For once, Elizabeth must have been glad to be away from this volatile world of the court, safe in the comparative haven of her household. She derived increasing comfort from her studies, which were progressing apace, thanks to the attentions of her governess. During the early years of their association, Kat had sole charge of Elizabeth's education. She taught her the alphabet and the rudiments of grammar, together with reading and writing skills.[20]

As well as learning her letters, Elizabeth would have been taught the typical accomplishments expected of a female royal, including needlework, music, dancing, and riding. Etiquette was another essential prerequisite and included table manners and forms of address. Such courtly pursuits were the closest that Elizabeth came to being trained for the throne. The role of royal daughters was to marry a king or a prince and beget heirs, so it was not considered necessary to give them the same education as that of a future king. Instead the focus was upon strengthening their character and maintaining a good reputation. It was believed that women were not just physically weaker than men but morally so. As a result, they needed to be given strict instruction in virtuous behavior, especially chastity. They should read not for pleasure but for instruction. Idle conversation and gossip were strongly discouraged; young girls should be seen and not heard.

When she reached the age of five, Elizabeth's education was expanded to include foreign languages and other subjects. Children would be taught to speak classical languages such as Greek and Latin before they could write them. Under the guidance of her governess, Elizabeth's capacity for languages increased considerably. Kat was also able to draw upon the benefits of her own education to inspire her pupil with an enthusiasm for the classics that would last throughout her life.

Elizabeth flourished under her governess's careful tutelage, and Kat won widespread praise among her contemporaries. In 1539, when Elizabeth was six years old, the courtier Sir Thomas Wriothesley, Earl of Southampton, visited the household and was very impressed by how advanced the young girl's intellect had become. "If she be no worse educated than she now appeareth to me," he reported back to court, "she will prove of no less honour to womanhood than shall beseem her father's daughter."[21]

Elizabeth's progress soon came to the attention of Roger Ascham, one of the leading educationalists of the day. He was full of praise for Kat and amazed at the results that she had achieved with her young charge. He was also rather taken aback by the pace of Elizabeth's learning and urged the governess not to push her too much. "If you pour much drink at once into a goblet, the most part will dash out and run over," he warned. "If you pour it softly, you may fill it even to the top, and so her Grace, I doubt not, by little and little may be increased in learning, that at length greater cannot be required."[22]

But Ascham underestimated the precocity of the young girl, who had a seemingly insatiable lust for learning and was delighted to find a tutor who could go at least some way toward satisfying it. So far from bewildering the young girl, Kat actually found herself struggling to keep up, and in 1542, after five years in charge of Elizabeth's education, she was forced to hand over the reins to Dr. Richard Coxe, who had been appointed tutor to Prince Edward. She did so with selfless alacrity, however, for she was aware of her own limitations and did not wish to impede the progress of her young charge.

The fact that Elizabeth responded with such enthusiasm to the guidance of her new tutor did nothing to diminish her bond with Kat, which seemed to grow stronger with every passing day. By now, the pair were inseparable. Having been effectively abandoned by her parents—her mother through death and her father through neglect—Elizabeth clung to her governess with a fierceness born of insecurity. Kat was a fixed point in an ever-changing world, and Elizabeth was determined never to lose her. To her governess, she confided everything. "I will know nothing but that she shall know it," she once remarked.[23] Kat returned her affection with no less zeal. Jealously protective of her young charge, she delighted in being the sole focus of her love. Some remarked that she had become a mother to Elizabeth, and to a certain extent this was true.

But their relationship went beyond the straightforward mother-daughter bond. Indeed, in some respects, Elizabeth often seemed the more mature of the two. Not only had she outstripped Kat intellectually, but she also showed far greater shrewdness and discretion than her affectionate governess. Although intelligent and accomplished, Kat was also somewhat naïve in the ways of the world, as well as impulsive and overly romantic. These were hardly commendable qualities in a woman who had been charged with raising Elizabeth to have the impeccable morals and manners expected of a king's daughter. But Elizabeth had long since ceased to be the subject of much attention at court, as all eyes were now focused on England's new heir, Prince Edward. The more worrying aspects of Mistress Champernowne's character were therefore apparently overlooked.

Another preoccupation at court was who the king would take as his next wife. In less than a decade, Henry had married Anne Boleyn, Jane Seymour, Anne of Cleves, and Katherine Howard. His wives had changed with a bewildering speed that reflected the ever-shifting balance of power at court, just as much as the king's own fickle affections. There were now a number of candidates for his new bride, and it was even rumored that he would remarry Anne of Cleves. She and Henry exchanged New Year's gifts in 1542, and Anne seemed keen to revive their association. Indeed, she may only have agreed to the annulment so readily because she hoped that Henry would soon change his mind. This hope had only increased by witnessing the rapid downfall of Katherine Howard. But the king made no indication of wishing to revive their union, and when Anne heard that he was rumored to be looking elsewhere for a new wife, she was bitterly disappointed.

Rumors about the king's next marriage were also circulating in Elizabeth's household, and her indiscreet governess no doubt failed to resist gossiping about it. Elizabeth would have had mixed feelings when contemplating the arrival of yet another stepmother. Her first, Jane Seymour, and last, Katherine Howard, had been rather distant figures, closeted away at a court that she was rarely permitted to visit. However, these women had still had an impact upon her status and position. Elizabeth had been able to observe their fates in a detached, almost clinical way, amassing precedents that would help to shape her future behavior. Even though she had been—indeed, still was—much closer to Anne of Cleves, her visits to her were all too brief.

Things would be entirely different with her next stepmother. If Elizabeth had known how great an impact she would have on her life, she would have welcomed the prospect with intense excitement. For Henry's sixth and last wife would not just influence Elizabeth's status at court—transforming her from a royal bastard to an heir to the throne—but would profoundly affect her character and outlook, helping to make her into the queen that she would one day become.

Born in 1512, Katherine Parr hailed from a long line of dominant, enlightened women. Her grandmother, Elizabeth Fitzhugh, had been a remarkable woman of considerable intellect. Her mother, Maud Parr, had lost her husband, Sir Thomas Parr, at a young age and had brought up her three children almost single-handedly, while serving Catherine of Aragon as a lady-in-waiting. Maud's eldest daughter, Katherine (who was named after the queen), stayed in this female-dominated household until she was in her late teens. During this time, she received an education from her mother that, by early-sixteenth-century standards, was extremely enlightened. Maud was an unusually articulate and independent woman, and set up a school in her household to educate her children. The program of studies that she devised was based on that favored by Sir Thomas More, who had taught his own daughters to believe that they were the intellectual equals of men.

By the time she reached adulthood, Katherine had grown into an assertive woman, keen to promote her intellectual ability when female voices were afforded little credence. She had a much greater sense of independence than most women of her generation and had learned to take an active role in shaping her destiny. This lesson would take her to the very heart of the royal court and would later be passed on to her future stepdaughter, Elizabeth.

For all her independence, Katherine knew that she was expected to take a husband as soon as a suitable one was offered. Her mother worked assiduously to bring this about, and in May 1529 she secured an agreement with the third Baron Borough of Gainsborough in Lincolnshire for Katherine, who was then just sixteen years old, to marry his son, Thomas. The match was not a happy one: Katherine's father-in-law was a notorious bully, and there were hints of insanity in his family. But she did not have to endure it for long, for in 1533 her young husband died. Katherine's mother had died two years before, so she now faced the first test of the independence and resourcefulness that she had learned during childhood. After taking refuge with some wealthy cousins, she soon found herself another husband, John Neville, third Baron Latimer, of Snape Castle in Yorkshire, whom she married in 1534. Thanks to his conservative religious principles, Lord Latimer became caught up in the Pilgrimage of Grace in 1536 and only narrowly escaped prosecution for treason. He was saved by his kinship with Katherine's family, who had opposed the rebels.

Shaken by the experience, Katherine and her husband left their estates in the north and moved southward to their manor of Wick, near
Pershore in Worcestershire. After nine years of marriage, Lord Latimer died. This does not seem to have come as a shock to Katherine,
who some months before, toward the end of 1542, had already secured
herself a place in the household of the king's elder daughter, Mary.
The princess was inclined to favor all those who had shown loyalty to
her late mother, so the appointment was no doubt thanks to Maud
Parr's royal service. The two women were of a similar age[24] and soon
established a close rapport, Katherine assuming the very personal task
of ordering the princess's clothes for her. Her service was to be of an
unexpectedly short duration, however, for it brought her to the attention of Mary's father, who, undeterred by his disastrous marital history, began to pay court to her.

By now thirty years of age, Katherine was a comely, if not exactly
beautiful, woman. The fact that there are no contemporary accounts of
her appearance suggests either that she was considered to be rather plain
or that her looks were unimportant. This has been compounded by subsequent portrayals of her as an aging widow with few attractions—certainly compared to the likes of Anne Boleyn and Katherine Howard. Yet
this is more the construct of filmmakers and romantic historians than of
fact. A portrait of her painted in 1545 shows a well-dressed young
woman with a pleasing, dignified appearance. She has rich auburn hair
and soft grey eyes, together with the pale, flawless skin that was fashionable at the time. However, it was her character that held the greatest attraction. Katherine was a witty and engaging woman who enjoyed lively
conversation, and who, thanks to her education, could speak knowl-
edgeably on a wide range of subjects. She was also well versed in the
more courtly accomplishments of music and dancing, and loved fine
clothes and jewelry. Her favorite color was crimson, which set off her
hair and pale skin to dramatic effect. She also paid a great deal of attention to her personal hygiene. In a court where pungent body odors were
prevalent among even the highest-ranking men and women, she was a
sweet-smelling anomaly. She indulged in milk baths and adorned her
body with expensive oils and perfumes such as rosewater. She even used
lozenges to sweeten her breath.[25]

Above all, though, Katherine had great presence and dignity and a sense of inner calm that drew men to her. Among them was one of the most handsome bachelors at court, Thomas Seymour, brother of the late queen, who was already trying to win her favor by the time she caught the king's eye. This defies the image of a plain spinster whose only attraction for Henry lay in her ability to nurse him through his final years. Moreover, the king had always had a weakness for pretty women, and the episode with Anne of Cleves had proved that he was not willing to marry for politics alone.

By February 1543, it was noted that Henry was paying more visits than usual to his daughter Mary's apartments. Courtiers soon suspected that his motives were inspired by more than just fatherly concern. This was a very different courtship than the ones that had gone before, however. The bitter experience of Katherine Howard's betrayal had dealt his confidence a severe blow, and rather than showering the new object of his affections with love tokens and grand gestures, he assumed a melancholy attitude whenever she was around, appearing "sad, pensive and sighing."[26]

Katherine was far from flattered by his attentions. Having already endured two loveless marriages, she had no wish to embark upon a third, particularly as she had already fallen deeply in love with Thomas Seymour. The king was plagued with jealousy when he learned of this, and in a repetition of the tactics he had used with Anne Boleyn and Henry Percy, he found an excuse to send Seymour away from court. This did little to allay Katherine's reluctance, however, and she remained out of his reach—not as part of a tactic, but as a true reflection of her feelings. But she knew that to gainsay the king was fraught with danger, and her reformist family no doubt put pressure on her to carry out God's will by accepting Henry's courtship. Eventually, therefore, she reluctantly put her own desires aside and consented to be the sixth wife of this much-married monarch.

The wedding took place on July 12, 1543, in the Queen's private apartments at Hampton Court Palace. There were just eighteen guests, including Princess Mary and her half sister, Elizabeth. Both girls had been invited to court a month before in order to meet their

new stepmother. Elizabeth was dispatched back to one of her estates
in the country soon after the wedding, but even in the brief time that
she and Katherine had spent together, they had apparently established
a close rapport. Katherine was no doubt charmed by the pretty, precocious nine-year-old, whose educational accomplishments were already
rivalling her own. For her part, Elizabeth seemed instantly to warm to
the dignified presence and lively humor of her father's latest wife.
Katherine had come into her life at precisely the right time. At nine
years old, Elizabeth was rapidly outstripping the intellectual capacity
of her governess, and the latter's waywardness left her in need of a stable female presence. What's more, unlike her other stepmothers, there
was little prospect of Katherine producing an heir to supplant Elizabeth and her siblings, for the king was incapacitated by age and infirmity. It was therefore very easy for her to like his new wife, and it must
have been a great disappointment to be summarily dismissed from
court when their acquaintance was still so new.

But for Katherine, it was not a case of out of sight, out of mind.
Having already had the experience of being a stepmother, thanks to
her second marriage, she was evidently fond of children and resolved
to befriend all of the king's children. She and Mary had grown close
during her service in the princess's household, and Katherine took pity
on her and her half sister. It was probably thanks to her suggestion that
the two daughters were included in a portrayal of Henry VIII and his
family, completed in or around 1545. Katherine seemed determined to
enhance her younger stepdaughter's position as the daughter of the
king—albeit an illegitimate one.

Katherine also cultivated the young Prince Edward's affection, and
he was pleased by her attention. Appreciating that he and his half sister
Elizabeth, young as they were, had a keen interest in scholarly pursuits,
Katherine took it upon herself to oversee their education. Thanks to
her influence, it was agreed that Elizabeth should share some of her
lessons with Edward. They were taught a range of different subjects as
part of a standard curriculum for royal children, including languages,
theology, history, rhetoric, logic, philosophy, arithmetic, literature,
geometry, and music. Sir John Cheke, Regius Professor of Greek at

Cambridge and a proponent of the reformist religion favored by Katherine, was appointed to lead their studies in the early part of 1544. Under his tutelage, the two children learned to question established orthodoxies and advocate the principles of the reformed religion, which they embraced enthusiastically.

Roger Ascham was a frequent visitor to the children's household and would take part in their studies. An expert in the italic hand, he inspired Elizabeth in this art, and her flamboyant signature, with its swirls and flourishes, testifies to the mastery she acquired. When she and her brother were tired of their books, Ascham would accompany them for walks on the grounds of Hatfield, Ashridge, or Hunsdon. His expertise extended to archery, and both children loved to practice this. It was thanks to Ascham that Elizabeth also developed a love of cockfighting and bearbaiting, for he took a keen interest in these brutal sports himself.

Katherine Parr soon realized that Elizabeth was far outstripping her younger brother, so she decided that the girl needed a tutor of her own. William Grindal was a scholar of St. John's College, Cambridge, and was renowned for his expertise in Latin and Greek. He had been recommended to Katherine by Elizabeth's governess, who was keen not to be completely sidelined in the girl's education. Having satisfied herself of his credentials, Katherine proceeded with the appointment. Elizabeth flourished under Grindal's charge and acquired a passion for the humanist principles so favored by her new stepmother.

Shortly after Grindal had been appointed as Elizabeth's tutor in 1544, he wrote to Kat Astley expressing his astonishment and gratitude for everything she had achieved with her protégée. "Would God my wit wist [knew] what words would express the thanks you have deserved of all true English hearts, for that noble imp [Elizabeth] by your labour and wisdom now flourishing in all goodly godliness, the fruit whereof doth even now redound to her Grace's high honour and profit. I wish her Grace to come to that end in perfectness with likelyhood of her wit, and painfulness in her study . . . which your diligent overseeing doth most constantly promise."[27] This was more than just the flattery of a grateful beneficiary of Kat's favor. Elizabeth's exceptional intellectual ability was by now widely talked of. "She was learned (her sex and times considered) beyond all common belief," another contemporary enthused.[28]

The rapid progression of Elizabeth's education proved to be a source of tension with her half sister. Mary had overseen her younger sister's studies after Anne Boleyn's death and had been proud of her precociousness, as evidenced by the fulsome praise that she relayed to her father. But Elizabeth had soon outstripped Mary's knowledge, just as she had her governess, and now she was being schooled in theological views that were directly opposed to Mary's own. Although she had acknowledged her father as supreme head of the Church, Mary remained a staunch traditionalist in matters of religion and severely disapproved of her siblings' studies. It would be one of the most dangerous causes of disagreement between them when Elizabeth reached adulthood. Furthermore, her younger sister was gaining a confidence and independence of spirit that jarred with Mary's reserved nature. She was no longer the sweet little girl who was endearingly dependent upon her elder sister. She had apparently already decided that she had little more to learn from Mary, and although she remained politely deferential in the letters she later wrote to her, there was a cool detachment in her tone.

This new sense of distance was not helped by the fact that Mary was spending more and more time away from Elizabeth's and Edward's households. She and her entourage were often at Richmond Palace or other royal residences farther from London. This pattern had begun during Katherine Howard's ascendancy, prompted by the reduction in Mary's household that she had ordered. No doubt the late queen's obvious favoritism of Elizabeth had not helped matters. Even after Katherine Howard's demise, Mary had chosen not to rejoin her siblings immediately. It is not clear whether this was by choice or by the king's command.

As a testament to everything she had learned under William Grindal, Elizabeth wrote a series of letters to her stepmother that were exceptionally articulate and accomplished for an eleven-year-old. These are the earliest surviving letters by the future queen, and the fact that they are all to her new stepmother suggests that she already held her in great esteem. The first one was written in July 1544, exactly a year since Elizabeth had last seen Katherine. She was then at St. James's Palace in London, while her two siblings were both with Katherine at Hampton Court. It is not clear why Elizabeth was separated from them, but it may have been due to the perennial fear of the plague. The letter is written in Italian, which was a compliment to Katherine, who had also learned this language—although perhaps enjoyed less mastery of it than her stepdaughter. The tone is confiding and respectful. Elizabeth begins by lamenting: "Unkind fortune, envious of all good and the continuous whirl of human affairs, has deprived me for a whole year of your most Illustrious presence." She goes on to say that this would be "unbearable" were it not for the hope that she would soon be reunited with her stepmother. Elizabeth was profoundly grateful for the care that Katherine had shown for her welfare, despite their separation. "In this my exile, I know well that in your kindness, your highness has had as much care and solicitude over my health as the King's Majesty. So that I am bound to serve you and revere you with a daughter's love."[29]

Much has been made of that word *exile*. The assumption is that Elizabeth was in deep disgrace with her father, particularly as she also refers to "not daring" to write to him herself. It is thought that Katherine was therefore trying to intervene with the king on his younger daughter's behalf and persuade him to allow her back into his favor. But this is unlikely, given that there is no record of what such a disgrace might have been. Men such as Chapuys, who knew everything about court affairs, did not mention it in their correspondence.

Moreover, a short while before Elizabeth wrote this letter, she and her half sister had been shown the greatest honor of all by their father, as he had restored them to the succession. The years of uncertainty and obscurity finally seemed to be at an end. Elizabeth was a princess once more, and as an heir to the throne, she had an immeasurably higher status than as the illegitimate daughter of a disgraced former queen. This transformation was at least in part, if not wholly, due to the benign influence of her new stepmother. Katherine knew that the chances of her bearing the king any children were slight, and therefore she focused her efforts upon uniting him with all of his offspring, not just the adored male heir, Edward. A key part of this strategy was to insure that Mary and Elizabeth were afforded their rightful place in the succession. Little wonder that the latter professed such love and esteem for Katherine in her first letter, assuring her that she was "Your most obedient daughter and most faithful servant." She later thanked her for having "not forgotten me every time you have written to the King's Majesty."[30]

Prior to Katherine's ascendancy, Elizabeth had been taught to look upon herself as the illegitimate daughter of a whore, with little prospect of inheriting the crown of England. Even though she was third in line to the throne, it is likely that from this moment she cherished the hope that one day she might be queen. According to the seventeenth-century historian Leti, Katherine encouraged her in this belief. "God has given you great qualities. Cultivate them always, and labour to improve them, for I believe that you are destined by Heaven to be Queen of England."[31] His account was mostly based upon evidence that has since been lost or destroyed, so it is far from reliable. But this apocryphal story has survived the centuries, and the influence that Katherine is known to have had upon Elizabeth's education and outlook does lend it some credibility.

Following their restoration to the succession in 1544, Elizabeth and Mary came to live at court for prolonged periods. Katherine was the first stepmother for whom the girls felt the same affection. Jane Seymour had favored Mary; Anne of Cleves and Katherine Howard had favored Elizabeth. If Katherine Parr also felt a stronger bond with her younger stepdaughter, however, she did not show it and treated both girls with equal love and respect.

Nevertheless, it was during her ascendancy that Elizabeth and Mary's relationship began to fall apart. Although an act of Parliament in 1544 had restored them both to the succession, Henry had made it clear that Mary's rights were strictly conditional upon her agreeing to abide by his religious reforms. If she did not, then Elizabeth would inherit "as though the said Lady Mary were then dead."[32] Mary bitterly resented this demonstration that her favor was still very tenuous and that her younger sister could at any time regain precedence over her. In making this stipulation, Henry had effectively pitted his daughters against each other. This revived the old jealousy that Mary had felt toward Elizabeth, and in the battle for precedence in their father's affections, she resorted to underhanded tactics.

From the moment of Elizabeth's birth, Mary had voiced doubts about her paternity. Having been present at Anne Boleyn's confinement, she had allegedly heard some whispered gossip among the other ladies about the Queen's promiscuity. According to one account, this "made her declare that she was sure the infant was not her sister."[33] In fact, as Elizabeth grew into adulthood, she came to look much more like Henry VIII than Mary did, which must have been galling to the latter. During Mary's own reign, the Venetian ambassador would report that "everybody" at court was saying "that she [Elizabeth] also resembles him more than the Queen does; and he therefore always liked her and had her brought up in the same way as the Queen."[34] Yet still Mary persisted in her belief that Elizabeth was not Henry's daughter. Toward the end of her life, she told a priest that the girl "was neither her sister nor the daughter of King Henry," and she used this as an excuse to delay naming Elizabeth as her successor. Even when it became obvious that she had no choice but to confirm her younger sister as heir, the prospect of seeing "the illegitimate child of a criminal who was punished as a public strumpet, on the point of inheriting the throne with better fortune than herself, whose descent is rightful, legitimate and regal" was said to be "bitter and odious" to her.[35] Perhaps Mary had to believe this for her own peace of mind; otherwise the jealousy caused by her father's apparently preferential treatment toward her younger sister would have been too much to bear.

In late July 1544, Elizabeth received the summons she was longing for, and she hastened to Hampton Court to join her stepmother. Edward and Mary were already there. The reunion was no doubt a joyful one for both Elizabeth and Katherine, and the former enjoyed her longest stay at court since her earliest infancy. It was also to be the most significant one of her young life, and she had cause to remember it for many years to come.

When Elizabeth arrived at Hampton Court, she found a household that was greatly changed from the last time she visited. Her father was not there. He was leading a military expedition to France, perhaps as much to recapture his youth as to win any strategic advantage. He had appointed his wife to reign as regent in his absence, with the full exercise of royal authority. Not since 1513, when he had left his first wife, Catherine of Aragon, in charge while he went on campaign to Scotland, had England been ruled by a queen. It was an entirely new experience for Elizabeth, who had only ever known her father to be the ruler. She looked on with awe as Katherine, who clearly relished her newfound power, proceeded to rule in the king's name.

Katherine clearly appreciated—and was determined to exercise—the full extent of her authority. She was afforded all the pomp of a reigning sovereign: she sat in state in the royal presence chamber, was served on bended knee, and was lauded by all who came to seek her favor. But it was more than just a symbolic role. Katherine assumed full powers in her husband's absence, presiding over the regency council, signing royal proclamations, and approving expenditure on additional troops for the French wars. It would have been a challenging enough task to rule during a period of peace and prosperity, but Katherine had assumed control over a country beset by war, plague, and religious division, and troubled by the ever-constant threat of conflict along the Scottish border. That she succeeded not just in avoiding catastrophe but in establishing herself as a figure of decisive authority is a testament to her enormous capability.

All this would have been witnessed by Elizabeth, who was constantly in attendance upon her stepmother. She looked on as courtiers and ambassadors paid court to the Queen with as much state as they had to Henry VIII. She would also have seen members of the council presenting matters of business for Katherine's advice or approval. And she would have been present at the banquets and other state occasions that were held in Katherine's honor. Elizabeth had been raised to be in awe of her father, the king, whom she learned was God's representative on earth. She had also been taught that women were the weaker sex, incapable of bearing the responsibility of monarchy. And yet here was Katherine presiding over the court with as much confidence and authority as any king. The sight of some of the most powerful men in the country bowing low before a woman was something that she would never forget.[36]

Elizabeth stayed with Katherine throughout her regency, accompanying her from palace to palace as the court travelled on its summer progress. During this time, she and her siblings came to enjoy a greater stability than they had ever had. They were united as a family under the auspices of a capable and benevolent stepmother who had proved to be genuinely committed to their happiness and welfare. In her letters to the king, Katherine wrote not just of state business, but of his children. "Your Majesties children are all thanks be to God in very good health," she told him in one.[37]

This was undoubtedly one of the happiest periods of Elizabeth's life. Freed from her "exile" in the country, she was now at the very heart of court life, and she loved it. She was learning more in this short time than she had during the many hours of lessons with William Grindal. Classics and languages were all very well, but this sojourn with her stepmother was providing her with something far more valuable: a role model for queenship.

It came to an end all too soon. Henry signalled his intention to return in September, and Elizabeth was dispatched to Ashridge with Edward before their father had even arrived back at court. She had much to contemplate on the way.

Later that year, Elizabeth decided to show her gratitude to Katherine by making her an exquisite New Year's gift. Determined to prove how much the experience had meant to her, she eschewed the customary trinkets, such as jewels and gold plate, and instead resolved to make this gift a very personal one. With great care, she set about translating "Le miroir de l'âme pêcheresse," or "Mirror of the sinful soul," a poem by Margaret of Angoulême, Queen of Navarre and the favorite sister of King Francis I. Her choice was significant. Katherine had read the poem to her during her stay at court that summer, and had introduced her to other writings of Margaret, who shared her religious sympathies and was a leading patroness of reform at the French court. As such, she was also an example of a powerful female figure whose learning and intellect was influencing the lives of many both at court and beyond.

One of the most important lessons that Elizabeth had taken from Margaret of Navarre was that women were essentially weak, inferior beings and that only by emulating the characteristics of men could a queen succeed in this world. As Margaret wrote to her brother, King Francis I, "All my life I wanted to serve you not as a sister but as a brother."[38] Elizabeth took this so much to heart that after she had ascended the throne, she more often referred to herself not as "Queen" but as "Prince" when addressing her subjects. Advocating her "male" characteristics of courage, authority, and shrewdness enabled her not just to survive but also to reign supreme over a society dominated by men. Strengthened by the teachings of powerful female figures such as Margaret of Navarre, together with the example that her stepmother had set as regent, Elizabeth came to believe what to most people—her father included—would have been an abhorrence: A woman could rule successfully in a man's world.

The theme of "Le miroir" is the inadequacy of the human soul, and the subtext is that only by faith can one be saved. This was a key facet of the new Protestant religion to which Katherine was so devoted, and she had shown her stepdaughter the poem in the hope that it would persuade her to believe in it herself. The fact that Elizabeth chose to translate the poem as a gift suggests that she meant Katherine to know that she had succeeded.

Elizabeth's choice may have had an additional inspiration. Anne Boleyn had been a favorite of Margaret of Navarre during the time she had spent at the French court in her youth. They later renewed their association in 1534–35, when Anne was queen, and it was probably at this time that Margaret presented her with the original manuscript of "Le miroir." Elizabeth's gift could therefore have been a covert symbol of her loyalty toward her late mother.

The translation had evidently taken Elizabeth a great deal of time and effort—more than she had expected, for the handwriting suggests that she had had to finish it in a hurry in order to get it to the Queen in time for New Year's. She prefaced it with an amusing note in which she begged her stepmother not to show the translation to others at court because it was "all unperfect and incorrect" and "nothing is done as it should be." But she comforted herself with the knowledge that "the file of your excellent wit and godly learning, in the reading of it . . . shall rub out, polish and mend . . . the words (or rather the order of my writing) the which I know in many places to be rude."[39]

As a finishing touch, Elizabeth embroidered a beautiful cover for the book, which was bound in exquisite blue cloth. She carefully stitched little forget-me-nots onto the spine and worked heartsease— a herb that signified domestic harmony—in violet, yellow, and green silk at the corners. On the front, she embroidered the initials KP in silver, mirroring the Queen's customary signature.

Katherine no doubt cherished the gift as a symbol not just of her stepdaughter's affection but also of the impact that her "godly learning" had had on the girl. The latter would be one of the most significant outcomes of Elizabeth's relationship with Katherine Parr. Chiming as it did with the beliefs favored by Anne Boleyn, Katherine's devout Protestantism would confirm and strengthen Elizabeth's religious views. While the girl may have felt an affinity to that religion out of loyalty to her late mother, under Katherine's careful tutelage her understanding of it was considerably increased, and she learned to appreciate it for its own merits.

With the new year came a surprising development in Elizabeth's household. Her old governess, Kat Champernowne, relinquished her spinsterhood and married John Astley, a courtier of good standing.[40] The couple may have been introduced by Roger Ascham, who was an acquaintance of both. They shared intellectual and religious interests, but perhaps the greatest attraction for Kat was the fact that John was a relative of her beloved Lady Elizabeth. Hailing from Norfolk gentry, he was first cousin to the late queen Anne Boleyn. This may have won him favor with Elizabeth, for there is no sign that his marrying her governess caused any friction between them. Indeed, John rapidly struck up a strong friendship with Elizabeth and became a valued member of her household. He later recalled the happy times that the three of them had shared in the various houses of Elizabeth's childhood: "Our friendly fellowship together at Cheshunt, Chelsea and Hatfield . . . our pleasant studies in reading together Aristotle's Rhetoric, Cicero and Livy; our free talk mingled always with honest mirth."[41]

This was one of the happiest periods of Elizabeth's youth. When she was living in her own household, she revelled in the company of Kat and her new husband, and she was also treated to regular visits to court, where she spent time with her stepmother. But she had by no means put her turbulent past behind her. The evidence suggests that Elizabeth developed a fascination with her late mother during her prepubescent years.

In 1545 Henry VIII commissioned a portrait of his family in order to reinforce the strength of the Tudor dynasty and confirm the succession. This was idealized, because it included Jane Seymour, who had died in childbed some eight years before. The other sitters were the king himself, along with his three children. Elizabeth and her sister, Mary, had been restored to the succession, if not to their father's affection, the year before, and this painting helped to emphasize their new status. Anne Boleyn's name had been banned from court ever since her execution, for the king hated to be reminded of her. Elizabeth would have known this and would also have been well aware that her return to favor at court was entirely dependent upon the king's notoriously fickle goodwill. And yet she took a colossal risk by deciding to demonstrate her loyalty to her late mother in this new painting. When she sat for the artist, she wore Anne's famous *A* pendant around her neck. This would have been clearly visible in the preliminary sketches, but is barely perceptible in the finished painting. Certainly it was discreet enough to escape the king's eye. The artist aside, Elizabeth alone would have known that it was there. She must have secretly triumphed every time she saw it.

As Elizabeth grew into adulthood, she became ever more her mother's daughter, both in appearance and character. With her red hair and long nose, she was every bit the Tudor, and while others questioned her paternity, Henry never had any doubt that she was his. Yet as she matured, it became increasingly obvious that she had inherited a number of her mother's features. Most strikingly, she had her dark, bewitching eyes that sparkled with something between intelligence, humor, and cunning. She also had Anne's high cheekbones, long, thin face, and pointed chin, together with her swarthy complexion—although in later life she would disguise this with makeup, in order to achieve the luminous white visage of the Virgin Queen.

All of these features can be seen in the portrait of Elizabeth that was painted around 1546, when she was thirteen years old. She has an aura of intelligence beyond her years, and her dark eyes appraise the viewer with almost uncomfortable perception. A prayer book is clasped in the long fingers of which she would later be so proud, and which bore no trace of the "sixth nail" for which Anne was notorious. The slender figure and small, pert breasts inherited from her mother are set off to great effect by a beautifully tailored dress. Elizabeth's red hair is swept neatly under a French hood, a fashion that Anne had introduced at court. Indeed, both women shared a sense of style that went beyond a love of fine clothes. Elizabeth, like her mother, realized that to act the part, she must dress the part. From Anne she inherited the instinct for image making, and would go on to exploit this to spectacular effect—much more so than her mother was ever able to.

Meanwhile, the devoutness of Elizabeth's new stepmother was being remarked upon by many at court. She was said to be "so formed for pious studies that she considered everything of small value compared to Christ."[42] Determined to share her passion for religious reform, Katherine gathered around her a group of intellectual women, each renowned for their "learning" and faith. They included her sister, Lady Herbert, and her stepdaughter Margaret Neville, together with the Duchess of Suffolk, Lady Lisle, and other notables. She would hold "conferences" with these ladies in her apartments, discussing matters of doctrine, hearing sermons, and offering prayers. She also spent many hours reading the scriptures and receiving instruction from chaplains and other learned men. Nicholas Udall, the scholar and playwright, praised her piety and that of her "circle" in a dedication to a translation that he made of a work by Erasmus in 1548: "When I consider, most gracious Quene Katerine, the greate noumber of noble woemen in this our time and countrye of England, not onelye geven to the studie of humaine sciences and of straunge tongues, but also so throughlie experte in holy scriptures, that they are hable to compare with the beste wryters as well in endictynge and pennynge of godlye and fruitful treatises to the enstruccion and edifynges of whole realmes in the knowledge of god, as also in translating good bokes oute of Latine or Greek into Englishe . . . I cannot but thynke and esteem the famous learned Antquitee . . . ferre behynde these tymes."[43]

Among these "noble woemen" was Elizabeth herself. She proved a conscientious student, receptive to everything her stepmother could teach her. The textbooks that she studied included works by Katherine herself. One of the most influential of these was *The Lamentation of a Sinner,* published in November 1547. This drew upon Lutheran and Calvinist teachings, but went much further by promoting the author's equality to men—a shocking concept for the time. Katherine adopted the persona of a ruler who chastised his subjects for their lack of faith. She claimed that Christ embodied the typically female qualities of meekness and humility, while she herself had a very manly ambition to "covet rule over my brethren."

Katherine was not promoting the rights of women over men; she was promoting her own rights, and in so doing, setting herself apart from the rest of the female sex as an exceptional example of learning and authority. It was the image of queenship that the young Elizabeth would take as her role model. Famously lamenting that she was but a "weak and feeble woman," she would assume a masculine stance in order to stamp her authority on a man's world. And as a queen regnant in her own right, she would be able to do so with much greater effect than the woman who had originally inspired her.

The influence that Katherine Parr had had upon Elizabeth's spiritual upbringing was celebrated in the New Year's gift that her stepdaughter gave to the king at the beginning of 1546. It was another translation, but this time the work was Katherine's own. *Prayers and Meditations* was published in May 1545 and included five original prayers written by the Queen. Elizabeth undertook an even more ambitious project than the previous year by translating the work into not one but three languages: French, Italian, and Latin. She used the accompanying letter to let her father know just how much she had been influenced by his sixth wife. Introducing the book as having been "compiled by the Queen your wife . . . [and] translated by your daughter," she lauded it as "a work of such piety, a work compiled in English by the pious industry of a glorious Queen and for that reason a work sought out by all!" Pointing out that theology was the "proper study of Kings," Elizabeth concluded that Katherine's *Prayers* must have been "by your Majesty highly esteemed."[44]

Intended both to signify her adoption of Katherine's teachings and to heighten her father's respect for his wife, Elizabeth's present backfired. Far from esteeming Katherine's work, it is unlikely that Henry had ever read it. When he learned how much it had influenced his impressionable young daughter, however, he was gravely concerned. Theology was too weighty a matter for women to trespass upon, and if Katherine was publishing her own works and gathering around her an ever-growing circle of supporters at court, then she was grossly exceeding her limited authority as queen. He had been browbeaten by Elizabeth's mother, and it had taught him a lesson he would never forget. Katherine would have to be slapped down.

The Queen's enemies at court seized this as an opportunity to get rid of her. Feeding the king's all-too-ready suspicion, a group of religious conservatives led by Stephen Gardiner, Bishop of Winchester, started to put about rumors of her demise. When the king confided in him that he resented his wife's outspokenness in matters of religion, Gardiner went at once to his fellow plotters, who included powerful men like Sir Richard Rich and the Lord Chancellor, Sir Thomas Wriothesley, and gathered enough evidence of Katherine's heresy to condemn her to

death. The men found a number of banned religious texts within her library, which enabled them to secure a warrant for her arrest.

By chance, the Queen heard of this before the men could reach her
and immediately took to her bed, claiming that she was mortally ill.
When the king rushed to see her, she cleverly told him that she was
sick with fear that she had displeased him. Henry admitted that he was
aggrieved at her for overstepping the mark in religious matters, and
Katherine proceeded to give a skillfully submissive defense of her actions. She professed that she had entered into spiritual debates with
him only in order to take his mind off his many ailments, as well as to
learn from his responses. Knowing that one of the chief causes of his
displeasure was that she, a mere woman, had meddled with affairs that
were rightly the preserve of men, she went on to plead the weakness
of her sex. "Your Majesty doth know right well, neither I my self am
ignorant what to great imperfection and weakness by our first creation
is allotted to us women, to be ordained and appointed as inferiour and
subject unto Man as our head from which head all our direction ought
to proceed." She then resorted to simple flattery, telling her husband
that he was "so excellent in giftes and ornaments of wisdom," whereas
she was "a simple poor woman so much inferiour in all respects of nature to you."[45] By the end of her speech, Katherine had talked herself
out of trouble. The king was instantly mollified and railed against Gardiner and all the others who had dared to question his wife's loyalty.
Katherine knew, though, that it had been a close call. She had narrowly
escaped with her life and would not risk it again.

Elizabeth was almost certainly at court when this controversy unfolded. Earlier that year, Katherine had succeeded in gaining the king's
consent that his two daughters might join her household on a more or
less permanent basis. Elizabeth and Mary duly headed the list of ladies-
in-waiting "accustomed to be lodged within the King's Majesty's
house." Witnessing the Queen's sudden fall from grace, Elizabeth
must have felt sick with terror at the thought of losing the dearest of
all her stepmothers, and perhaps also guilty at the part she had unwit-
tingly played in it by her ill-advised choice of New Year's gift. But
Katherine's submission, like that of Anne of Cleves six years earlier,

had taught Elizabeth the wisdom of pragmatism. Much as she might have shared Katherine's stance on her natural equality with men, she realized that this was not a belief to be defended in a court that was still ruled by men. Only when she herself became queen, with no male consort to limit her powers, did she reignite this notion and exercise it to staggering effect.

The controversy of Katherine's near arrest and condemnation had clearly shaken her, and thenceforth she played a much more low-key role at court. Fortunately for Elizabeth, her father's suspicions had been sufficiently allayed for her to continue spending time with her stepmother. When the court was at Greenwich or Whitehall, Katherine insured that the girl's apartments were situated next to her own: an honor that singled her out as the Queen's favorite stepchild. "The affection that you have testified in wishing that I should be with you in the court, and requesting this of the king my father, with so much earnestness, is a proof of your goodness," Elizabeth enthused in a letter to her stepmother. "So great a mark of your tenderness for me obliges me to examine myself a little, to see if I can find anything in me that can merit it, but I can find nothing but a great zeal and devotion to the service of your Majesty."[46] Things seemed to be settling into a comfortable pattern, and Elizabeth was enjoying her prolonged stay at court. She attended the Christmas celebrations at Greenwich in 1546 and was pleased to see her stepmother fully restored to favor, as the king talked and feasted with her. But a little over a month later, everything had changed. Henry VIII, England's seemingly invincible king, was dead.

CHAPTER 5

Governess

Elizabeth was with her brother, Edward, at Ashridge when they received the news that their father had died. The children were said to have clung to each other and wept piteously. Henry's widow, Katherine, was also bereft, for if she had never truly been in love with her husband, she had at least come to care for him deeply. Her grief was intensified by disappointment when her hopes of assuming the regency during Edward VI's minority were dashed. Having distinguished herself when she had taken the reins three years earlier, she confidently expected to do so again. Her late husband had apparently never shared her confidence, however, and three days after his death, Edward Seymour, brother of the late queen Jane, was appointed Lord Protector of England. Under his leadership, the Privy Council would rule until the nine-year-old Edward reached his majority. Katherine was afforded no role at all in the government of the realm.

The king's death heightened the tension between his two daughters. Mary was generously provided for in her father's will, which also reaffirmed her place in the order of succession. She inherited substantial estates in East Anglia, as well as her favorite residences of Hunsdon in Hertfordshire and New Hall in Essex. This was only fitting, as she was the elder of the two daughters and higher up in the order of succession, so naturally she took precedence over her half sister in the division of Henry's spoils. But Elizabeth may have had cause for some resentment, given that she had been raised in a household that was structured around her needs above those of her half sister.

Mary and Elizabeth, along with their brother, were taken back to court in the immediate aftermath of Henry's death. However, they could not remain there for long because the new king had no queen and it was not considered fitting for unmarried ladies to be present without a female household in which to serve. A few days after Henry's death, they therefore joined the entourage of their widowed stepmother. This was only intended as a temporary arrangement, because it was expected that the two daughters would soon move to their own estates.

Thomas Seymour, the Lord Admiral and brother of the Lord Protector, now began to renew his advances to the dowager queen. Theirs was an attachment of some duration, and it is likely that they would have married some years before had Katherine not come to the attention of King Henry. But Seymour was far from being a one-woman man, and his name had been attached to various other ladies of standing at court. Among them were the king's own daughters.

Seymour had once paid court to Mary but had been swiftly rebuffed. Kat Astley was of the firm belief that Henry had been on the point of offering the Lord Admiral his younger daughter's hand in marriage when death had robbed him of the chance. She had little to base this upon, apart from the rumors that were always virulent at court, together with the fact that Henry had enhanced Seymour's standing by appointing him a Privy Councillor five days before he died. As Kat tried to interfere in the matter, the indiscretion and naivety of her character appeared in sharp relief. It seems that she was driven not merely by the best interests of her young charge, but by a kind of vicarious pleasure. She was undoubtedly attracted to the handsome Seymour and anticipated a union between him and Elizabeth with almost as much excitement as if she herself had been involved. She later admitted that she had tried to further the matter while King Henry was still alive by telling Seymour that she believed he and Elizabeth were well matched. Coming across him in St. James's Park one day, she had remarked: "I had heard it said he should have married my lady." Seymour had denied it at once, jesting that "he loved not his life to lose a wife," and that the council would never allow it.[1]

Meanwhile, Katherine's feelings toward Seymour had clearly never gone away, for within as little as three months of Henry's death, she had married her former suitor. The wedding was conducted in such secrecy that even today it is still not known exactly when it took place, but it is likely to have been in either April or May 1547. Such an impetuous action was at odds with the calm good sense for which Katherine was so well known. But perhaps this was the point: After three political marriages, she had had enough of acting out of duty. Now it was time to take something for herself. At thirty-four—an advanced age for marriage in Tudor times—she may have felt it was her last chance of happiness. There may also have been an element of defiance in her decision to marry the Lord Protector's brother so swiftly. The late king had denied her a place in the government of the country, so why should she show him any loyalty now?

The unseemly haste with which the widowed queen married again brought upon her the condemnation of the council, whose permission she had not sought. It also caused a scandal at court. Edward VI noted in his journal: "The Lord Seymour of Sudeley maried the Quene whose nam was Katerine, with wich mariag the Lord Protectour was much offended."[2] The dowager queen, still angry at being ousted from the government of the country, bitterly resented his disapproval. Shortly after the wedding, she wrote indignantly to her new husband: "my Lord your Brother hathe thys Afternone a lyttell made me warme. Yt was fortunate we war so muche dystant, for I suppose els I schulde have bytten hym." Such a show of passion was rare for a woman who was renowned for her calmness and good sense. In her fury, she was determined to go see the young King Edward himself, "wher I intend to utter all my coler to my Lord your Brother," but assured Seymour that she would obey him if he advised her not to do so.[3] It is not clear whether she made good her threat.

When she heard of Katherine's marriage, Princess Mary was outraged at what she perceived to be a blatant show of disrespect to her late father. Seymour had sought her help in securing Katherine's consent to marry him and had met with a sharp rebuke. "Consyderyng whose wyef her grace was of late . . . I ame nothyng able to perswade her to forget the losse of hyme, who is as yet very rype in myne owne remembrance," she wrote.[4] By the time this letter arrived, Seymour had already married Katherine. Close though she had been to her last stepmother, Mary cut off all contact and refused to have anything more to do with her.

Elizabeth's response was rather more measured. Although she paid lip service to the same disbelief that was felt by the rest of the royal family, it was with telling alacrity that she accepted Katherine's invitation to live with her at Chelsea in west London, delighted at the prospect of being reunited with her stepmother. Edward VI's Privy Council had soon established a routine of rigid ceremonial, and the staid, almost puritanical atmosphere formed a stark contrast to the glittering world of his father's court. Little wonder that Elizabeth showed no hesitation when Katherine's letter arrived.

When Mary heard of this, she was horrified. The only way she could comprehend it was to assume that Elizabeth had felt she had nowhere else to go. She therefore wrote at once to the girl, urging her not to associate with the dowager queen, considering the "scarcely cold body of the King our father so shamefully dishonoured by the Queen our stepmother."[5] She went on to offer Elizabeth a place in her own household so that the two sisters might show a united front in their disapproval of their stepmother's actions. Elizabeth's reply was full of deference and respect, assuring Mary of her loyalty and obedience. But she clearly had no intention of passing up the opportunity of being reunited with her stepmother, and therefore told Mary that they must "submit with patience to that which could not be cured." She added that although Katherine's behavior might not have been entirely proper, "the Queen having shown me so great affection, and done me so many kind offices, I must use much tact in manoeuvring with her, for fear of appearing ungrateful for her benefits."[6]

Mary left in disgust for her estate at New Hall in Essex. Although her relationship with Elizabeth had been difficult for some time, this was the first show of defiance on the part of her younger sister. She made it clear that she seriously disapproved of Elizabeth's choice and predicted (rightly, as it turned out) that it would damage the girl's reputation.

Kat Astley was similarly aghast at her protégée's decision to live with Katherine Parr. She was still bitterly disappointed that her hopes for a marriage between Elizabeth and Thomas Seymour had come to nothing, and was also jealous of the woman to whom Elizabeth had become increasingly attached. Her jealousy may account for her subsequent actions, which would run counter to practically every behavior expected of a royal guardian.

Elizabeth heartily approved of her stepmother's choice of husband— but perhaps not for entirely selfless reasons. Seymour was lively, audacious, and handsome; a magnet for the ladies of the court. Fully aware of his charms, he had a seductive self-confidence that many found irresistible—Elizabeth included. If she felt any jealousy or regret when she heard that he had married her stepmother, she did not show it. Perhaps she was glad of the change that he had brought in Katherine, who was clearly besotted with him and overjoyed to have finally married for love. In any case, she showed no compunction in joining the household of her stepmother and her new husband, even if he was her own former suitor. She did not do so lightly; she realized that by thus associating herself with Katherine, she would probably be denied the court. Although this had lost its appeal, it was still the center of political power, and Elizabeth knew that she was risking her position by forsaking it. It is a testament to the esteem in which she held her step-mother that this pragmatic young girl chose Katherine over her own political welfare.

Together with her tutor, William Grindal, and a select number of her household staff—including, of course, her governess, Kat Astley—Elizabeth joined Katherine and Seymour's household at the pretty manor house of Chelsea in the summer of 1547. Her joy at being re-united with the most beloved of her stepmothers was not tempered by fear that she could be sent away at any time, as had happened so often when her father was alive. "Madam Elizabeth . . . will remain always in her company," noted the Imperial ambassador.[7]

The house was situated on a beautiful stretch of land close to the river Thames, surrounded by picturesque gardens and woodland. It was a handsome building of red brick, two stories high and furnished with well-appointed rooms. It also had many amenities that would have been considered a luxury at the time, including piped water from a nearby spring. In short, it was a fitting place for a dowager queen's re-tirement, and Elizabeth looked set to enjoy a very comfortable life there. As well as her stepmother, she also had another well-born young woman for company: Lady Jane Grey, her second cousin. The ambi-tious Seymour had struck a deal with Jane's father, the Duke of Suf-folk, whereby he bought her wardship and marriage rights for the sum of £2,000.[8] The Duke later claimed that Seymour had promised to se-cure Edward VI as Jane's husband, although Seymour denied this.

Chelsea proved a stimulating place for Elizabeth's precocious mind. Together with Lady Jane Grey, who also had a considerable intellect, she enjoyed many hours of lessons with William Grindal, under the overall supervision of her stepmother. This happy arrangement would soon come to an end, however, for Grindal's health was declining, and he died a few months after joining the dowager queen's household, in January 1548. This was a blow to Elizabeth, who had flourished under his tutelage for the past four years. Much as she regretted his death, however, she was anxious to continue her education, and therefore asked her stepmother if Roger Ascham could take his place. Katherine had planned to offer the post to Francis Goldsmith, a loyal retainer, but she was well acquainted with Ascham and realized that he would be an even more suitable replacement. She therefore readily agreed to her stepdaughter's request. Her confidence was well placed. Elizabeth thrived under Ascham's careful tutorship and eagerly absorbed all the learning he could convey. Within a short time of his appointment, Ascham was writing in awe of his new pupil: "Her ears are so well practiced in discriminating all these things and her judgement is so good, that in all Greek, Latin and English compositions there is nothing so loose on the one hand or so concise on the other which she does not immediately attend to, and either reject with disgust or receive with pleasure as the case may be." He went on to pay her an even higher compliment by the standards of the time, claiming: "Her mind has no womanly weakness . . . her perseverance is equal to that of a man."[9]

As well as enhancing her studies, living with her stepmother brought Elizabeth other advantages. It was during this time that she first became acquainted with the man who was to play a hugely important role in her life. William Cecil, an able and ambitious young courtier, came to pay his respects to Elizabeth at Chelsea, possibly at Katherine's invitation. The pair apparently got on well straightaway, and Elizabeth trusted this wise and assiduous man more quickly than was her custom. Cecil agreed to take on the management of her estates and revenues, and proved so effective in this task that Elizabeth entrusted him with other matters. She quickly came to rely upon his advice and guidance, particularly in affairs of state. It was this early acquaintance that laid the foundations for the partnership that was to be one of the most successful in British monarchical history.

Meanwhile, the geographical separation of Elizabeth and her sister Mary had increased the emotional distance between them. With Elizabeth living at Chelsea and Mary at New Hall in Essex, they were literally on opposite sides of the capital. On the surface, relations between them remained cordial, and they exchanged courteous letters. Mary was by far the more faithful correspondent, however, and it seemed that Elizabeth was enjoying the diversions of her new life at Chelsea too much to bother replying to all of her elder sister's sober letters. Nevertheless, when in autumn 1547 Elizabeth learned that Mary had

been unwell, she wrote to express her concern. "Good Sistar as to hire
of your siknes is unpleasant to me, so is it nothinge fearful, for that I
understande it is your olde gest that is wont oft to viset you, whose
cominge, though it be oft, yet it is never welcome."[10] She went on to
thank her for "your oft sendinge to me." Elizabeth's gratitude rings a
little false, however, for she then declined a request to send Jane Russell, one of her servants, to attend Mary in her illness. Claiming that
she was prevented from doing so by Jane's husband, William, who had
apparently refused to spare his wife even for a short while, she expressed her regret that she could not help her sister. Even if this were
true, surely if Elizabeth had been genuinely concerned for Mary's welfare, as William's employer she could have ordered him to release his
wife.

But Elizabeth was too preoccupied with her new life at Chelsea to
give much thought to her elder sister. She was clearly thriving under
her stepmother's care, and the first few months that she spent in her
household were the happiest and most stable she had known since that
first summer with Katherine during the latter's regency in 1544. Life
was full of stimulation and excitement as she pursued her studies with
one of England's finest scholars and, during her leisure hours, enjoyed
the lively company and entertainments on offer in the household. Her
stepmother loved music and was acquainted with some of the greatest
musicians of the age, who may have accompanied the regular dances
at Chelsea.

Elizabeth was developing sexually as well as socially and intellectually, and was blossoming into a striking young woman, with the fine
features and bewitching eyes of her mother, and an abundance of the
trademark red hair of her Tudor forebears. She was an alluring
prospect to any suitor, and a fatally irresistible one to her stepfather,
Thomas Seymour.

According to Kat's later testimony, early one morning, before the
household had risen, she and her charge were shocked by the sudden
appearance of the Lord Admiral in Elizabeth's bedchamber. All innocence, he smilingly told them that he had simply come to bid his stepdaughter good morrow.[11] The two women no doubt soon recovered

themselves and laughed about it together, but both had been affected. Elizabeth already seemed infatuated with Lord Seymour and was seen to blush whenever he was spoken of. Kat, meanwhile, was evidently excited by such intimacy and gave little thought to its inappropriateness. When her husband, who was evidently a good deal more conscious of the potential gravity of the situation, warned her not to encourage Seymour's advances, she chose to ignore him.

Before long, Seymour's morning visits to Elizabeth's bedchamber had become a regular habit. And the more often he came, the more outrageous his behavior grew. Kat herself recounted: "He wold come many Mornyngs into the said Lady Elizabeth's Chamber, before she was redy, and sometyme before she did rise. And if she were up, he wold bid her good Morrow, and ax [ask] how she did, and strike hir upon the Bak or on the Buttocks famylearly." Things soon got out of hand. "If she were in hir Bed," Kat continued, "he wold put open the Curteyns, and bid hir good Morrow, and make as though he wold come at hir: And she wold go further in the Bed, so that he could not come at hir." By now, even the indiscreet Kat was alarmed, and when upon the occasion of his next visit, Seymour "strave to have kissed hir in hir Bed," she admonished him, telling him to "go away for shame."[12]

Kat had realized, too late, the dangers this flirtation could bring to her fourteen-year-old protégée. In vain she tried to restrain the Lord Admiral as he came again and again to Elizabeth's bedchamber, dressed only in his nightgown. She reprimanded him that it was "an unsemly Sight to come so bare leggid to a Maydens Chambre," at which he cried: "What do I do? I would they all saw it!" and stormed out in anger.[13] Kat knew that he would be back, however, and, driven on by the fierce protectiveness that she had always felt toward Elizabeth, she decided to tell Lady Katherine of Seymour's shocking behavior.

At first, Katherine dismissed Mistress Astley's concerns, assuring her that Seymour's antics amounted to little more than a light-hearted prank—a sign of his affection for his new stepdaughter. As if to humor the distressed governess, however, she offered to accompany her husband on his morning visits to Elizabeth in future.

Katherine's reluctance to believe Kat has been viewed as a gross misjudgment on her part. But this is based upon the knowledge of what happened afterward. Should Katherine have shown greater caution at this stage? Perhaps not. She was renowned for her calm good sense and not accustomed to acting rashly. And she had little more to go on than the testimony of a woman who was known to be impetuous, indiscreet, and overly fond of gossip. Moreover, the evidence suggests that Katherine had scant regard for Kat Astley. The two women had been rivals for Elizabeth's affection ever since Katherine's marriage to Henry VIII, and each had tried to fulfil the role of mother figure during this time. It cannot have been a comfortable arrangement with them both now under the same roof. Katherine may therefore have believed that Mistress Astley was trying to stir up trouble between herself and her stepdaughter.

Katherine's skeptical reaction to the concerns of the governess may also have been due to the closeness that existed between herself and Elizabeth. The latter's many expressions of love and devotion to her stepmother were proof that she appreciated just how much she owed her. By the time of Henry's death, the trust that the two women shared was seemingly unimpeachable. It is therefore to Katherine's credit that she refused to believe that her stepdaughter would so betray her as to accept—or even encourage—the advances of her husband. Besides, she did not suspect that there was anything improper in Seymour's behavior; she held him in too great an esteem to believe him capable of infidelity at so early a stage of their marriage.

However, when Katherine started to keep a closer eye upon her husband, she saw more than enough to corroborate Kat Astley's suspicions. Seymour constantly sought out Elizabeth's company and was increasingly physical in his contact with her. Rather than restraining him, Katherine became an accomplice to his "jests"—on one occasion pinning back Elizabeth's arms as her husband cut the girl's dress "into a hundred pieces."[14] What on earth had possessed this otherwise astute and sensible woman? By this time, Katherine was pregnant with Seymour's child, so perhaps the hormonal disruption had clouded her accustomed judgment. Or perhaps she was so blinded by her love for her new husband that she could not see that he was deceiving her.

There is another explanation. Katherine may have known full well that Seymour's attentions toward Elizabeth were motivated by lust, not fatherly affection as he claimed. She also knew her husband's temperament. He was not the sort of man to be gainsaid or restrained: that would be the surest way of losing his love. She knew, too, that he lusted most after that which remained out of his grasp. Far from giving up after losing her to the king, he had remained on the sidelines, determined to take his chance if it should ever be offered again—as it was four years later. If she had insisted that he leave Elizabeth well alone, then she would only have increased the girl's appeal in his eyes. She may therefore have judged that the best way to cure him of his temporary infatuation was to allow him the prize that he sought. Of course, she had no intention of permitting him to gain the ultimate victory of sleeping with Elizabeth, but she was apparently willing to let him taste just enough of her charms to sate his appetite.

If such it was, then her plan backfired spectacularly. Far from being satisfied by the fleeting touches and caresses that his romps with Elizabeth allowed him, these merely served to stoke his desire still further. Perhaps his wife's condition intensified his lust, for it was believed to endanger the life of the unborn child if the mother indulged in sexual intercourse during pregnancy. Elizabeth therefore became the sole focus of his attention.

By the time the household moved to Hanworth around spring 1548, Seymour's behavior had got so out of hand that his wife was compelled to act. She summoned Kat Astley and instructed her to "take more heed, and be as it were in watch betwixt the Lady Elizabeth and the Admiral." It was with a certain satisfaction that the governess subsequently confided to Elizabeth's cofferer (or principal officer of the household), Thomas Parry, that "the Quene was jelowse on hir and him."[15]

Things finally came to a head when Seymour alleged that he had happened to look through a window of the house and had seen Elizabeth "cast hir Armes about a Man's Neck."[16] This was almost certainly a fabrication on his part, designed to cover his own guilt in the matter. A short while later, Katherine, whose suspicions were by now strongly aroused, went in search of her husband and stepdaughter, and "cam sodenly upon them, wher they were all alone, (he having her in his Armes)." Outraged at being so deceived, she ordered Elizabeth to leave her house at once.[17] Rather than accepting this meekly, Elizabeth doggedly insisted upon her innocence and railed against the injustice of her punishment. Such a blatant attempt to deny the truth in the face of apparently incontrovertible evidence angered Katherine even more, and the two women proceeded to have a furious row.

They were barely on speaking terms by the time Elizabeth left the household a few days later in June 1548, hardly a year since her arrival. But despite the hurt and anger at discovering her stepdaughter's disloyalty, Katherine could not entirely relinquish the love that she felt for her. As Elizabeth, still resentfully silent, took her formal leave, Katherine impulsively told her that she would send warning if she heard that any rumors about the affair had got out. This would enable the girl to defend her reputation, which her stepmother knew was one of the most precious things she had.

Elizabeth's dismissal from Katherine's household had a profound effect upon the girl. The shock of it brought her immediately to her senses and made her realize the foolishness of her actions. Even if she had not committed the ultimate betrayal, she had still put her reputation in danger by allowing a married man of Seymour's rank and profile to flirt with her so openly. As she and her servants made their way to Cheshunt, the Hertfordshire home of Kat Astley's brother-in-law, Sir Anthony Denny, Elizabeth had cause to reflect upon the events of the previous few months. The realization of how close she had come to danger made her resolve never to be so reckless in the future.

Cheshunt provided Elizabeth with much-needed privacy while the scandal in the dowager queen's household began to die down. Fortunately, it had not spread much further, so it seemed that Elizabeth had escaped with her reputation intact—no thanks to her governess. Kat might well reproach herself for her conduct in the whole affair. She had been carried along by her romantic sentiments toward Seymour, a notorious rogue, and had failed to appreciate the danger that his indecorous behavior had put her charge in, even when warned about this by her own husband. When this fact had finally dawned upon her, she had done her best to salvage the situation, but by then it was too late. Elizabeth was clearly besotted with the Lord Admiral, and her reputation had been sullied by her apparent connivance in those morning romps in her bedchamber.

But what if Kat's failure to control her charge had done much worse harm than this? There has been speculation that by the time Elizabeth left Katherine's household, she too was pregnant with Seymour's child. She was kept in seclusion at Cheshunt, and Kat reported that she was sick. The nature of her illness was not specified, but she was laid low with it for some time. Shortly afterward, a rumor was circulated by a local midwife, who claimed that she had been called upon to assist a lady in a "great house," and taken there blindfolded so that she would not know the identity of the family. She was ushered into the house and attended a young lady who was in labor. A short while later, the girl was delivered of a stillborn baby. The midwife was returned to her home, still none the wiser about who she had attended. But it did not take long for her to surmise that it must have been a lady of some importance to necessitate such secrecy. And was not the Lady Elizabeth reported to be ill at nearby Cheshunt? Surely the coincidence was too great.

But stories such as this were forever being put forward where members of the royal family were concerned. Without other evidence to corroborate it, it is at least equally possible that Elizabeth's illness that summer was due to the stress of recent events. However far things had really gone with Lord Seymour, in a sense the damage had already been done: it was enough that people were merely speculating that Elizabeth had been defiled. In this first real test of her abilities as a guardian, Kat Astley had failed miserably.

For a while, it looked as if the whole sorry affair would blow over, and Kat would be able to repair the damage done to her protégée. Elizabeth had evidently spent a good deal of time reflecting upon her stepmother's parting words. No doubt humbled by Katherine's benevolence and, for all her defiance, appreciating that she had instead deserved her censure, she resolved to do everything she could to make amends. Upon reaching Cheshunt, she wrote at once to her stepmother, thanking her for her kindness and assuring her that she had been "replete with sorowe to departe from your highnis." She went on to excuse her sullenness, saying: "although I answered little, I wayed it more dipper [deeper] whan you sayd you wolde warne me of al euelles [evils] that you shulde hire of me, for if your grace had not a good opinion of me you wolde not have offered frindeship to me that way." She concluded the note with the heartfelt words "thanke God for providinge suche frendes to me."[18]

Elizabeth's letter had the desired effect. Katherine was eager to forgive her stepdaughter's behavior, which she no doubt put down to little more than youthful indiscretion. She may well have been right. Elizabeth's furious protestation of her innocence, together with her subsequent resentment toward the woman who had suspected her of adultery, suggests that while she may have allowed Seymour to overstep the mark, she never conceded to his desire for a full sexual relationship. It is likely that a combination of loyalty to Katherine and a keen sense of the danger of such a relationship to her own position, prevented her from giving in to this, no matter how much she might have wanted to.

The speed at which Katherine and Elizabeth overcame this first serious challenge to their friendship is a testament to the depth of their love for each other. There is no trace of any lingering hostility in the letters they exchanged after their separation in June 1548. Katherine wrote several times to Cheshunt telling her stepdaughter how much she was missing her. Elizabeth replied to each of these with pledges of loyalty and affection. "My humbel thankes that your grace wisshed me with you til I ware wery of that cuntrye . . . although hit were in the worst soil in the wor [world] your presence wolde make it pleasant," she wrote a month after their separation.[19] After a time, Katherine even allowed her husband to write to Elizabeth, and the latter repaid her trust by sending a politely detached reply, thanking him for his attention but clearly putting a stop to anything more.

Although the affair quickly died down, it had taken its toll on Katherine. Heavily pregnant with her first child at the comparatively advanced age of thirty-six, the stress of it all had seriously damaged her health. Elizabeth expressed her anxiety that her stepmother had seemed very ill when the two women had parted at Hanworth, and sent regular inquiries after her health. Even though Katherine assured her that she was much recovered, she was clearly still unwell. In addition to the increasing exhaustion and discomfort of her pregnancy, she had injured her wrist so badly that she was barely able to reply to her stepdaughter's letters. "Although your hithnys letters be most joyfull to me in absens, yet consyderinge what paine hit ys to you to write your grace beinge so great with childe, and so sikely [sickly] your comendacyon wer ynough in my Lordes lettar," Elizabeth assured her.[20]

Katherine derived some comfort from removing to her husband's beautiful castle of Sudeley in Gloucestershire shortly after Elizabeth's departure, where she awaited the birth of her child. From there, she wrote to her husband at court, assuring him that his "little knave" was in good health and kicking her boisterously to prove it. Seymour replied that she should keep the child lean by feeding it a good diet and taking regular exercise, so that "he may be small enough to creep out of a mousehole."[21]

A few weeks after issuing this wry piece of advice, Seymour joined his wife at Sudeley. Her pains began shortly afterward, and on August 30, she was delivered of a girl—not the "little knave" they had confidently expected. The baby was christened Mary in honor of her eldest stepdaughter, whose attitude toward Katherine had softened upon hearing that she was with child. Their joy was short lived, however, for Katherine soon fell prey to that most dreaded of childbed illnesses, puerperal fever. In her delirium, all the suppressed pain and humiliation caused by her husband's betrayal burst forth, and she ranted against him as he stood by her bedside with her ladies, desperately trying to comfort her. Seizing her husband's hand, she appealed to Lady Tyrwhit, one of the ladies present, crying: "I am not wel handelyd, for thos that be abowt me caryth not for me, but standyth lawghyng at my Gref; and the moor Good I wyl to them, the les Good they wyl to me." In vain, Seymour tried to calm her with the assurance that he had meant her no harm. As Lady Tyrwhit later recalled, "she saed to hym agayn alowd, no my Lord, I thinke so; you have geven me many shrowd tauntes." The more her husband tried to pacify her, the more Katherine dealt with him "rowndly and shortly."[22]

Eventually, after six days of lying in this state, racked by pain and sorrow, Katherine died on September 5. She was buried the same day at Sudeley Castle. Lady Jane Grey, who had replaced Elizabeth as her protégée, was the chief mourner. Lord Seymour appeared to be genuinely grieved at the sudden loss of his wife, for whom, despite his indiscretions, he had cherished a real affection. He was reported to be "the heaviest man in the world."[23] It was he who sent news of her death to Elizabeth at Cheshunt. There is no record of her reaction, but the fact that she was shortly afterward reported to be gravely ill suggests that the news had affected her badly. From the time that Elizabeth was nine years old until two days before her fifteenth birthday, Katherine had been one of the most influential women in her life. Grief at the loss of her stepmother was no doubt mingled with guilt at the thought of the pain she had caused her. This guilt would have been intensified when she heard of Katherine's "unquyettydd" mind as she ranted against Elizabeth and Seymour for so betraying her.

In the immediate aftermath of Katherine's death, Elizabeth seemed determined to honor her stepmother's memory. A number of disputes had arisen over the late dowager queen's estates. Among the claimants was the Duchess of Somerset, wife of the Lord Protector, who tried to get her hands on one of Katherine's London residences, Durham Place. Perhaps using this as an excuse to reestablish close contact with Elizabeth, Seymour wrote to ask her to intervene with the duchess on his behalf. He received a curt response. "In faith I will not come there, nor begin to flatter now," Elizabeth told him.[24]

Her governess showed a good deal less fidelity to the late queen. Apparently having learned precious little from the scandal, she immediately started plotting to marry Seymour to Elizabeth. With unseemly haste, she told her charge that "her old husband, appointed at the king's death, was free again, and she might have him if she wished." This is where Elizabeth's greater maturity, at fifteen years old, than that of her governess, by now well into her forties, is revealed. The girl immediately dismissed Kat's foolish proposition, saying that her marriage was a matter for the king and the council to decide. She knew—as Kat should have done—that for a royal to marry without the sovereign's permission was treason. Yet still Kat persisted. Surely Elizabeth would not refuse him if the council did give its permission? In the meantime, she ought to write to the Lord Admiral, assuring him of her continued affection. Again it was Elizabeth who had her wits about her. She refused to write, "lest she be thought to woo him." Undeterred, her governess asked for leave to go to London and pass on Elizabeth's good wishes—and no doubt a good deal more besides. Elizabeth would have none of it.

If she had left the matter there, Kat would have avoided a great deal of future trouble. But her preoccupation with Lord Seymour bordered upon infatuation, and she was determined to have the vicarious pleasure of seeing him married to Elizabeth. She therefore continued to wheedle and plead, praising Seymour's qualities incessantly and insisting that Elizabeth had been the real object of his affections all along. Gradually, with nobody else to advise her, Elizabeth began to succumb to the alluring fantasy that her governess was concocting. While she still refused to write to the Lord Admiral herself, she agreed that Kat could do so.

Triumphant, Kat could hardly refrain from talking about the matter and evidently told her husband all about it. Perhaps not surprisingly, this resulted in a bitter row between them, and John Astley forbade her to meddle any further in the matter, warning her with some considerable foresight that "the admiral's suitors would come to an evil end."[25] Furious at her husband's obstinacy, Kat left for London, no doubt intending to seek out Lord Seymour, despite Elizabeth's instructions to the contrary. Seymour evidently knew of Kat's weakness for him and played on it, sending her a flirtatious message to inquire "whither her great Buttocks were grown eny les or no?"[26]

Kat stayed in London for some time but evidently failed to see Lord Seymour herself, so she sent a message to him via the cofferer Thomas Parry, a rather self-important busybody. Kat instructed Parry to assure Seymour of her own goodwill and friendship, to which he replied:

"Oh, I know she is my Frend." When Parry went on to relay Kat's message that "she wold her Grace were your Wief of any Man's lyvyng," Seymour sensibly retorted that his brother, the Lord Protector, would never agree to it.[27] He did, though, make a suggestion that would have been considered most improper at the time. He said that he could pay a visit to Elizabeth on his way to Sudeley Castle. Parry wrote to tell Mistress Astley this, and she, for once, showed a measure of decorum, replying that "he shuld in no weye come hether for feyr of Suspicyon." She then went to tell Elizabeth all about it, who was "mych offendyd with her," telling her that she should not have committed the matter to paper because then there would be proof that she knew of the proposal.[28]

But this was a mere show of propriety on the young girl's part, for when Parry returned to Hatfield, she immediately besieged him, desperate to hear everything that had passed between him and the Lord Admiral. She listened, enraptured, as he told her of Seymour's "Gentlenes and kind offeres," and urged him to go and tell Kat what he had just told her, no doubt keen to compare notes with her governess later. Kat was every bit as excited as her protégée, if not more so.

A short while later, Kat dined with Thomas Parry and his wife in their chambers. After supper, their talk turned to Lord Seymour and the prospect of his marrying their mistress. According to Parry's account, he expressed concern about this, saying that Seymour was "not onely a very covetouse Man, and an Oppressor, but also an evill jelouse Man." He went on to describe "how cruelly, how dishonestly, and how jelowsly he hadd used the Quene." Kat was having none of it. "Tushe, tushe," she said, "that is no Matier, I know him better then ye do." She affirmed that she "wold wishe her his Wife of all Men lyving." Warming to her theme, she confided to Parry that the dowager queen had once found Elizabeth in Lord Seymour's arms, and that this had been the cause of their peremptory departure from her household. Registering the shock on Parry's face, she knew that she had gone too far. She begged him not to repeat what she had told him, and Parry assured her, "I had rather be pulled with Horses." With one final plea for discretion, Kat took her leave.[29]

Parry was not true to his word. By Christmas 1548, the rumor that Lord Seymour was on the verge of taking Elizabeth as his wife was the talk of the whole court. It was said that Seymour had retained his late wife's ladies so that they could wait upon his new bride, Elizabeth. Thanks to Kat, there was more evidence to support these rumors. She had apparently engineered clandestine visits between Elizabeth and Seymour, flouting the restraints to which an innocent young lady should have been subject—particularly one of royal blood. Anne Seymour, Duchess of Somerset, wife of the Lord Protector, was horrified to learn that Mistress Astley had allowed her charge to go on a romantic boat ride on the Thames at night with only Seymour for company. She exclaimed that Kat "was not worthy to have the Governance of a King's Daughter," and said that "another shuld have her Place, seeing that she bare to much Affection to my Lord Admyrall."[30]

Things now began to unravel with alarming speed. In mid-January 1549, Thomas Seymour was arrested on the charge of high treason and committed to the Tower. A key part of the evidence against him was that he had conspired to marry the king's sister without the council's permission, which in itself would have been enough to send him to the block. Elizabeth was deeply shocked when she heard of it, but how much greater must her dismay have been when she was soon afterward told that her beloved Kat had also been arrested and taken to the Tower, along with Thomas Parry. "She was marvelous abashede, and ded weype very tenderly a long Tyme," reported Sir Robert Tyrwhit, the formidable official who had been appointed to interrogate the princess at Hatfield. Elizabeth would have been fully aware of just how much danger her governess was in. Only seven years earlier, Jane, Viscountess Rochford, had proved how fatal complicity in royal affairs could be when she had been executed for helping to arrange Katherine Howard's secret meetings with Thomas Culpepper. Was Elizabeth's beloved Kat now to meet the same end?

The council had decided that the best way of uncovering the truth would be to interrogate Elizabeth separately from Mistress Astley and Thomas Parry so that the notes could be compared afterward and any discrepancies found out. This was the greatest ordeal Elizabeth had

faced since her mother's execution, and as she was still only an adoles-
cent, it must have been terrifying indeed. But if she had been deprived
of Kat Astley, then she could at least rely upon her old servant Blanche
Parry for support. The records suggest that the latter was with Elizabeth throughout this anxious time. Tyrwhit even used her to relay
messages to the princess, which must have helped soften the blow of
what they contained.

At first, all three of the accused stood by their story that nothing
improper had ever passed between Elizabeth and Lord Seymour, that
there had been no offers of marriage, and that in any case Elizabeth
would certainly not have consented to such without the council's permission. "I never secretly moved [Elizabeth's] affections to [Seymour]
or any other, but always counselled her to keep her mind safe and at
the council's appointment," Kat insisted. "I told her not to set her mind
on it, seeing its unlikelihood."[31]

Elizabeth was just as firm, and her answers to Sir Robert Tyrwhit
show the strength of her loyalty toward her beloved governess. She insisted that Kat had not spoken to her about marrying Seymour, much
to the frustration of Sir Robert, who clearly did not believe her. "She
will no more accuse Ashley than herself," he told his master, the Lord
Protector, "and cannot now abide anybody who disapproves of her doings."[32] When he tried to cajole her into confessing her servants' guilt
on the basis that "all the Evyll and Shayme shuld be ascrybyd to them,
and her Yowth consedered," he was given short shrift. "I do parsav
[perceive] as yet, she wyll abyd mo [more] Stormys, or [before] she ack-
ews Mestrys Aschlay," he complained.[33] He had got no further a week
later. Having read the transcripts of Mistress Astley's and Parry's interrogations and compared them with his own, he was convinced that the
three must have conferred beforehand. "They all synge onne Songe,"
he lamented, "and so I thynke they wuld not do, unles they had sett the
Nott befor: For surly they wold confesse; or ells they could not so well
agree."[34]

Getting nowhere with their prisoners, the interrogators at the
Tower changed tack. They moved Kat Astley to one of the darkest,
most uncomfortable cells in the entire prison. Let her see how long she

would continue to stand by her mistress under those conditions. Kat was wretched with terror and discomfort, as the full implications of what she had done seemed at last to sink in. Miserably she lamented her "great folly in speaking of marriage to such a person as she," and promised that if she were allowed to return to Elizabeth, she would never commit such a transgression again. "I have suffered punishment and shame," she went on, pleading that the council be lenient toward "this first fault." When they showed no signs of relenting, she begged them to move her to a different cell: "Pity me . . . and let me change my prison, for it is so cold that I cannot sleep, and so dark that I cannot see by day, for I stop the window with straw as there is no glass." It is a testament to her love for Elizabeth that despite the wretchedness of her condition, she still would not betray what had really happened with Lord Seymour. As her interrogators continued to press her, she claimed that she could not remember all the events because "My memory is never good, as my lady, fellows and husband can tell, and this sorrow has made it worse."[35]

With Elizabeth still remaining tight lipped at Hatfield, she and Kat might well have weathered the storm. But their resolve was not matched by that of Thomas Parry. In the middle of February, a month after their arrest, he tried to save his own skin by telling his interrogators everything they wanted to know about Seymour's relationship with Elizabeth. The half-dressed romps in her bedchamber, the young girl's blushes every time he came into the room, their moonlit ride along the Thames, her being caught in his arms—everything was revealed in all its sordid detail. At last the interrogators had what they wanted. Confronted with Parry's confession, the half-truths and falsehoods of Elizabeth and her governess would collapse like a house of cards.

"False wretch!" wailed Kat upon hearing of Parry's betrayal. So much for being "pulled by horses": it had taken only verbal threats to get him to tell everything. Knowing there was now no choice, she reluctantly uncovered all the details she had so firmly held back before. Yes, she admitted, Seymour had "come at" Elizabeth in her bedchamber on many occasions; he had tickled her and kissed her and cut her gown to shreds. And yes, she and her royal mistress had talked of the marriage "diverse Tymes," and Kat had "wishid both openly and priv/ely, that thei two were maried together."[36]

When Mistress Astley had miserably confessed the whole sorry story, a messenger was dispatched to Hatfield, where a triumphant Sir Robert Tyrwhit received the news that would finally defeat his stubborn opponent. With barely concealed satisfaction, he showed Elizabeth her governess's confession, at which "she was much abashed, and halffe Brethles," horrified that all the lurid details of her relationship with Lord Seymour had been (literally) laid bare. After she had recovered herself, she showed a courage that was remarkable for a fifteen-year-old, refusing to implicate either Kat or Parry in the affair. "In no weye she wyll confesse, that owr Mestrys Aschlay, or Pary, wylled her to eny Practys with my Lord Admyrall, ether by Message or Wryttynge," a dismayed Tyrwhit wrote to his master. Much to his disappointment, neither would she corroborate or deny the salacious details that they had revealed about her romps with Seymour. All that she did admit was that she had talked of the Lord Admiral "manye Times" with her governess, and the latter had tried to persuade her that he intended to marry her. She was careful to add, however, that she had told Kat that nothing could be done without the council's consent.[37]

This latter fact, which was backed up by Kat Astley's and Thomas Parry's confessions, was what saved Elizabeth. However improper her relations with Seymour had been, she had never given any indication that she would be so foolish as to marry him without the council's permission. It was therefore impossible for them to convict her of treason. Her relief was short lived, however, for the council subsequently decreed that Mistress Astley was "far unmeete . . . to se to the good Education and Government of your Parson." Kat was hauled before them and told "rowndeley" that "she hath not shewed herself so moche atendant to her Office in this Part, as we looked for at her Hands."[38] She was therefore dismissed from her position as governess and replaced by Lady Tyrwhit, the wife of Elizabeth's interrogator.

When Elizabeth heard of this, she was devastated. "She took the

Matter so hevely, that she wepte all that Nyght, and lowred all the next Day," reported Sir Robert Tyrwhit, who was dismayed that the girl could so mourn the loss of one who had been such a bad influence upon her. "The Love that she beryth her [Kat Astley] ys to be wondert at," he told Edward Seymour, Lord Somerset and Lord Protector of England. When Elizabeth eventually emerged from her bedchamber, red eyed and pale faced, she defiantly told him that "Mestrys Aschlay was her Mestrys, and that she had not so demened her selffe, that the Counsell shuld now need to put eny mo Mestressys unto her . . . the Worlde wold nott [note] her to be a great Offender, havyng so hastely a Governor appointid to her." Refusing to submit to Lady Tyrwhit's authority, she stubbornly clung to the conviction that she would be able to "recover her old Mestrys agayne."[39]

But as time wore on, it became clear that this would not be the case, and Elizabeth had no choice but to accept Lady Tyrwhit as her new governess. Her reluctance to do so was driven by more than her loyalty to Kat, strong though that was. Lady Tyrwhit had a reputation as a religious firebrand, and her fiercely puritanical views would have been anathema to Elizabeth, who favored a much more pragmatic approach. Even Sir Robert admitted that his wife was "not sane in divination." How such a sternly pious woman must have viewed the disgraceful impropriety of Seymour's relationship with Elizabeth can well be imagined. But in Elizabeth she had met her match. Far from being bowed by Lady Tyrwhit's self-righteous condemnations, Elizabeth would not hear a word spoken against Lord Seymour, and if ever her new governess criticized her predecessor, Mistress Astley, the girl would be "very redy to make answer veamently."[40]

In early March, shortly after Kat's dismissal from her service, Elizabeth received the news that Lord Seymour had been found guilty of treason and condemned to death. Quite apart from any feelings she might have had toward him, Elizabeth was distraught at the prospect that her beloved Kat and Thomas Parry, who were still prisoners in the Tower, could meet the same fate. Putting aside her grief for Seymour, she therefore resolved to write to the Lord Protector and plead for Kat's release. It was the most important letter she had ever written: the life of her beloved governess—the woman who had been a mother to her for more than twelve years and had shared all of her grief, joys, hopes, and fears—now depended upon every word she wrote.

Elizabeth began with a show of humility, thanking Somerset for issuing a proclamation of her innocence, as she had requested him to do. She then turned to the main subject of the letter. Shrewdly realizing that she would get nowhere by simply insisting on Kat's innocence, as she had done so many times in the past, she acknowledged that her governess had been somewhat at fault. "I do not favor her in any iuel [evil]," she wrote, "(for that I wolde be sorye to do), but for thes consideracions wiche folowe." Numbering these "consideracions" in turn, Elizabeth proceeded to put the case for Mistress Astley with as much skill and eloquence as the most highly trained lawyer. First, she asked the Lord Protector to consider Kat's service to her: "she hathe bene with me a longe time, and many years, and hathe taken great labor, and paine, in brinkinge of me up in lerninge and honestie." Second, she pointed out that whatever Kat did to further Lord Seymour's cause with Elizabeth, she did on the assumption that he would have kept the council fully informed of it, being a member of it himself. Finally, she claimed that if they continued to detain her former governess, "it shal and doth make men thinke that I am not clere of the dide [deed] myselfe, but that it is pardoned in me bicause of my youthe, bicause that she I loved so wel is in suche a place."[41]

In writing all of this, Elizabeth was putting herself at great risk. She had not been charged with treason because there had been insufficient evidence that she had acted without the council's knowledge and consent. But it had been a close call, and she knew that her favor with the Lord Protector and his councillors was tenuous at best. She had already succeeded in persuading them to proclaim her innocence, but all the signs were that this had exhausted their scant reserves of goodwill toward her. In trying to persuade them to now pardon and release the woman who had been unquestionably at fault in the whole sordid affair, Protector Somerset could well have decided that Elizabeth was pushing her luck too far and ordered her arrest. But the gamble paid off. To Elizabeth's joy and relief, Kat was released from the Tower, along with her partner in crime, Thomas Parry. Her triumph was not quite complete, for upon one matter the council would not relent: Mistress Astley would never be restored to her old post of governess during its watch.

For the time being, though, it was enough for Elizabeth to know that Kat was safe. She therefore settled into a quiet and orderly routine at Hatfield under the auspices of Lady Tyrwhit. Relations between the two women gradually began to thaw as Elizabeth resigned herself to this new authority. Although she lacked Kat's warmth and humor, Lady Tyrwhit was not without sympathy, and she gave her charge much-needed stability after the turmoil of the preceding months. Elizabeth even grew to appreciate the strength of her religious convictions, and she benefited from the routine of prayers, meditations, and devotions that her new governess introduced in the household. She learned from Lady Tyrwhit's own "godly sentiments" that she had written down as a guide of conduct for daily life. One of her favorite maxims—"Be always one"—so inspired Elizabeth that she translated it into Latin as *"Semper eadem"* and adopted it as her own motto.

For all her goodwill toward Lady Tyrwhit, Elizabeth had by no means forgotten her old governess, and she continued to work toward her reinstatement. Eventually, some two years after her release from the Tower, Kat was restored to Elizabeth's household, along with her husband, John.[42] The reunion between Kat and Elizabeth was an emotional one: After spending almost every day together for twelve years, the separation had seemed endless. The closeness of their relationship was restored immediately, and Kat remained at Elizabeth's side throughout the remainder of Edward's reign.

Once more, Elizabeth's life seemed to have settled down into a more comfortable pattern. She was even able to welcome a new addition to her household. Lady Katherine Howard was the daughter of Henry Carey, later first Baron Hunsdon, the son of Mary Boleyn.[43] The date of Katherine's birth is uncertain, and could have been anywhere between her parents' marriage in 1545 and 1550. Katherine was raised in Elizabeth's household from a very early age, and she quickly formed a close attachment to her. Philip II of Spain's envoy, the Count de Feria, would later remark that the young girl "was brought up with her [Elizabeth] and is devoted to her."[44] Like Elizabeth Fitzgerald, who had also spent time with Elizabeth as a child, Katherine grew up to become one of her most trusted ladies at court when Elizabeth became queen.

Elizabeth also paid regular visits to her former stepmother, Anne of Cleves. Anne's status had diminished upon the death of Henry VIII. Edward's council viewed her as an irrelevance, not to mention a drain on its resources, and confiscated two of the manors that Henry had given her—Richmond and Bletchingly. Forever the pragmatist, Anne resolved to make the most of the life that she had left. She established her house at Hever as a lively social center—a kind of miniature court, where she could receive esteemed guests from across the kingdom. Through these guests, she kept abreast of events at court and solicited invitations to visit it herself. But her favorite guest by far was Elizabeth. Anne provided the young girl with a much-needed mother figure away from the confines of her household. She may not have rivalled Kat Astley's place in Elizabeth's heart, but she was a more sensible and level-headed role model. Through her visits to Hever, Elizabeth also became better acquainted with her own mother's history. There would certainly have been mementos of Anne Boleyn at the castle.

Now that Elizabeth was established in her own household and Katherine Parr was no longer a factor in her life, there seemed to be the chance of a reconciliation with her elder sister, Mary. In fact, the situation became worse. While Elizabeth suffered from a tainted reputation, Mary, pure as ever, gained the moral high ground and exploited it to full advantage. She no doubt derived some satisfaction from being proved right about the unsuitability of Elizabeth's choice of guardian. Her younger sister's behavior might also have reminded her of Anne Boleyn, whose notorious flirtatiousness had landed her in similar trouble. Like mother, like daughter—not a thought to render Elizabeth more pleasing to Mary than she had been before their separation the previous year. Their growing hostility was noted by the Venetian ambassador, who reported that Mary "demonstrated by very clear signs" that she no longer loved her half sister.[45]

Although Elizabeth had displeased her brother the king by going to live with their disgraced stepmother and by her involvement in the Seymour scandal, she was soon back in favor. By contrast, Edward had little time for their elder sister. Even before he had become king, he had come to resent Mary's presence at court. On one occasion, he had complained about her to Katherine Parr, asking that she should "attend no longer to foreign dances and merriments" because it was "not becoming a Christian princess."[46]

By the time of his accession, Edward had become a strictly devout Protestant and reformer, and deeply resented Mary's intransigence on matters of religion. She had resisted the Lord Protector's Act of Uniformity, continuing to celebrate Mass with great ceremony at her residences. This was an open defiance of Edward's authority, and he was not prepared to tolerate it. He ordered that she was no longer to hear Mass, even in the privacy of her own house, and made her unwelcome at court. "Your near relationship to us, your exalted rank, the conditions of the times, all magnify your offence. It is a scandalous thing that so high a personage should deny our sovereignty; that our sister should be less to us than any of our other subjects is an unnatural example," he raged. "Truly, sister, I will not say more and worse things because my duty would compel me to use harsher and angrier words. But this I will say with certain intention, that I will see my laws strictly obeyed, and those who break them shall be watched and denounced."[47] Accordingly, it was noted that "the care they [the king and the council] had of her [Mary] is decreasing daily, and is principally shown in making her move from one house to another."[48]

This treatment, together with her brother's reprimand, enraged Mary and sparked a show of that same stubbornness that had caused her such trouble during her father's reign. Although she claimed that Edward's letter had "caused me more suffering than any illness even unto death," she continued to defy his commands, and as a result became a figurehead for opposition to his regime.[49] Religious opinion was still very much divided in England, and while many welcomed the reforming zeal of the new king, others regretted the demise of the monasteries and longed for the country to return to the Roman Catholic fold. In courting the support of the religious conservatives,

Mary was playing a dangerous game. But she was driven more by a genuine passion for that faith than by any sense of political advantage, and she refused to give way. She had withstood the terrifying wrath of Henry VIII; in her eyes, the displeasure of Edward VI was probably no more than a little brother's tantrum.

During the remaining years of Edward's reign, Elizabeth and Mary saw little of each other. They continued to exchange letters, but they were seldom at court together, and there is no evidence that they visited each other's houses. With her dogged conservatism, Mary increasingly represented the past, while Elizabeth and Edward represented the future. Although the latter referred to Mary as "our nearest sister," he was furious at her stubborn nonconformity and increasingly demoted her to second place behind Elizabeth, whose views and beliefs were so much closer to his own. As the reign progressed, the younger of his two half sisters was treated with ever-greater honor. Upon one of her visits to court, she rode into London "with a great suite of ladies and gentlemen," and was received by an impressive delegation of councillors and noblemen. This was a studied gesture on the part of the young king to "show the people how much glory belongs to her who has embraced the new religion and is become a very great lady." It was subsequently noted that Elizabeth was "continually with the King," and that he and his councillors "have a higher opinion of her for conforming with the others and observing the new decrees, than the Lady Mary, who remains constant in the Catholic faith and stays at her house 28 miles from here without being either summoned or visited by the Council."[50]

By 1553, when Elizabeth was nineteen, everything augured well for her future. She had survived the crisis of the Seymour scandal and was now in great favor with the king. Knowing of his growing disapproval toward their half sister, she may even have hoped that he would alter the succession to give herself precedence. Any such hopes were incidental, however, for there appeared every chance that Edward would go on to marry and produce a long line of male heirs to succeed him. He was still only fifteen years old and had always enjoyed good health. But if her life thus far had taught Elizabeth anything, it was how quickly fate could turn against her. In the spring of that year, her young brother suddenly fell ill. An apparently minor chest infection had turned into something altogether more serious, and within weeks it was clear that the boy was dying. The cause is not certain, although the symptoms were consistent with tuberculosis. Knowing that Mary would undo all the reforms for which he and his council had worked so hard, Edward was loath to leave his throne to her. Persuaded by ministers, he therefore altered the succession, not—as Elizabeth might have hoped—in her favor, but in that of his young cousin, Lady Jane Grey.

This was not just unexpected, it was illegal, and Mary knew it. When Edward died on July 6, she immediately sent the council letters claiming her right to the throne. For once, her characteristic stubbornness won her the day. Convinced of the justice of her cause, she mustered considerable forces in East Anglia, where she had fled upon hearing that her brother was close to death, knowing that the council in London was against her. Setting up camp at Framlingham Castle, the ancient fortress of the Dukes of Norfolk, she attracted ever-greater numbers to her cause, most of whom were staunchly traditional men opposed to both the religious reforms of Edward's reign and the attempts to overthrow the rightful Tudor succession.

Meanwhile, Elizabeth remained at Hatfield, hedging her bets by refraining from showing support either to Mary or to Jane, her former companion at Chelsea. She did not have long to wait. On July 19, barely two weeks after Edward's death, Elizabeth received the news that Mary had triumphed. Terrified by rumors of the forces that were gathering in ever-greater numbers to support her, the council had capitulated and abandoned its plot to place Lady Jane upon the throne. It was with mixed feelings that Elizabeth contemplated Mary's new status. Her half sister, with whom she had never enjoyed an easy relationship, was now her queen.

CHAPTER 6

Sister

M ary Tudor ascended the throne on a wave of popular rejoicing. There were street parties across the capital and lively celebrations throughout the realm. A woman she might be, but she was also a Tudor, and as such the only true heir in the eyes of most Englishmen. Elizabeth wrote at once to offer her congratulations. Showing all due deference, she also humbly craved Mary's advice as to whether she ought to appear in mourning clothes out of respect for their brother, Edward, or something more festive.[1] Although relations between them had soured during Edward's reign and Elizabeth had failed to throw her support behind Mary's campaign for the throne after his death, Mary was apparently prepared to be magnanimous in her triumph. She therefore invited Elizabeth to accompany her to London.

On July 29, flanked by a magnificent retinue that included the two thousand soldiers whom she had refrained from offering to her half sister in her time of need, Elizabeth rode through the city of London.

Mary had evidently advised her against wearing mourning, for she sported the Tudor colors of green and white. Having taken up residence at Somerset House on the Strand, she and her company set out the next day to meet Mary at Wanstead, to the east of the capital. Her reception was so cordial that any uninformed observer would have thought that the pair had long been affectionate sisters. Mary embraced her half sister warmly and proceeded to kiss each of her ladies in turn. She subsequently presented them with gifts of jewels and gave Elizabeth an exquisite necklace of white coral beads trimmed with gold, together with a ruby-and-diamond brooch. In the celebrations that followed, the new queen accorded her half sister first place after herself and appeared anxious to keep her by her side at all times. It seemed that Mary's accession had healed the old wounds between the two sisters and that thenceforth they would enjoy mutual affection and harmony.

In fact, this was to be a fleeting high point in their relationship. Even as Mary rode through the streets of London in celebration of her triumph, accompanied by Elizabeth as a demonstration of the unity of the Tudor dynasty, the seeds of discord were already being sown. Naturally introspective and lacking her father's ability to charm and enthrall the crowds, Mary progressed through them, responding awkwardly to their cheers and appearing distant and aloof. When a group of poor children sang a verse in her honor, it was noted with disapproval that she "said nothing to them in reply."[2] By contrast, Elizabeth, who had inherited Henry VIII's gift for public relations in abundance, attracted the most attention as she gracefully inclined her head and waved her hand, making every member of the crowds that thronged the streets feel that she had saluted him or her personally. As the procession gained ground, she drew loud cheers and used her instinct for the theatrical to attract more still. "Her Grace, by holding up her hands and merry countenance to such as stood far off, and most tender and gentle language to those that stood nigh . . . did declare herself thankfully to receive her people's good will," remarked one bystander.[3]

Elizabeth's popularity was enhanced by her appearance. With her "comely" face, long, flowing red hair, "fine eyes," and youthful exuberance, she enjoyed by far the greater share of beauty between the two sisters. The Venetian ambassador remarked how much she resembled her mother.[4] Like Anne, Elizabeth was not recognized as a conventional beauty, but she knew how to make the best of herself, and she had that same indefinable allure that drew men to her. She was also taller than her sister, who was described as being "of low rather than of middling stature." Although she was only in her midthirties, Mary appeared much older. The turmoil and sadness of her youth had aged her prematurely, and the sombre, tight-lipped expression she wore did nothing to lift her lined face. "At present, with the exception of some wrinkles, caused more by anxieties than by age, which make her appear some years older, her aspect, for the rest, is very grave," remarked the Venetian ambassador. Her appearance was not helped by the fact that she had lost nearly all her teeth in her twenties. Among her most noticeable features were her eyes, which were so piercing that they "inspire, not only respect, but fear, in those on whom she fixes them." In fact, Mary's tendency to stare intently at people was due more to her severe shortsightedness than an intention to intimidate, but her gruff voice, which was "rough and loud, almost like a man's," did not make her any more appealing.[5]

Although she loved fine clothes and paid a great deal of attention to her wardrobe, Mary lacked the sense of style that came so naturally to her half sister. She would adorn herself in richly decorated gowns of bright colors that clashed with her red hair rather than complementing it. Even the Spanish ambassador was forced to admit that if she dressed more stylishly, then "she would not look so old and flabby."[6] By contrast, Elizabeth dressed with understated elegance, favoring simple gowns of white or green that set off her coloring to perfection. While Mary was embarrassed by her sexuality and preferred to hide her emaciated figure in heavy, high-necked gowns, Elizabeth flaunted hers with a knowingness beyond her years, exuding a sex appeal that many men at court found irresistible.

Despite their differences, the sisters made a convincing show of unity in these critical early days of the new reign. With every prospect that Mary would rule for a long time, and perhaps even beget an heir, Elizabeth had reason to court her favor. She also sought to benefit from her half sister's experience as the first queen regnant for more than four hundred years. Not since Matilda had claimed the throne in 1141 had England seen a woman rule in her own right. That episode had spelled disaster, plunging the country into a prolonged civil war. The subsequent examples of Isabella, wife of Edward II, and Henry IV's queen, Margaret of Anjou—both of whom had tried to seize power—provided further proof of how disastrous it was for the country when a woman was at the helm. Little wonder that Henry VIII had been so desperate in his pursuit of a male heir. When this heir himself had lain dying and his councillors had tried to justify preventing Mary from inheriting the throne, they had protested "the inferiority of the female sex," which was a flimsy argument, given that they had subsequently named another woman as heir.[7]

Most of Edward VI's subjects shared these prejudices. Even though there was no law forbidding a woman to reign, in contrast to France, it was not seen as desirable. Quite apart from Matilda's unfortunate example, there was the fact that women were generally regarded as the weaker sex, entirely dependent upon and subservient to fathers, husbands, and brothers. They could manage affairs of a domestic nature, but not affairs of state. Queens were there to produce heirs, not to rule. Even those who had proved themselves capable of understanding and dealing with greater matters faced the seemingly impenetrable barrier of male prejudice. "A woman is never feared or respected as a man is, whatever her rank," explained Mary, queen dowager of Hungary, to Charles V, when resigning her regency.

To many men, the very concept of a woman ruling over them was not just abhorrent, it was unnatural. "To promote a woman to beare rule, superioritie, dominion or empire above any realme, nation or citie, is repugnant to nature, contumelie to God, a thing most contrarious to his reveled will and approved ordinance, and finalie it is the subversion of good order, of all equitie and justice," railed the Protestant preacher John Knoxe. Women, he argued, were "weake, fraile, impacient, feble and foolish: and experience hath declared them to be unconstant, variable, cruell and lacking the spirit of counsel and regiment."[8]

Despite the almost universal hostility to female rulers, Mary had at least ascended the throne on a wave of popularity. Also in her favor was the example provided by her maternal grandmother, the formidable Isabella of Castile, who had been one of the greatest rulers in Spanish history, combining the apparently conflicting duties of military leadership and producing heirs with staggering success. Mary's mother, Catherine of Aragon, had been an immensely popular queen, albeit a traditional consort rather than a ruler in her own right. If Mary had wished, she could have drawn upon these strengths and shattered the perceptions of female rule by setting herself up as a powerful figurehead, providing decisive leadership for her country. But in truth, she was every bit as conservative in her attitude toward women's place in society as her male courtiers. Therefore, although she undoubtedly had the courage and capacity for hard work to make a success of her new role, she lacked the vital ingredient of political confidence, convinced that her sex was a fatal impediment to her ruling effectively.

Nevertheless, she went through all the due ceremonies inherent in establishing a new sovereign's position. First was the task of establishing her government and household. Her sex caused further complications here. Traditionally, officials in both the king's privy chamber and his council had been of equal importance, and in some cases interchangeable. With the accession of a queen regnant, who must be attended by female servants, this could no longer continue. All the much-coveted positions in Mary's privy chamber were awarded to women. At a stroke, this effectively neutralized the household department that had long been the most politically important, because, unlike their male counterparts, women could not serve their sovereign in both a personal and a political capacity.

Mary set the tone of court life by appointing ladies of irreproachable character, such as Susan Clarencieux and Jane Dormer, who served her with the utmost loyalty and had no ambition to interfere in matters of government. All the Queen's ladies were staunch Catholics, and coupled with their high morals, they made her privy chamber a sober and devout retreat, which was very much to the taste of their royal mistress. The later contrast to Elizabeth's glittering court filled with glamorous attendants could not have been greater. Upon first arriving in England, a member of Philip II of Spain's entourage would remark: "The Queen is well served with . . . many ladies, most of whom are so far from beautiful as to be downright ugly," complained one of Philip II's entourage, "though I know not why this should be so, for outside the palace I have seen plenty of beautiful women with lovely faces."[9]

The presence of the thirty or so women who surrounded the new queen was a constant in Mary's life. Very much a woman's woman, she revelled in their company and valued their advice, and was so jealous of their affection that she disapproved of their leaving her service to marry. Neither did she wish them to meddle in matters of government, and she made it clear that they should restrict their activities to household affairs. In both these respects, she anticipated the attitude shown by her half sister—much more extremely—when she became queen.

As well as her household attendants, Mary also had to appoint her councillors. This was no easy task, for there were two opposing forces in government, and she could not risk alienating either. The first was her inner circle of supporters, who had proved loyal during the uncertain years leading up to her accession. The second comprised the men who had held power during her brother's reign, and who had been responsible for the coup that had placed Lady Jane Grey upon the throne. The latter had apologized profusely to Mary, assuring her of their allegiance. Although she still did not trust them, she could not afford to dismiss them because most had considerable influence in their localities. She therefore created a council which, compared to those that had served her father and brother, was large and unwieldy, bringing together staunch Catholics and reformers. Although there were inevitably tensions among such a large and diverse body of men, all were united by their loyalty to the Tudor regime.

Mary's government and household thus established, the next priority was to organize the ceremony of her coronation. This was not a simple task, for there was no guidance for the anointing of a woman as a ruler in her own right. The only precedents to draw upon were those for the coronation of a queen consort. In the event, it was decided to proceed with all of the ceremonials that would have been accorded a male ruler. Thus Mary eschewed the traditional white worn by queen consorts to their coronations, and donned the blue of kings. She also wore the customary gold trelliswork cap and gold garland, both embellished with jewels and pearls, which her male predecessors had worn—although these proved too heavy for her delicate frame, and she had to hold them up with her hands as she rode beneath the canopy of state.

An impressive array of knights, councillors, peers, and ecclesiastics accompanied her procession, together with an unprecedented number of ladies—one report estimated as many as seventy. Elizabeth was given a place of honor, along with her former stepmother, Anne of Cleves. The two women shared an open chariot richly arrayed with crimson velvet and cloth of silver. It was pulled by six horses bedecked in the same gorgeous material. Elizabeth and Anne were also given new dresses made from a similarly rich silver material, so the whole sight must have been quite spectacular to the onlookers who thronged the processional route. When Elizabeth and her former stepmother alighted at the Abbey, they walked together directly behind the new queen. Once inside, everything proceeded according to the council's carefully laid plans. Mary received the full anointing of a male sovereign, "on the shoulders, on the breast, on the forehead and on the temples."[10] She was now England's first crowned queen regnant.[11]

Even though her coronation had been an undoubted success, strengthening Mary's image as queen, there remained some complex legislative problems as a result of her sex. All the monarchical powers were based upon a male ruler, and it was not clear whether they could or should apply to a woman. This uncertainty could not continue, so in April 1554, Parliament passed "An Acte for declaring that the Regall power of thys realme is in the Quenes Maiestie as fully and absolutely as ever it was in anye her mooste noble progenytours kynges of thys Realme."[12] Put simply, the act determined that there would be no distinction between male and female with regard to the powers of the crown. This would potentially benefit not just Mary but every subsequent queen regnant. Mary herself, though, failed to exploit the powers it gave her, and it was only when Elizabeth ascended the throne that its full potential was realized.

But Mary was no pushover. One contemporary observer described her as having "a terrible and obstinate nature," and at least initially, she set about the task of governing with determination and conscientiousness.[13] Keen to get to grips with matters of state, she would spend long hours forgoing both food and sleep as she pored over official documents, deliberated over appointments, and tried to solve problems that would have tested the most experienced politician. She also displayed a fierce resolve to return England to the Roman Catholic fold, and would pursue this ambition with increasingly blind fanaticism over the ensuing years. But although her upbringing had given her this sense of purpose and courage, it had not given her the skill of leadership. Mary had none of the guile and shrewdness necessary to succeed in the fickle world of Tudor politics, and she ceded much of her authority to her councillors. "Respecting the government and public business, she is compelled (being of a sex which cannot becomingly take more than a moderate part in them), according to the custom of other sovereigns, to refer many matters to her councillors and ministers," observed the Venetian ambassador.[14] She was also dangerously single minded, governed by principle rather than pragmatism—the exact opposite of her half sister.

If Mary was blind to these faults, she was painfully aware of what she perceived as her essential weakness as a woman. Within weeks of her accession, she told the Imperial ambassador that she "knew not how to make herself safe and arrange her affairs."[15] As a result, her response to the challenge of carving out a position for herself as England's first queen regnant for centuries was to resort to the conventionally female strategy of seeking a husband. Desperate to relieve herself of at least some of the burden of government and to relinquish, as Ambassador Renard put it, "those duties which were not the province of ladies,"[16] she made it clear that her first priority was to marry. Her councillors assumed that she would consult them on such a weighty matter, but Mary had already made up her mind. The bond she had forged with her cousin Charles V during childhood had been strengthened by the natural affinity she felt for Spain because of her late mother. Although Charles had broken their early betrothal by marrying his cousin, Isabella of Portugal, he now had a son of marriageable age.

At the time of Mary Tudor's accession, Philip of Spain was twenty-six years old, some eleven years her junior. He had already been married once, at the age of sixteen, to Maria Manuela, Princess of Portugal, who had died giving birth to his son, Don Carlos, two years later. Already inclined to favor Philip because of his connection with her mother's homeland, and having precious little experience with men, Mary is said to have fallen desperately in love with him upon first seeing his portrait. She was now even more determined that this was the man she would marry, and would brook no opposition from her council. Neither—fatally—did she appreciate the strength of feeling among her xenophobic people, who had enough to deal with in reconciling themselves to a female ruler, let alone one who had allied herself to a foreigner. "The English . . . are most hostile by their nature to foreigners," remarked the Venetian ambassador.[17] Clearly this official, who had spent little time in England compared to Mary, understood it better than she did herself.

Her councillors were also well aware of the anti-Spanish feeling, but realizing that it was futile to attempt to dissuade the Queen from her choice, they determined to carve out a marriage settlement that would significantly restrict her intended husband's powers. Although it was agreed that Philip "should have and enjoy jointly together with the said most noble Queen his wife the style, honour and kingly name of the Realm," his role was confined to simply "aiding" Mary, rather than ruling over her, as would be the expected prerogative of any other husband. A number of specific limitations to his authority were then spelled out, including the inability to control appointments to office and to take his wife out of the realm without her express consent. Neither was he to drag England into his father's wars with France. Even her royal jewels were denied him. In short, from the council's perspective, it was imperative to safeguard Mary's identity as queen regnant from her position as Philip's wife.

Philip himself and his Spanish advisers clearly saw the situation

very differently. They expected him to provide the masculine leadership so lacking in Mary's sovereignty. He would not just advise her; he
would rule for her and "make up for other matters which are impertinent to women."[18] It seems that Mary herself believed that this would
be the case, for she made no secret of her desire for guidance from her
new husband. She also hoped above all that the marriage would provide her with that other essential prerequisite of sovereignty: an heir.

Once the marriage settlement had been agreed to in January 1554,
the council published it in order to allay the suspicions of the people,
who feared that England would become a mere hand puppet to the
might of the Spanish empire. But these fears remained: a fact of
which, in her excitement at the prospect of Philip's arrival, Mary was
blissfully unaware. Her councillors were acutely conscious of them,
however, and put further safeguards in place by including terms specific to the forthcoming royal marriage in an act, which also confirmed the equality of male and female power. Although this allowed
Philip to "ayde your highnes, being his wife, in the happye adminis-
tracion" of her realms, it went on: "that youre maiestye as our onely
Quene, shal and may, solye and as a sole quene use, have, and enioye
the Crowne and Soverayntye of, and over your Realmes, Dominions
and Subiectes . . . as your grace hath had . . . before the solemniza-
tion of the sayde mariage."[19]

Mary's superiority over her husband was further emphasized during
the marriage ceremony, in which she took precedence, being always on
Philip's right, in the exact opposite of the traditional arrangements for
royal weddings. But almost as soon as the ceremony was over, Mary
made it clear that she intended to give her new husband a good deal
more authority than the settlement had prescribed. She wrote to the
Lord Privy Seal, instructing him "First to tell the kyng the whole state
of the Realme . . . Second to obey hys commandment in all thynge[s]."
She added that Philip should be permitted to declare his opinion on any
matter he chose.[20] As if to confirm what she saw as their joint sovereignty, Mary ordered a new coin to be struck in honor of their marriage. This showed the couple facing each other, and above their heads
was a crown, placed equally between them.

With her council and most of her people fiercely opposed to Philip's authority, and Mary equally determined that he should exert it, a clash was inevitable.

The rise of anti-Spanish feeling was rapid and, in terms of Mary's authority as queen, disastrous. Within a few months of the wedding, there was already fierce hostility between the English and Spanish at court. "It will be very difficult to make the Spaniards get on well with the Englishmen," Renard predicted shrewdly. "There is the obstacle of language; and then, as I have often explained in my letters, the English hate strangers and have never seen so many of them together at once. Several Spaniards have already been robbed while landing or on their way hither; and they are given bad and insufficient lodgings." A member of Philip's entourage who was on the receiving end of such treatment complained: "We Spaniards move among the English as if they were animals, trying not to notice them; and they do the same to us. They refuse to crown our Prince, though he is their King, for they do not recognise him as such or as in any way their superior, but merely as one who has come to act as governor of the realm and get the Queen with child. When she has had children of him, they say, he may go home to Spain. Would to God it might happen at once!"[21]

Coupled with her increasingly fervent attempts to reestablish Roman Catholicism, Mary's Spanish marriage ensured that within just a year of her accession, she had already alienated a dangerously large number of her subjects. As far as her role was concerned, she was providing her half sister with an example of what not to do.

Elizabeth had evidently foreseen the trouble that the Spanish marriage would cause. Upon hearing of Mary's betrothal to Philip, she had written a cautious letter of congratulation. "I have no doubt it will redound to the glory of God, the repose of your Majesty and the safety and preservation of your Kingdoms," she began. However, she then went on to say that it was "a deep and weighty matter" and warned her half sister: "I doubt not that your will shall be made the instrument of His." Urging her to insure that she clarify her position before signing the marriage treaty, she concluded: "For a house built on a sound foundation can only stand firm, whereas one built on the sand may soon be wrecked by winds and sudden tempests."[22]

Reading this letter, one could be forgiven for assuming that of the two sisters, it was Elizabeth, not Mary, who had the benefit of greater age and experience. Despite being just twenty years old, she had evidently gained a far greater insight into both the politics of female rule and the tenor of public feeling. But was her attempt to warn her half sister of the dangers inherent in her forthcoming marriage entirely selfless, a demonstration of female solidarity in a world dominated by men? Perhaps not. Tellingly, Elizabeth had not warned Mary against the marriage on the grounds that Philip was a foreigner, which was, after all, the greatest drawback to the match. She had cautioned her against marriage full stop. She was all too well aware that her position as next in line to the throne was highly tenuous. If Mary were to marry and beget an heir, then she might well have to give up any hope of one day becoming queen. If she could therefore dissuade her half sister from entering that state a little longer, then it could only be to her own benefit. By the standards of the day, at thirty-eight, Mary was very old to be trying to conceive for the first time. Her childbearing days were numbered, and Elizabeth knew it.

Regardless of Elizabeth's motives in seeking to discourage her half sister from taking a husband, they worked little effect. The harmony between the two women had rapidly disintegrated as the euphoria of Mary's accession subsided. Ambassadors at court wasted no time in stirring up trouble between them. The Imperial representative, Simon Renard, dropped poisonous words about Elizabeth in Mary's ear, claiming that she was a traitor to the new queen and intended her downfall. "The Lady Elizabeth . . . might, out of ambition, or being persuaded thereto, conceive some dangerous design and put it to execution," he counselled, adding that the girl was "clever and sly."[23] Meanwhile, the French ambassador Noialles courted Elizabeth's favor and encouraged her to set her sights on the throne. He also put about a rumor that there was a "misunderstanding between the Queen and the Lady Elizabeth."[24]

For her part, Mary was all too easily persuaded of the evil intentions of her half sister. Renard's comments helped to reawaken the old
hostility toward this young upstart whose very existence had for a time
threatened to destroy her own. "She still resents the injuries inflicted
on Queen Catherine, her lady mother, by the machinations of Anne
Boleyn, mother of Elizabeth," noted Renard.[25] His satisfaction was
even greater when, shortly afterward, Mary confided to him that she
thought Elizabeth would grow to be like her mother, a woman "who
had caused great trouble in the Kingdom."[26]

Relations between the two women took a turn for the worse during
the first Parliament of Mary's reign, which passed an act declaring
Catherine of Aragon's annulment void and thereby confirming Mary's
legitimacy. However, it left unrepealed the clauses relating to Elizabeth's bastardy—they could not both be legitimate daughters of
Henry VIII, because if his first marriage had been valid, then his second had not, and vice versa. Nevertheless, Mary's refusal to alter Elizabeth's status was still a cause of some resentment to the latter.[27] But
the deliberations surrounding the 1554 act seemed to have revived in
the new queen bitter memories of the torment that she and her
mother had suffered at the hands of Anne Boleyn, which in turn provoked her antipathy toward the daughter of the "Great Whore." She
therefore proposed to go still further, for although her younger sister's
illegitimacy had been confirmed, she was still the heir to the throne according to the statute passed in Henry VIII's time. Mary was therefore
determined to repeal this and instead make her cousin Lady Margaret
Douglas heir.

Lady Margaret, Countess of Lennox, was the daughter of Henry
VIII's elder sister, Margaret, and as such had a strong claim to the
throne, even though Henry had barred his older sister's descendants
from the succession. She and Mary also enjoyed a close affinity, being
of a similar age and having shared the same household in their youth.
Margaret Douglas had first come to England as a baby in March 1516,
when her mother had been forced to give up the regency of Scotland
and seek refuge at Henry VIII's court. The king had ordered that Margaret be installed in the royal nursery at Greenwich Palace, joining her

young cousin, Mary, who had been born the previous month. That stay had been brief, but Margaret had returned to England in 1530 and had again been invited to join Mary's household, which was then in residence at the castle of Beaulieu in Hampshire. The two girls immediately struck up a close and lasting friendship, and a contemporary would later recall the "special love" that existed between them.[28] Margaret had also become a firm favorite of the English king, and he indulged her with lavish gifts, such as the ermine-trimmed gowns that he sent to Beaulieu "for the use of our dearest niece."[29]

Margaret Douglas's subsequent career at court had been marked by extreme highs and lows. A shrewd politician, she had succeeded in winning favor with each of Henry's wives, whom she had served in turn. She was probably in attendance at the birth of Princess Elizabeth, and was subsequently named first lady of honor to her new cousin, which made her superior in status to the king's elder daughter. Yet Margaret had been reckless in her personal life. Fully Tudor in appearance, she had the characteristic red hair and long nose, but also striking heavy-lidded eyes and a vivaciousness that charmed many members of the court. The French ambassador enthused that she was both "beautiful and highly esteemed."[30] Among her admirers was Thomas Howard, a cousin of Anne Boleyn, with whom she was secretly betrothed in the spring of 1536. This earned her a prolonged spell in prison. The king was so incensed that he ordered a clause to be added to the 1536 Act of Succession making it treason to "espouse, marry or deflower being unmarried" any of the king's female relations.[31] The ramifications of this act would be felt for centuries. A year later, Margaret was declared illegitimate by the king because her father, Archibald Douglas, had divorced her mother after finding evidence of a precontract on her part. He had successfully argued that this rendered their marriage unlawful, and their daughter had been openly reputed a bastard in Scotland.

Perhaps by way of rebellion, in 1540 Margaret began another clandestine courtship. Her suitor this time was Katherine Howard's brother. When this was discovered, Margaret was once more incarcerated. She was allowed back to court three years later and was sufficiently restored to the king's good graces to be appointed a bridesmaid at his wedding to Katherine Parr. Realizing that his twenty-seven-year-old niece was overly ripe for marriage, Henry arranged a match for her himself. The husband he chose was Matthew Stewart, fourth Earl of Lennox. His reasons were entirely political. In 1543 the Scots had reneged upon an agreement made some time before to marry Henry's heir, Edward, to the young Mary, Queen of Scots. The English king therefore needed an alternative Scottish alliance, and Lennox was an ideal candidate for his niece. But he was also cultured, well educated, and "a strong man, of personage well shaped . . . fair and pleasant faced, with a good and manly countenance."[32] Margaret was instantly smitten, and they were married on July 6, 1544, in St. James's Palace. Henry's two daughters were invited to join in the banquet held to mark the occasion.

The match would be a fruitful one, producing two male heirs, Henry and Charles. But although the king had allegedly told his niece that he "should be right glad if heirs of her body succeeded to the Crown," he cut her out of his will in 1547 following a bitter quarrel over her attachment to the Roman Catholic faith.[33] The fact that he had already declared her to be illegitimate a decade before meant that there was now no prospect of her inheriting the English throne.

With Mary's accession, Margaret's fortunes had changed once more. The two women were united by their faith, and Mary saw her cousin as one of her few genuine allies. She had wasted no time in inviting Margaret back to court, and arranged for her to be accommodated in some of the finest apartments in the Palace of Westminster. A suite of luxurious furniture was ordered for her, together with rich tapestries, new dresses, and precious jewels. In addition, Margaret was granted a generous annual allowance, and she and her household were provided with food and drink at no cost. Not since Henry VIII had considered making her his heir had the countess been treated with such preferment.

If the statute confirming Elizabeth's status as next in line to the throne could be repealed, then so could Henry VIII's attempts to cut Margaret out of the succession. Mary duly began to make much of her cousin, according her precedence at court. Thus she took her place ahead of Elizabeth at all the state banquets and receptions, much to the chagrin of Henry's younger daughter. Margaret was also in attendance at Mary's marriage to Philip of Spain. The countess relished her newfound status. It flattered her natural vanity and ambition, and she soon came to act as if she were the Queen's rightful heir. Indeed, Mary increasingly treated her as such, and it was widely expected that she would soon officially name her as successor.

Even though he had no great love for Elizabeth, William Paget, one of Mary's most trusted advisers, urged caution, all too conscious of the danger of alienating her popular half sister. Mary therefore reluctantly abandoned her plans to disinherit Elizabeth—for now. Nevertheless, the act, together with Mary's obvious favor toward Margaret Douglas, had badly damaged Elizabeth's status. Renard dismissed her as having "too doubtful lineage on her mother's side," and the Venetian ambassador, Michiel, went further still, describing her as "the illegitimate child of a criminal who was punished as a public strumpet."[34] Rather than being bowed by such treatment, Elizabeth displayed the same resilience that her mother had shown when faced with hostility at court and among the people. Adopting an attitude that the Spanish described as "proud and haughty," she asserted that she was every bit as legitimate as her sister the Queen and also made a rare reference to her mother in support of her claim. Michiel noted with some distaste: "Although she knows she was born of such a mother, she nevertheless does not consider herself of inferior degree to the Queen, whom she equals in self-esteem; neither does she believe herself less legitimate than her Majesty, alleging in her own favour that her mother would never cohabit with the King unless by way of marriage, with the authority of the church . . . so that even if deceived, having as a subject acted with good faith, the fact cannot have invalidated her mother's marriage, nor her own birth."[35]

The atmosphere between the two sisters had become unbearably strained. Just two weeks after her accession, Mary confided to Renard that she intended to send Elizabeth from court. Although she apparently changed her mind, her resentment continued. She tried to conceal her "evil disposition" toward Elizabeth under a mask of affection, forcing herself to converse only on "agreeable subjects."[36] Elizabeth also took care to display the utmost courtesy and respect, although there was little real feeling behind it. But there was one issue that threatened to cause their simmering resentment to break out into open hostility. Upon her accession, Mary had made it clear that her most urgent priority was to restore England to the papal fold. Driven on by an evangelizing zeal, she ordered increasingly drastic measures in order to achieve this, not flinching from burning those subjects who persisted in their "heretical" beliefs.

Closer to home, she was determined to bring her younger sister to heel on religious matters. Elizabeth had tactfully declined to attend Mass with Mary at court. The first occasion on which this was raised was the funeral of their brother, Edward. The new queen at first ordered that he should be buried as a Catholic but eventually gave way and agreed that he could be afforded Protestant rites in accordance with the beliefs that he had cherished in life. She insisted, though, upon holding a Mass for his soul at the Tower of London. Elizabeth pointedly failed to attend this, along with their former stepmother, Anne of Cleves, who had been welcomed back to court by Mary. Perhaps hoping that this had set a precedent, the younger sister absented herself from all subsequent Masses at court for the following two months. But Mary was far from content to let this pass. Again Renard intervened, convincing her that there was something more sinister behind Elizabeth's defiance than a mere difference of doctrinal opinion. He urged that "it would appear wise in your Majesty not to be too ready to trust the Lady Elizabeth, and to reflect that she now sees no hope of coming to the throne, and has been unwilling to yield about religion, though it might be expected of her out of respect for your Majesty and gratitude for the kindnesses you have shown her, even if she had only done so to accompany you. Moreover, it will appear that she is only clinging to the new religion out of policy, in order to win over and make use of its adepts in case she decided to plot."[37]

Mary needed little convincing of her half sister's evil intent, and with the approach of one of the most important dates in the religious calendar, she decided to force Elizabeth's hand. She demanded that her half sister attend Mass to celebrate the Feast of the Nativity of the Virgin, on the eve of which Elizabeth had been born some twenty years earlier. By now, Elizabeth had evidently caught wind of the rumors that were whispered at court about her disloyalty to the Queen and had also noticed that Mary "did not show her as kindly a countenance as she could wish."[38] Afraid of where this loss of favor might lead, she beseeched her half sister to grant her a private audience so that she might plead her loyalty. Mary did not immediately accede to her request, but instead kept her in suspense for two days. She then summoned Elizabeth to a secluded gallery, each of them accompanied by one of their ladies. Upon seeing the Queen, Elizabeth immediately threw herself to her knees and, weeping, begged for forgiveness at having so offended her in the matter of religion. She excused herself on the basis that "she had been brought up in the way she held, and had never been taught the doctrine of the ancient religion," and therefore entreated Mary to send her books from which she might learn the true faith, cleverly adding: "so that having read them she might know if her conscience would allow her to be persuaded."[39]

This was a masterly performance on Elizabeth's part. At a stroke, she had convinced her half sister that she genuinely repented her perceived disobedience in religion without actually committing herself to adopting the Catholic faith. Mary was apparently taken in immediately and expressed herself "exceedingly glad to see her turn to such good resolves," assuring her that she would at once arrange for her to be instructed in the old religion. But Elizabeth's mask soon slipped. Having agreed to attend Mass on the Feast day of September 8, she made it as far as the chapel before suddenly complaining of a violent stomachache and begging to be excused. Mary, though, was having none of it and insisted that she abide by her original promise. Elizabeth therefore reluctantly went ahead, although "complained loudly all the way" and adopted "a suffering air" throughout the service.[40]

Less than two weeks after faithfully promising to learn the ways of the Catholic faith, Elizabeth was failing to attend Mass once more. The Queen noted with disgust "that she has half-turned already from the good road upon which she had begun to travel." She now openly declared her distrust for her half sister and demanded to know "if she firmly believed what the catholics now believed," telling her plainly that she suspected her of hypocrisy. Elizabeth put on another show of humility, appearing "timid" and "trembling" as she assured the Queen that she acted out of conscience, not policy. Renard dismissed this as pretense, remarking: "we interpreted her answer and trembling rather differently."[41]

Elizabeth found a welcome supporter in Anne of Cleves. Although the coronation had given Anne cause to hope that her status would be enhanced under Mary, it soon became clear that her religious views set her at odds with the new regime. Anne's position was compromised by the renewed Anglo-Imperial alliance, for she had first been brought to England as part of a coalition against the emperor, Charles V. Soon there were rumors that she was conspiring with Elizabeth against the Queen. Simon Renard, the Imperial ambassador, claimed that Anne had approved of plots to prevent Mary's marriage to Philip II and that she was conspiring with her brother, the Duke of Cleves, to further the dynastic ambitions of Elizabeth. Such rumors were extremely unlikely to be true, given Anne's sensible pragmatism, but they were nevertheless damaging to both women at a time of intense suspicion and uncertainty. Mary evidently remembered enough of her former affection for Anne of Cleves not to act against her, however, and she remained at court for the time being. Elizabeth was glad of this, for she found Anne a much-needed ally as she came under increasing pressure to conform to Roman Catholicism. It was noted that both women failed to attend any Masses at court, even though there were six or seven every day. "My Ladies of Cleves and Elizabeth have not been present yet," the Spanish ambassador observed disapprovingly.[42]

For all her skill as an actress, Elizabeth could not conceal her distaste for the Catholic religion from Mary, whose resentment against her now reached fever pitch. The Queen told Renard that she heartily wished to remove her half sister from the succession "because of her heretical opinions, illegitimacy and characteristics in which she resembled her mother," adding that "as her mother had caused great trouble in the kingdom, the Queen feared that Elizabeth might do the same."[43] For all her defiance on religion, it is unlikely that Elizabeth was so foolhardy as to plot against her half sister. But the fact that people suspected her of doing so was enough to place her in grave danger. Meanwhile, tension at court was mounting. In late November, when the Queen was on her way to chapel, there was a cry of "Treason!" When Elizabeth heard this, she was gripped with terror, suspecting that the cry was against her. One of Mary's own ladies was obliged to comfort her and rub her stomach until she calmed down.[44]

By now, it was clear that the situation could not continue. Elizabeth begged Mary to let her leave court and take up residence at her country estate of Ashridge in Hertfordshire, well away from the plots and intrigues that were beginning to surround her. This threw Mary into a quandary. While at heart she wanted nothing more than to be rid of her half sister's irksome presence, she also considered the old adage of keeping one's enemies close and did not know if she could risk Elizabeth being at a distance from which she could more easily plot against her. As was increasingly her policy, she went to seek Renard's advice. He immediately cautioned that Elizabeth's request was "suspect," arguing that "as it was known that the heretics were building upon her, it would be better to keep her here." He concluded that the Queen had only two choices: either to keep her half sister at court and maintain a show of harmony, or to "shut her up in the Tower." This latter course was too drastic for Mary, but the first was little more appealing. At length, therefore, she assented to Elizabeth's request to leave court.

Eager though she was to get away, Elizabeth was also anxious that her enemies would use her absence to turn the Queen against her once and for all. She therefore petitioned Mary not to believe any "evil reports" of her until she had first given her a chance "of proving the false and malicious nature of such slanders." Mary duly promised to give her a fair hearing, but at heart she was still consumed by resentment against her younger sister, and "recalls the trouble and unpleasantness before and since her accession, unrest and disagreeable occurrences to which Elizabeth has given rise." Nevertheless, when the two sisters took their leave of each other, they gave one last show of affection.

Elizabeth was "very courteous" toward the Queen, who "dissembled well" herself and gave her younger sister a rich hood of sable, which was a thoughtful gift given that she was about to embark upon a long journey in the cold December air.[45] But Renard noted afterward that it had taken a great effort on Mary's part to appear so civil. As Elizabeth's entourage set off on the thirty-mile journey to Ashridge, the Queen was heartily glad to see the back of her.

Elizabeth was no doubt equally relieved to be away from court, but it was by no means a complete escape, for Mary had insured that she would remain under close scrutiny at Ashridge. Her cousin, Charles V, had urged her to take this precaution, and he ordered his ambassador, Renard, to "now and then remind the Queen to have the Lady Elizabeth watched." Mary needed no reminder. Even though her half sister reiterated her loyalty and affection toward her upon reaching Ashridge, Mary refused to be won over by fair words, and dispatched some trusted men "to watch what takes place in her house."[46]

They found little cause for suspicion. Having been greatly alarmed by the Queen's displeasure toward her, Elizabeth was careful to avoid any cause for further complaint. And yet her very position made her the natural focus for plots and conspiracies against the regime. Although outwardly loyal to her half sister, she represented everything that Mary was not. She was Protestant at a time when Mary's increasingly severe religious policies were turning people against the Catholic faith. She was also entirely English, while her half-Spanish sister had caused fierce resentment by marrying a foreign prince. One pamphleteer described Elizabeth as "a prince of no mingled blood, of Spaniard or stranger, but born mere English here amongst us."[47] It was one of her greatest strengths, and she knew it. In the years to come, she would repeat the proud boast that she was "meer English," and use this to bolster not just her own popularity but a sense of national pride.

Within months of Mary's ascending the throne, there was already widespread opposition to her administration. This went right to the heart of government. In January 1554, a group of MPs (members of Parliament) plotted to prevent the Queen's marriage by coordinating four separate risings in Kent, Hereford, Devon, and Leicestershire.

This plot was soon discovered by Mary's council, and the conspirators in three of the counties rapidly dispersed. The atmosphere at court was growing increasingly tense, however, and the Queen's closest advisers urged her to rid herself of any rival claimants. Elizabeth aside, the principal threat seemed to be Lady Jane Grey, who was still imprisoned in the Tower following her father's abortive attempt to thrust her onto the throne. Even though Lady Jane had been a most unwilling pawn in his political power games, she remained a dangerous figurehead for opposition to the regime. Under pressure from her council, Mary therefore reluctantly ordered her execution and that of Guilford Dudley in February 1554.

This sent a stark warning to Elizabeth, whose sorrow for her former companion at Chelsea mingled with fear for her own life now that she was the main figurehead for opposition. A plot to place her on the throne was already gathering ground in Kent. Its leader was Sir Thomas Wyatt, who in January 1554 raised a considerable band of men (one report said as many as three thousand) and marched toward London.[48]

The threat from Wyatt became serious when the band of trained soldiers led by the Duke of Norfolk, who had been ordered to head off the rebels before they reached the capital, suddenly deserted. This left Mary and her council virtually undefended. She bravely refused to flee, however, and instead placed her trust in the citizens of London. For the first time, she displayed the decisive, authoritative behavior that might be expected of a daughter of Henry VIII. Rallying the loyal troops that had gathered at the Guildhall in the heart of London, she delivered the greatest speech of her reign. "I cannot tell how naturally the mother loveth the child, for I was never mother of any," she began. "But certainly, if a prince and governor may as naturally and earnestly love her subjects, as the mother doth the child, then assure yourselves, that I being your lady and mistress, do as earnestly and tenderly love and favour you."[49] For this one fleeting occasion, Mary seemed entirely at ease as a female sovereign, allowing her sex to lend her a maternal role over her subjects, while also displaying the majesty and leadership to inspire respect and awe among everyone who heard her. Perhaps

Elizabeth learned of her sister's speech, for she herself would draw upon the same theme of motherhood when addressing her people as queen, and it would become one of her most effective ways of insuring their loyalty.

Mary's address was a decisive factor in defeating the rebellion. But for Elizabeth, the danger was only just beginning. Even though she was almost certainly innocent of any involvement in the plot, Wyatt claimed that he had written to her to tell her of his plans. This was supported by a fellow conspirator, who alleged that he had delivered the letter himself. Their testimonies gave Elizabeth's enemies at court enough ammunition to do away with her for good. For months they had tried to persuade the Queen to throw her half sister into the Tower and find an excuse to condemn her as a traitor. Now they had the perfect justification. "I will do all I can to obtain that result," Renard assured his master.[50] Until now, Mary's conscience had not allowed her to agree to their requests, but even she appreciated that Elizabeth could no longer be kept at liberty. She therefore wrote to summon her to court. Her letter was, on the surface, full of cordiality. Referring to Elizabeth as her "Right dearly and entirely beloved sister," she invited her to come to court for "the surety of your person."[51] Meanwhile, she dispatched some of her councillors to accompany her half sister back to London.

As Elizabeth waited anxiously at Ashridge, the stress of the situation became too much, and she collapsed with nervous exhaustion. Throughout her life, Elizabeth would suffer from attacks of sickness at times of extreme stress, and this was undoubtedly the most stressful situation that she had faced since her interrogation over the Seymour affair some five years before. Kat Astley was frantic with worry and tended her at her bedside throughout the desperate hours and days of her beloved girl's illness. Then, late into the night of February 10, the delegation of officials arrived from court, charged with arresting the Lady Elizabeth and bringing her to London for questioning. Legend has it that they burst into her bedchamber, brushing aside Mistress Astley, who begged them to show mercy to the sick girl. But they were resolute in carrying out their orders and conveyed a weak, protesting Elizabeth away from the safety of Ashridge.

Attended by a handful of her most trusted ladies, among them Kat Astley and Blanche Parry, Elizabeth and her entourage made their way slowly to the capital. On arrival, she was installed at Whitehall Palace in cramped apartments and told to wait there until the Queen saw fit to have her incarcerated in the Tower. Sick with worry, Elizabeth endured long days holed up in her rooms with no news from the court. Her discomfort was heightened by the noise and smells that were constantly emitted from the chamber above. This belonged to Margaret Douglas, Countess of Lennox, the woman whom Mary looked set to name as heir. Margaret had no liking for Elizabeth and resolved to emphasize her own superiority. She gave orders that her apartment be converted into a kitchen so that her young cousin would be plagued by the commotion attendant upon the preparation of her meals. The countess would live to regret this petty act of triumph.

When the councillors whom Mary had appointed to convey her half sister to the Tower arrived at Whitehall, Elizabeth succeeded in delaying them by insisting that she first be allowed to write to the Queen. The longer she wrote, the less time they had to catch the tide, and she knew it. By the time she had finished the letter (thereafter known as the "tide letter"), they had missed it. Elizabeth had won only a temporary reprieve, but the words she wrote to her sister may have won her a good deal more.

"If ever I did try this old saying, that a king's word was more than another man's oath, I beseech your majesty to verify it in me, and to remember your last promise and my last demand that I be not condemned without answer and proof; which it seems now I am," she began. "For without cause proved I am by your council from you commanded to go to the Tower. I know I deserved it not, yet it appears proved. I protest before God I never practised, counselled or consented to anything prejudicial to you or dangerous to the state." Convinced that if she were allowed to see her half sister, she would be able to persuade her of her innocence (as she had over the matter of religion), Elizabeth begged an audience. "Pardon my boldness. I have heard of many cast away for want of coming to their prince," she assured the Queen. "I pray God as [that] evil persuasions persuade not one sister against the other." Because she had finished the letter at the top of the second page, she carefully scored lines across the rest of it so that none of these "evil persons" could add defamatory words afterward.[52]

The tide letter was to be one of the most famous that Elizabeth would write. It is a testament to her great presence of mind that, sick with panic and fearing for her very life, she was able to write such an eloquent, well-reasoned defense of her actions. There is no record of the effect it had upon her sister, who chose not to reply. Neither did she accede to Elizabeth's request for an audience. Instead she dispatched her to the Tower without delay, along with Kat Astley and Blanche Parry. For Elizabeth, this was her worst nightmare. She now faced the same fate that her mother had met some eighteen years before. As she mounted the steps to the fortress, she looked up to the heavens and exclaimed: "Ohe Lorde! I never thought to have come in here as a prisoner; and I praie you all, goode frendes and fellowes, bere me wytnes, that I come yn no traytour, but as true woman to the quenes majesty as eny is nowe lyving."[53]

A group of interrogators was sent to question the new prisoner closely about her alleged involvement in the conspiracy, and it was noted that "her fate should depend on her answers."[54] Although still very sick and terrified, Elizabeth defended herself as stoically as her mother had done when faced with accusations of treason. Again and again she proclaimed her innocence, insisting that she was the Queen's loyal subject. When this was reported to Mary, she dismissed it, saying that "Elizabeth's character was just what she had always believed it to be."[55] By now, Elizabeth was frantic with worry. Her mother had once said of Mary: "She is my death, or I am hers." Elizabeth could have said the same. The relationship between the two sisters, so opposite in character, appearance, and outlook, was little short of a battleground. Elizabeth was so convinced that she was going to die that she even contemplated asking if she could be executed with a sword rather than an axe, as her mother had been, because she had heard that it was quicker.

A few days later, it seemed that Elizabeth's worst fears would be realized. The lieutenant of the Tower, Master Bridges, received a warrant for her execution. Thankfully, despite her openly voiced hostility toward her half sister, he had the presence of mind to doubt the warrant's validity and went with all haste to the Queen. She immediately denied that she had issued it, and when it was discovered that it had been the work of her lord chancellor, Gardiner, and his party, she railed against them "for their inhuman usage of her sister."[56] Perhaps she was mindful of public opinion, which was firmly in favor of Elizabeth. Or perhaps, despite all the hostilities and resentments, the bond between the two sisters was stronger than it appeared. Even while Elizabeth was still a prisoner in the Tower, Mary started to refer to her as "my sister" again and ordered that her portrait be reinstalled in its former position in the royal gallery.

It soon became clear that Mary would not sanction Elizabeth's execution—at least not unless some definitive proof of her involvement in Wyatt's rebellion came to light. So, as Renard neatly put it: "The question of the day is: what shall be done with her?"[57] Two months after her imprisonment in the Tower, Elizabeth was released— ironically, on the very same day that her mother had been executed. But she was not set at liberty: instead she was conveyed to the gloomy palace of Woodstock in Oxfordshire to be kept there under house arrest. Most of her old servants were replaced by those appointed by the council. This included Kat, who, like Elizabeth, had been released from the Tower but immediately placed in custody. Only Blanche Parry was permitted to remain with Elizabeth, who appointed her chief gentlewoman later that year in recognition of her loyalty.

As Elizabeth made her way through the city, accompanied by a considerable armed guard of some four hundred men, the people cheered when they saw her, assuming that she had been set at liberty. A group of merchants "shot off three cannons as a sign of joy," much to the disapproval of the Queen when she heard of it.

Although she could rejoice at having narrowly escaped her mother's fate, Elizabeth had been deeply affected by the whole episode. She would remember it for years to come, and it would have a profound impact upon her own development as a ruler. Years later, she would recall: "I stode in dangere of my lyffe, my systere was so ensenst [incensed] ageynst me."[58] The experience had taught her two valuable lessons: first, the need to strenuously safeguard her reputation, and, second, the

strength of her own ability to talk—and write—her way out of danger.
The Seymour episode had given her a taste of the latter. Now the Wyatt
conspiracy fostered within her a new sense of confidence that would
grow ever greater as Mary's reign progressed, equipping her for the task
of one day taking the throne herself.

Elizabeth would be at Woodstock from May 1554 until April 1555.
While there, her every move was watched, and she was not allowed to
correspond with anyone unless it was by the council's sanction. This
included her old governess, Kat, who was under the custody of Sir
Roger Cholmley at his house in Highgate, north London. However, in
April 1555, a warrant was issued to Sir Roger, ordering him to "set at
libertie Katheryne Assheley who hath of long tyme remained in his
custodie."[59] By June of that year, Kat was reinstated in Elizabeth's service, much to both women's joy and relief. The terror of Kat Astley's
second brush with the Tower, and the months of imprisonment that
followed, might reasonably be expected to have encouraged her to
keep a low profile from then onward. The council was well aware of
the closeness of her relationship with Elizabeth, and as far as they were
concerned, she was tainted by association. For as long as Elizabeth remained under suspicion, so would her governess.

Whether or not it had been her intention, Mary's decision to keep
her half sister under close confinement protected Elizabeth from being
implicated in any future plots. Perhaps conscious of this, and buoyed by
the confidence she had gained from escaping condemnation for the
Wyatt rebellion, Elizabeth requested permission to write to the Queen.
This was granted, and although her subsequent letter does not survive,
Renard saw it and immediately related its contents to his master. Apparently Elizabeth had dispensed with her former deference and had instead shown a distinct lack of respect toward her half sister, addressing
her not as "Highness" or "Majesty," but simply "You." The Queen was
furious at this slight, and wrote at once to Elizabeth's custodian, saying:
"Our pleasure is that we shall not hereafter be molested any more with
her disguised and colourable letters." Unrepentant, Elizabeth angrily
retorted: "I must say for myself that it was the plain truth."[60]

The balance of power between the sisters might have undergone a subtle shift, but Elizabeth still had to be careful to maintain an un-blemished reputation with regard to her loyalty to the Queen. As she and Kat tried to reestablish some semblance of normality in the household, a fresh plot against the Queen was already gathering ground.

With the number of Protestant martyrs lost to the flames growing ever higher and anti-Spanish feelings intensifying across the country, plots against Mary were never far away. In May 1555, another one was brought to light. Henry Dudley, a distant relative of the late Duke of Northumberland, had served in France under Edward VI and boasted a number of powerful henchmen among his entourage, including John Throgmorton. Together they conspired to place Elizabeth on the throne. Their outlandish plan comprised three key elements: robbing the Exchequer, seizing the Tower, and killing the Queen. It was soon discovered, and yet again the finger of suspicion pointed at Elizabeth, who was then residing at Hatfield.

Elizabeth immediately denied all knowledge of the ridiculous plot and wrote at once to the Queen, protesting her innocence and condemning the conspirators for not showing "the reverent fear of Romans to their Senate." She went on to give an eloquent assurance of her loyalty, saying: "And among earthly things I chiefly wish this one: that there were as good surgeons for making anatomies of hearts that might show my thoughts to your majesty as there are expert physicians of the bodies, able to express the inward griefs of their maladies to their patient. For then I doubt not but know well that whatsoever other should suggest by malice, yet your majesty should be sure by knowledge, so that the more such misty clouds obfuscates the clear light of my truth, the more my tried thoughts should glister to the dimming of their hidden malice." She signed the letter "Your Majesties obedient subiect and humble sistar."[61]

On this occasion, Mary was willing to believe Elizabeth's protestations, and she sent her a ring as an assurance of her goodwill, together with a message that she knew that Elizabeth was too "wise and prudent" to do anything prejudicial against her queen. But this did not prevent Elizabeth's name from being connected with almost every subsequent plot and conspiracy. "It unfortunately appears that never is a conspiracy discovered in which either justly or unjustly she or some of her servants are not mentioned," reported the Venetian ambassador. He added, though, that "having no suitable cause to proceed against her she [Mary] dissembles her hatred and anger as much as she can, and endeavours when they are together in public to receive her with every sort of graciousness and honour, nor does she ever converse with her about any but agreeable subjects."[62]

But Mary's officials were determined to find evidence of Elizabeth's involvement, and they searched both Hatfield and her London residence, Somerset House. They found nothing to implicate her, but Kat Astley was not so fortunate. In Somerset House, a number of "seditious" books and pamphlets were discovered, together with a small cabinet containing papers, paintings, "and other defamatory libels" against Queen Mary, her husband, Philip, and the Catholic religion. These items were found among Kat Astley's possessions, which was sufficient evidence to have the hapless governess arrested once more.

In late May 1556, a delegation of officials arrived at Hatfield to arrest Mistress Astley and two other members of Elizabeth's household who had been implicated by the search at Somerset House: Elizabeth's Italian master, Battista Castiglione, and a servant, Francis Verney. This caused "great general vexation" amidst the household, and Elizabeth herself was said to be "distressed and dejected."[63] Even though Mary had assured her sister that she did not believe her to be personally implicated, Elizabeth was still terrified at the thought of what might happen to Kat.

Her faithful servant arrived at the Tower a few days later. Familiarity did not make this imposing fortress any less fearful, and the chances of her escaping death a third time seemed unlikely. "What talk have you had with any person touching this conspiracy?" her interrogators demanded. "When, where, with whom and how often?" In vain, Mistress Astley protested that she had talked to nobody about it, and that the first she had heard of it was when Throgmorton was arrested. She insisted that she was the Queen's loyal subject, vowing: "If I cannot be true to my queen, especially one of whose virtues I have had long experience, it were pity the earth should bear me." She also used the opportunity to defend Elizabeth, adding: "If her I serve (whose love to her highness I have known from her youth) should prove me corrupt but in thought to her highness, I am sure she would never see me again."[64]

In the course of her spirited defense, Kat had committed an error that should have been enough to seal her fate. The fact that she admitted to knowing about the conspiracy, albeit not until after the arrest of its leaders, was sufficient evidence to convict her in those dangerous, paranoid times. Worse still, her fellow servants had also confessed that they had heard of it, but, like Mistress Astley, had not done what they should have and informed the authorities immediately. When the Venetian ambassador heard this, he reported to his master: "I am told that they have all already confessed to having known about the conspiracy; so not having revealed it, were there nothing else against them, they may probably not quit the Tower alive, this alone subjecting them to capital punishment."[65]

Once their confessions had been extracted, there followed an agonizing wait for Kat and the two other servants as to what sentence would be passed. It looked almost certain to be death. The days dragged on interminably, and still no news came. Speculation about their impending fate was rife at court. Michiel surmised that their execution was being deferred "for the purpose of adding to their numbers." Elizabeth was frantic with worry. With her beloved Kat incarcerated in the Tower, she had nobody to turn to. Mary cheerfully told her that Mistress Astley's arrest was an opportunity to review her entourage. She had taken a dim view of the "licentious life led, especially in matters of religion, by her household," which she believed had left Elizabeth "clandestinely exposed to the manifest risk of infamy and ruin." She therefore devised a plan for "remodelling her household in another form, and with a different sort of persons to those now in her service." All of Elizabeth's closest attendants were to be replaced with "such as are entirely dependant upon her Majesty," with the result that Elizabeth's own "proceedings," and those of her household, "will be most narrowly scanned." This would insure that the Queen's sister "keep so much the more to her duty, and together with her attendants behave the more cautiously."[66]

This was cold comfort indeed for Elizabeth. Not only had she lost her governess—perhaps forever—but she now faced the prospect of having a cast of sober matrons forced upon her. The lively conversation, music, dancing, and other diversions that she had enjoyed at Hatfield looked set to end. Sure enough, in mid-June 1556, Mary sent the first of her chosen servants to Elizabeth, the morally upright Sir Thomas Pope and "a widow gentlewoman" as governess. As the Venetian ambassador shrewdly observed: "So that at present having none but the Queen's dependants about her person, she herself may be also said to be in ward and custody."[67]

The summer came to a close, and still there was no word of Kat. Did Mary intend to keep her holed up in the Tower forever, reluctant to order either her execution or her release? The prospect was as awful for Elizabeth as it was for Kat. But then at last, on October 19, the news that neither had dared hope for was received: Mistress Astley was to be set at liberty. Disbelief and joy at cheating death a third time must have mingled with anticipation at being reunited with Elizabeth. The latter prospect was soon shattered, for the Queen ordered that Kat be dismissed from her office as governess and, worse, "forbidden ever again to go to her ladyship."[68] This was a cruel punishment for the woman who had proved Elizabeth's most loyal and devoted servant. It was also a crushing blow for Elizabeth herself, who was desolate at the thought of never seeing Kat again.

In a sense, there had been little other option for Mary, apart from ridding herself of this troublesome woman for good. Moreover, she seemed genuinely concerned to protect her sister's reputation: until she herself produced an heir, Elizabeth remained her most likely successor. Mistress Astley had already proved herself a most unfit person to take charge of a royal lady's upbringing, and Mary could not risk another scandal. As well as the flaws in her character, the former governess was also a heretic, and the Queen would stand no chance of bringing her sister back to the Catholic fold if she was surrounded by such people. The seditious books and pamphlets that had been found among Kat's possessions made it clear that she could not be trusted. Indeed, she had made little secret of her firmly Protestant beliefs, which, at a time when the Queen and her council were trying to eradicate this religion from England, was a reckless and foolhardy policy. It also made Kat a liability to Elizabeth, who was desperately trying to avoid implication in the many Protestant plots that were springing up against her sister. In a sense, therefore, Mary had done her a favor by dismissing Kat from her service—although Elizabeth might not have appreciated this at the time.[69]

Kat Astley had undoubtedly borne the brunt of the blame for the Dudley conspiracy. By contrast, Elizabeth had benefited from her half sister's goodwill on this occasion. Not only had Mary accepted her protestations of innocence; she also refrained from following up any further rumors of her involvement in plots and conspiracies. It is thought that this was less to do with sisterly affection than with the influence of her husband, Philip. Although Mary no doubt wished to have her half sister brought to the Tower and questioned, Philip insisted that instead she should send her a kind message "to show her that she is neither neglected nor hated, but loved and esteemed by Her Majesty."[70] Perhaps also thanks to Philip's intervention, Mary went on to invite Elizabeth to court, although on this occasion the latter politely declined. Apparently being kept under scrutiny in the country was preferable to having to put on an act of devotion—to both her half sister and the Roman Catholic faith—at court.

Philip might seem an unlikely ally to Elizabeth, given that Spain had traditionally viewed her as a dangerous heretic, but the Venetian ambassador affirmed that it was due to his influence with the Queen (which was considerable) that her half sister had been spared. "There is no doubt whatever but that had not her Majesty been restrained by the King, and by the fear of some insurrection she for any trifling cause would gladly have inflicted every sort of punishment on her."[71]

Philip was no doubt acting out of political, rather than personal, motives. He knew that Mary risked alienating her subjects further if Elizabeth was ill treated. Even so, the fact that he continued to defend her half sister was enough to invoke Mary's jealousy. From the very beginning, she had been utterly besotted with her husband. Philip had only to exist for her to adore him unreservedly. Philip himself was a good deal less enamored. Upon first meeting the Queen, he remarked that she was rather older than he had been made to expect. Shortly after the wedding night, he confided to an attendant that his new wife had been lacking in sexual prowess. Mary certainly knew precious little of the ways of men and was innocent about the bawdy talk that went on at court. Her love for Philip was of a much more romantic nature.

A month after the wedding, she wrote to Charles V, full of praise for his son. Proudly referring to Philip as "The King, my lord and husband" (overlooking the fact that he had not been crowned), she enthused: "This marriage and alliance, which renders me happier than I can say, as I daily discover in the King my husband and your son, so many virtues and perfections that I constantly pray God to grant me grace to please him and behave in all things as befits one who is so deeply embounden to him."[72] The object of her affections, meanwhile, was trying his best to act the part of a doting husband. His favorite attendant, Ruy Gómez de Silva, remarked: "The Queen is very happy with the King, and the King with her; and he strives to give her every possible proof of it in order to omit no part of his duty," adding in another dispatch: "His Highness is so tactful and attentive to her that I am sure they will be very happy."[73] In private, Philip complained of the marital duties that he had to perform in order to satisfy this unappealing wife, but he continued to maintain the pretense. "He treats the queen very kindly and well knows how to pass over the fact that she is no good from the point of view of sensuality," reported de Silva. "He makes her so happy that the other day when they were alone she almost talked love-talk to him, and he replied in the same vein."[74]

Philip's efforts were due to more than just courtesy. It was vital that he made the marriage a success in order to firmly establish the Habsburg dynasty in England. In the sixteenth century, marital success was defined by the number of heirs produced. Within weeks of the wedding, it seemed that he would be rewarded for his pains, for it was announced that the Queen was with child. In November 1554, Parliament

gave thanks to God for "the Quen's grace qwyckenyng," and the following month, Mary wrote joyfully to Charles V, assuring him: "As for
that which I carry in my belly, I declare it to be alive, and with great humility thank God for His great goodness shown to me, praying Him so
to guide the fruit of my womb that it may contribute to His glory and
honour, and give happiness to the King, my Lord and your son."[75]

The fact that Mary felt it necessary to confirm that she was carrying
a healthy child suggests that even at this early stage there had been rumors that the pregnancy was false. It was well known at court that
Mary had suffered menstrual problems since puberty. The Venetian
ambassador, Giovanni Michiel, observed that the Queen had a tendency "to a very deep melancholy, much greater than that to which
she is constitutionally liable, from menstruous retention and suffoca-
tion of the matrix to which, for many years, she has often been subject." She was thought to have what was known as "strangulation of
the womb," and her symptoms included periods that were not just irregular, but often entirely absent, as well as swelling in the abdominal
area and frequent bouts of nausea. Her physicians regularly "bloodied"
her "from the foot or elsewhere" in the mistaken belief that this would
help. In fact, it weakened her still further, and Michiel noted that she
was "always pale and emaciated."[76] Given her history and her age,
many doubted that her condition was real. It certainly suited the anti-
Spanish party at court to believe that it was not, and they began to
spread rumors that it was all a Spanish plot to pass off another infant
as the Queen's.

Nevertheless, Mary displayed all the symptoms of early pregnancy.
A Spanish envoy reported that she had been feeling nauseous and
added that her doctor had "given me positive assurance" that there
were other signs. Questions were evidently still being raised a few
months later, but Renard affirmed: "One cannot doubt that she is with
child. A certain sign of this is the state of the breasts, and that the child
moves. Then there is the increase of girth, the hardening of the breasts
and the fact that they distill."[77]

In Mary's mind, there was no doubt. She was carrying the prince
who would build upon all of her and Philip's plans for Roman Catholi-

cism, and would be a living embodiment of the Anglo-Spanish alliance—one that could never be broken. Beside herself with joy, she ordered that everything be put in place for her confinement. Sumptuous cradles and baby clothes were commissioned, and midwives and rockers were appointed. Hampton Court Palace was chosen as the place for the birth, and Mary duly made her way there in early April 1555, a little over a month before the baby was due. By now, her stomach was so swollen that it was thought the little prince might arrive any day.

The excitement of the impending birth made Mary feel more benevolent toward her half sister. After all, when she was delivered of a male heir, Elizabeth would be rendered almost obsolete. She no longer seemed a threat to Mary, and the latter was therefore willing to be generous and invite her to attend upon her during her confinement. She had played second fiddle to her alluring, intelligent younger sister for years, and now here she was, on the brink of the ultimate triumph. Elizabeth could do nothing to hurt her now.

Although the idea was no doubt distasteful to her, Elizabeth could not refuse the Queen's invitation without causing great offense. Besides, if she had heard the rumors about the pregnancy being false, she would have been curious to examine her half sister's condition for herself. She arrived at Hampton Court shortly after receiving Mary's summons. Happy though she was in her condition, Mary had by no means forgotten the resentment and suspicion that she felt toward Elizabeth. Determined to make a point, she took her time in granting her half sister an audience. Eventually, after three weeks of waiting, Elizabeth was summoned late one night to the royal presence. Accompanied only by Mary's trusted servant, Susan Clarencieux, she was led by torchlight to the Queen's bedchamber. Having been ushered into her half sister's presence, Elizabeth reenacted the scene of almost two years before and humbly protested her innocence and devotion. This time Mary was not to be fooled, however, and she curtly dismissed her younger sister's performance, chiding her for so stubbornly refusing to admit her guilt. Taken aback by the Queen's harshness, Elizabeth continued to plead her loyalty. After a while, Mary grudgingly uttered "a few comfortable words" to her before abruptly sending her back to her apartments. According to Foxe's account, Philip had been hiding behind an arras the whole time, eager to spy upon this exchange between his aging wife and his bewitching young sister-in-law.[78] There are no other accounts to corroborate this, but Philip certainly showed a keen interest in Elizabeth thenceforth.

As Mary's confinement progressed and the due date of May 9 drew closer, there was a fever pitch of anticipation both at court and across the country. Ambassadors overseas also waited anxiously for news. Any scrap of information was seized upon, and rumors swiftly became fact. Little wonder, then, that at the beginning of May, it was falsely reported that the Queen had delivered a son. The English ambassador in Brussels wrote to his agent at court, urgently seeking confirmation. But before he received it, the news had spread throughout the Netherlands, and bells were rung in celebration. Even the English merchants on the channel heard of it and set off cannons from their ships as a sign of joy.[79] Their celebrations were brought to an abrupt halt when the rumors were confirmed to be false.

The date that the baby should have been born came and went. Mary remained confident, however: after all, first babies were often late. A little over three weeks later, on June 1, she experienced some pains, and her ladies hastily gathered around, assuming that the labor had begun. Letters were duly prepared, announcing the birth of a prince, to the Pope, Charles V, Henry II, and all the potentates of Europe.[80] But it came to nothing, and Mary and her attendants resumed their waiting. The physicians must have miscalculated, they reasoned. "Her doctors and ladies have proved to be out in their calculations by about 2 months," Renard reported to Charles V on June 24, "and it now appears that she will not be delivered before 8 or 10 days from now." He added, with a note of anxiety: "Everything in this kingdom depends on the Queen's safe deliverance."[81] Meanwhile, those opposed to the regime spread counter-rumors, claiming that the pregnancy and confinement had been nothing but an elaborate sham. To explain the Queen's swollen girth, the French ambassador reported that she had been "delivered of a mole or lump of flesh, and was in great peril of death."[82]

For months, Mary had been surrounded by her ladies, who had constantly assured her that everything was progressing as normal with the pregnancy. All too willing to believe them, she had been whipped up into an ever-greater fervor of anticipation. Only one of them, Mrs. Frideswide Strelley, had doubted Her Majesty's condition from the start. In the early days of her pregnancy, Mary had joyfully summoned Mistress Strelley to feel the baby move. "Feel you not the child stir?" she had asked. "My fortune is not so great," had come the reply. Mary had dismissed her concerns at the time, but now, with the waiting dragging on inexorably, she again sent for Frideswide. "Ah, Strelley, Strelly, I see they be all but flatterers and none trewe to me but thou."[83] At last, a year after first believing herself to be pregnant, Mary gave up and admitted that in spite of all the signs to the contrary, she had been mistaken. "There is no longer any hope of her being with child," wrote Charles V to his ambassador in Portugal on September 14.[84]

By the time the Queen finally admitted defeat, Hampton Court was thronging with noblemen and women from across the country, all eager to see England's new heir. One suspects that those who arrived toward the end of the confinement wished rather to take part in what was rapidly becoming the greatest scandal of the reign so far. The humiliation that Mary felt can only be imagined. It would have been disappointing enough if she had given birth to a girl. The fact that she gave birth to nothing at all was mortifying.

Elizabeth, watching from the sidelines, had seen not only Mary's personal anguish but the damage that the whole episode had done to her public reputation. In just a few months, Mary had been transformed from the warrior Queen who had rallied her troops to defeat Wyatt's rebellion to the subject of ridicule and derision, as rumors abounded that she had tried to pass off another child as her own. Her subjects had been shockingly disrespectful, pinning satirical posters to the palace doors and even throwing abusive pamphlets into her own rooms. It was a serious blow to her authority, for Mary had failed in her duty not only as a queen but also as a woman. All the fears that had been voiced about the accession of a female sovereign seemed justified. If Elizabeth herself was to be a queen regnant, then she must learn from this harsh but invaluable lesson.

Witnessing Mary's phantom pregnancy might also have strengthened Elizabeth's growing conviction that she herself was incapable of
bearing children in any case. This conviction had been generated by
her mother's tragic example, together with that of her aunt, Mary Boleyn, as well as by the fact that, like her half sister, Mary, she herself
had experienced menstrual problems from early puberty. Quite apart
from her antipathy toward marriage, such a belief in her infertility
might have been enough to decide Elizabeth once and for all to remain
a virgin.

She had further evidence to support this view during those months
spent at Hampton Court, for she witnessed Mary's humiliation as a result not just of the false pregnancy but also of the inequality of her
marriage. The Queen made no secret of her adoration for her husband, and despite exceeding him in age and rank, she seemed meekly
to defer to him in all things. Philip, meanwhile, was struggling to keep
up the pretense of affection toward this prematurely aged, sexually inadequate wife, and the whole farce of her phantom pregnancy had
made him increasingly restless to be rid of her—and her hostile country. To his relief, in July 1555, as Mary's confinement at Hampton
Court dragged on, Charles V dispatched a summons for his son to attend to Imperial business in the Netherlands. Whether this was as a result of a private plea on Philip's part is not known. When Mary heard
of it, she was aghast and wrote at once to Charles, begging him to
withdraw his request. "I assure you, Sire, that there is nothing in this
world that I set so much store by as the King's presence," she wrote,
and claimed that "quite apart from my own feelings, his presence in
this kingdom has done much good and is of great importance for the
good governance of this country." Her appeals fell on deaf ears, however, and when it became clear that no child would result from those
long months of waiting at Hampton Court, Philip prepared to embark
for his father's dominions.

In the meantime, he found a great deal of diversion in the company
of his sister-in-law. Both having time on their hands in the long hours
of waiting for news from the Queen's birthing chamber, they became
better acquainted with each other. Philip found the Lady Elizabeth a
captivating and charming companion, and she seemed to go out of her

way to court his favor. "At the time of the Queen's pregnancy, Lady Elizabeth, when made to come to the court, contrived so to ingratiate herself with all the Spaniards, and especially the King, that ever since no one has favoured her more than he does," observed the Venetian ambassador.[85]

Before long, it was rumored that the Spanish king was so captivated by his half sister's youthful beauty, which contrasted sharply with his wife's faded looks, that he determined to have her for himself. Any subsequent favor that he showed toward her was therefore interpreted not as policy but lust. Elizabeth had herself been suggested as a bride for Philip at the age of just thirteen, as part of Henry VIII's schemes to forge an Imperial alliance. But such betrothals were made and broken all the time, so there is no reason to suppose that it was particularly remembered by either party. In fact, the only evidence to suggest that Philip was in love with Elizabeth was her own testament, made years later when she was queen. By then, the two sovereigns had long been at war, and Elizabeth was fond of boasting that their relationship had begun with love, at least on Philip's part.

There is little to support her contention. It seems more likely that Philip, aware of his wife's poor health and her continuing failure to produce an heir, had an eye to the future. If this heretical girl came to the throne, the best way to curb her activities would be either to marry her or to forge some other alliance that would secure her friendship. After all, England played a critical part in the ongoing power struggle between the might of France and Spain, and if she was loyal to the latter only for as long as Mary lived, then Philip needed to plan further ahead. What was more, the alternative claimant to the English throne was Mary, Queen of Scots, who had been raised in France and was betrothed to the Dauphin. From Philip's perspective, Elizabeth was therefore very much the lesser of two evils. As a result, he resolved to show her every courtesy during the weeks leading up to his departure from Hampton Court.

This was anathema to Mary, who by now hated her sister as much as she loved her husband. Philip II's envoy later reported that one of the principal causes of her resentment toward Elizabeth was "her fear

Elizabeth as a princess (ca. 1546–47).

Anne Boleyn, Elizabeth's mother,
at the height of her powers
(late sixteenth-century copy of
a lost original).

Locket ring owned by Elizabeth I
(ca. 1575), showing portraits of
herself and Anne Boleyn.

A recently discovered portrait of Henry VIII and his children, with his jester, Will Somers, in the background (seventeenth-century copy of a lost original from ca. 1545–50). Elizabeth is on the extreme right of the portrait.

The family of Henry VIII (ca. 1543–47). Elizabeth (shown on the right) is wearing Anne Boleyn's famous "A" pendant as a sign of loyalty toward her disgraced mother.

Blanche Parry's memorial at Bacton Church, Herefordshire. Blanche is shown on the left, kneeling in adoration before her royal mistress.

Katherine Astley, Elizabeth's long-standing governess and confidante.

ANNO DNE   1544

LADI MARI   DOVGHTER TO
THE   MOST   VERTVOVS PRINCE
KINGE HENRI   THE EIGHT

THE AGE OF   XXVIII YERES

Elizabeth's sister, Mary, in her youth (ca. 1544).

Jane Seymour (ca. 1536–37),
beloved third wife of Henry VIII.

Anne of Cleves (ca. 1539),
whose marriage to Henry VIII
lasted for just six months.

Katherine Howard (ca. 1541), Henry VIII's youngest—and shortest lived—bride.

Katherine Parr (ca. 1545), the last wife of Henry VIII and Elizabeth's most influential stepmother.

KATHARINE PARRE

Queen Mary Tudor (1554), whose relationship with her half sister,
Elizabeth, became increasingly stormy.

that if she died your Majesty would marry her." Having just failed to produce a child, and feeling increasingly old and barren, Mary was taunted by the thought that her half sister, who "is more likely to have children on account of her age and temperament," would bear Philip many sons if they married.[86]

Much to the Queen's resentment, her husband requested that Elizabeth should be among the party who travelled to Greenwich to bid him farewell as he sailed for the Netherlands. By now openly suspicious of her half sister and jealous of the esteem in which she was held by both her husband and her people, she assented only on condition that Elizabeth should be conveyed to Greenwich discreetly by barge, rather than through the streets of London, where she could be cheered by the crowds. She herself was loath to be seen by these crowds, painfully aware that the last time had been when, some five months before, she had made her way in expectant triumph to her confinement at Hampton Court. Now she must face her people again, her belly no longer distended, knowing that she had failed to give them the prince of whom she had been so confident. She endured the ordeal bravely, however, maintaining a proud bearing and gratefully acknowledging the cheers and good wishes that her kinder subjects cried out. But having got through this trial, she then had to face the greater one of bidding farewell to her adored husband. As she watched his ship embark from Greenwich, she suffered more intense misery than she had yet in her troubled life. Now she not only had no child, she had no husband to comfort her.

"To say the truth the Queen's face has lost flesh greatly since I was last with her," observed the Venetian ambassador, Michiel. "The extreme need she has of her Consort's presence harassing her, as she told me, she having also within the last few days in great part lost her sleep."[87] While it caused his neglected wife a great deal of grief, however, Philip's departure sparked something of a reconciliation between her and Elizabeth. Perhaps it had removed one of the causes of jealousy on Mary's part, or perhaps she was now more willing than ever to honor the commands of her husband, as it made her feel closer to him in his absence. His parting words had included a commendation of the

Lady Elizabeth, and he had reiterated this in a letter to the Queen written shortly after his arrival in Brussels. Mary duly allowed her half sister to return to her favorite country residence of Hatfield, and although she remained under surveillance, it felt a good deal less like a prison than Woodstock had done. In late 1556, she invited Elizabeth to court, although she stopped short of including her in the Christmas celebrations, and she was sent back to Hertfordshire at the beginning of December.

One of the reasons for Elizabeth's peremptory dismissal from court could have been her refusal to accept Philip's attempts to marry her to Emmanuel Philibert, the Duke of Savoy. Initially, Mary had heartily concurred with this plan. Marrying her heretical sister to a foreign Catholic would go a good way toward neutralizing the threat she still posed to Mary's authority. She therefore summoned Elizabeth to her presence and relayed her husband's scheme to her. She was taken aback by her sister's reaction. Elizabeth pleaded with the Queen not to make her enter such a match, crying that "the afflictions suffered by her were such that they had not only ridded her of any wish for a husband, but that they had induced her to desire nothing but death."[88] This was the clearest indication that Elizabeth had ever given of the profound impact that her early experiences had had upon her view of marriage. It had proved the death of her mother and two of her stepmothers, and had done irreparable damage to her half sister's authority as queen. She would not endure the same fate herself. Besides, with her hope of succeeding Mary growing ever stronger, leaving England in order to marry a foreigner—and a Catholic one at that—was entirely at odds with her plans.

In an increasingly hysterical plea to her half sister, Elizabeth burst into "a flood of tears," and by the end of their meeting, Mary herself was crying. But her tears seemed more the result of anger than desperation, and she immediately dismissed Elizabeth not just from her presence but also from the court. So outraged was she by her younger sister's defiance that she talked of summoning Parliament and having her removed from the succession.[89] Camden attested that she showed such "inveterate hatred to the Lady Elizabeth [and] did so boyle with anger, that shee loaded her with checks and taunts, and stucke not ever and anon to affirme, that Mary Queen of Scots was the certaine and undoubted heire to the Crowne of England next after her selfe."[90]

Rather surprisingly, though, Mary subsequently relented and told her husband that his plans for the Savoy marriage must be dropped. Quite why she decided to defend her half sister is not clear. Had Elizabeth's tearful plea moved her after all? Another explanation is that Mary continued to doubt Elizabeth's legitimacy as a princess of royal blood. She had often remarked in private that she suspected that Elizabeth "had the face and countenance of Mark Smeaton," one of the men with whom Anne Boleyn was accused of adultery. This would explain why her first reaction to Elizabeth's refusal to marry Savoy was to confirm her bastardy and remove her from the succession. This would also conveniently remove the wily young woman as a threat to her own queenship and leave her free to choose a more malleable successor.

Elizabeth had been glad to leave court in December 1556. From the haven of Hatfield, she was gradually building support from those who shared her religious views and hoped for the day when she might be queen. Among them was Katherine Knollys, the daughter of Anne Boleyn's sister, Mary. As Elizabeth's first cousin, Katherine was her closest female relative on her mother's side. She was some ten years older than Elizabeth but had spent time with her as a child. Elizabeth was drawn to Katherine because she had known Anne Boleyn and was said to have been among the women who attended her during her final days in the Tower, even though Katherine would have only been a young girl at the time.

Elizabeth and Katherine were also united by their Protestant beliefs and sought solace in each other's company during the dangerously uncertain years of Mary's reign. By 1556, they had become so close that when Katherine and her husband were obliged to flee to exile on the Continent in order to escape the religious purges, Elizabeth was heartbroken. Perhaps as much to reassure herself as her cousin, she wrote

Katherine a comforting letter as she and her husband were preparing to depart. "Relieve your sorrow for your far journey with joy of your short return," she began, "and think this pilgrimage rather a proof of your friends, than a leaving of your country." She went on to assure Katherine that she would remain as true a friend to her in her absence as she had been when they were together, and would do everything possible to protect her interests. "The length of time and distance of place, separates not the love of friends, nor deprives not the show of good will . . . when your need shall be most you shall find my friendship greatest. Let others promise, and I will do, in words not more in deeds as much. My power but small, my love as great as them whose gifts may tell their friendship's tale, let will supply all other want, and oft sending take the lieu of often sights. Your messenger shall not return empty, nor yet your desires unaccomplished . . . And to conclude, a word that hardly I can say, I am driven by need to write, farewell, it is which in one way I wish, the other way I grieve." She signed the letter: "Your loving cousin and ready friend, Cor Rotto [Broken Heart]."[91]

Elizabeth was ever one for fair words and well-turned phrases, but her letter to Katherine has a ring of truth about it. Her promise to help her cousin was not lightly made, for she would have known how dangerous it would be to antagonize Mary still further by advocating the cause of Protestant exiles. Furthermore, her actions when she herself became queen would prove the sincerity of her words. Katherine and Francis Knollys remained on the Continent for the rest of Mary's reign and became prominent among the increasing number of English exiles there. As they travelled from country to country, drumming up support for the Protestant cause, they kept an anxious eye on events back in England, praying for the day when they might return.

Meanwhile, another of Elizabeth's female allies, Anne of Cleves, had fallen gravely ill. She had left court soon after Elizabeth had departed for Ashridge during the early days of Mary's reign and had sensibly remained in the background ever since. Mary had permitted Elizabeth to visit her at Hever from time to time. Elizabeth valued—and learned from—Anne's pragmatic approach, and the two women had remained close. It was therefore with great sadness that Elizabeth learned that Anne had died on July 16, 1557, while in residence at

Chelsea Manor. Anne's will attests to the affection that she felt for Henry's daughters, for she bequeathed them her best jewels. She was buried with great ceremony at Westminster Abbey. Although she had been just short of forty-two years of age when she died, she had won the dubious honor of being the longest lived of all Henry's wives. But her death had come all too soon for Elizabeth, who had once more been deprived of a much-needed mother figure.

Meanwhile, events at court were moving apace. Philip II had returned from the Netherlands the previous year, and Mary was naturally overjoyed to have him back. He was a good deal less so, and was growing increasingly impatient with his cloyingly affectionate wife and her hostile, xenophobic subjects. No sooner had he returned than he was already looking for an excuse to leave again. The following summer, he set sail once more. Utterly wretched at being abandoned for a second time, Mary retreated to her privy chamber, taking all her meals there alone and refusing to appear in public. Desperate for consolation, she found it in the belief—real or imagined—that she was pregnant once more. Again she displayed certain symptoms to suggest that she was in this happy condition, including "the swelling of the paps and their emission of milk," as well as a growing stomach.[92] She wrote at once to her husband, telling him the joyous news that their child would be born the following March. So confident was she, despite her earlier experience, that she also made a will, "thinking myself to be with child in lawful marriage."[93]

This time, though, few people believed that the Queen was truly pregnant. No preparations were made for the birth, and her ministers and ladies merely humored her. Deluded though she was, Mary nevertheless realized what people were saying. "She the more distresses herself, perceiving daily that no one believes in the possibility of her having progeny, so that day by day she sees her authority and the respect induced by it diminish," remarked one observer at court.[94] "They [her subjects] had small hope of issue by the Queene, being now 40 yeeres old, dry, and sickly," wrote the seventeenth-century chronicler William Camden, with the same lack of sympathy that Mary's courtiers had displayed.[95]

Defying all of them, the Queen doggedly went ahead with her

plans for the birth. In February 1558, Elizabeth visited her at Richmond, ostensibly to offer her good wishes for a safe delivery, but in reality to judge for herself whether this time Mary really was pregnant. Although she can have little believed it after witnessing the farce of the first occasion, she went through the charade of presenting her elder sister with some baby clothes that she herself had made. She stayed just a week, which was no doubt long enough to satisfy herself that this pregnancy, like the first, was a figment of the Queen's imagination, and that her own position as heir was therefore secure.

The Queen entered her confinement shortly after her younger sister's departure. This time there was little sense of anticipation. There were considerably fewer lords and ladies gathered to await the birth, and those who attended Mary resigned themselves to the tedium of sitting it out until she herself came to the realization that there would be no child. As the weeks dragged on, she became increasingly despondent, weighed down not just by grief but by what would turn out to be a fatal illness. It is possible that the swelling in her stomach was due not to pregnancy but to cancer. Even during her first confinement, it had been rumored that "she was deceived by a Tympanie [tumor] or some other like disease, to think herself with child."[96] Now, as the Queen grew weaker, it became certain that this tumor, if such it was, was slowly killing her.

In April 1558, Mary rallied sufficiently to pay a visit to her younger sister at Hatfield. Elizabeth showed her all due honor, arranging lavish entertainments such as feasting, singing, and bearbaiting. The visit was such a success that in the summer, Mary invited Elizabeth to Richmond so that she could repay the compliment. She ordered a sumptuous pavilion to be constructed, from which she could receive her half sister. Bedecked with gold and crimson cloth, this was made to represent a mythical castle and would have presented an impressive sight to Elizabeth as her barge approached the palace. A host of festivities followed, including a splendid banquet and dancing accompanied by minstrels. It seemed that the Lady Elizabeth was higher in favor than she had been since the Queen's accession. Did Mary know she was dying and therefore wish to reconcile with her half sister, putting their difficult past behind them? If she did, then it was more likely to be for reasons of politics than sentiment. Besieged by urgent requests from her councillors formally to name her heir—something she had hitherto refrained from doing—it seemed that in her generous hospitality toward Elizabeth, Mary was making a public statement.

As it became clear to everyone at court that the Queen was dying, still abandoned by her husband, ambitious politicians and noblemen hastened to Hatfield, eager to ingratiate themselves with her successor. "Many persons of the kingdom flocked to the house of Miladi Elizabeth, the crowd constantly increasing with great frequency," reported Michiel, the Venetian ambassador.[97] Even while she still lived, Mary was becoming obsolete. Elizabeth would remember this when, years later, her own council urged her to name her successor, and it gave her the strength to resist them until almost her last breath. "I know the inconstancy of the people of England," she would tell the Scottish ambassador when under pressure to name his queen as her heir, "how they ever mislike the present government and has [sic] their eyes fixed upon that person that is next to succeed . . . I have good experience of myself in my sister's [time] how desirous men were that I should be in place, and earnest to set me up."[98]

Among Elizabeth's entourage was a woman with whom she had become acquainted as a child. Elizabeth Fitzgerald, now Lady Fiennes de Clinton, had recently arrived at Hatfield, and the two women had rapidly reestablished their former intimacy. In early November, Philip II's envoy, the Count de Feria, visited Elizabeth at Brocket Hall, the home of one of her tenants at nearby Hatfield. He was invited to dine with her and Lady Clinton. He was no doubt already acquainted with the latter because she was a friend of his fiancée, Jane Dormer, the favorite attendant of Queen Mary.

The three enjoyed a pleasant repast, as Feria reported to his master: "During the meal we laughed and enjoyed ourselves a great deal." But tensions were simmering beneath the surface, for the count's real purpose was to find out Elizabeth's intentions toward Spain and the Catholic religion when she finally inherited the throne, which looked set to be any day now, as Queen Mary's life was fast slipping away. Elizabeth and Lady Clinton knew this only too well, and the latter was every bit as skilled in the art of dissimulation as her royal mistress.

Lady Clinton assumed a placid and indifferent air as Elizabeth rose from the dinner table and told Feria that he might speak to her in Spanish on business matters. She assured him that although Lady Clinton and their two female attendants would remain in the room, they "could speak no other language than English."[99] Given that Lady Clinton had shared Elizabeth's education, which had included tuition in several languages, and had a number of acquaintances in Mary's Spanish-dominated court, it is likely that she understood a good deal more of the ensuing conversation than Elizabeth led Feria to believe. In any case, she had certainly conferred with her friend about the items that would be discussed, for in the course of her conversation with Feria, Elizabeth praised Lord Clinton and declared that he should never have lost his office of Lord High Admiral under Mary (albeit temporarily), "for the deprivation was unlawful as it was contrary to the patent he held, which he had deposited with [William] Paget and which Paget had kept for him ever since."[100]

Although their exchange had been civil enough, Feria left Hatfield with a less than favorable impression of the heir apparent. He reported to Philip II that Elizabeth was becoming increasingly arrogant, as she now firmly believed that nothing could come between her and the throne of England. She had told him "that the queen had lost the affection of the people of this realm because she had married a foreigner." Elizabeth, by contrast, had gone out of her way to court the goodwill of the English people and confidently told Feria "that they are all on her side"—which, he admitted, "is certainly true."[101] One of Elizabeth's greatest strengths as queen would be the ability to judge the mood of her subjects and never to lose sight of the need to cultivate their favor. This talent for public relations may have been inherited from her parents, but its necessity was learned during her half sister's reign.

"Madam Elizabeth already sees herself as the next Queen," Feria wrote to his master, "and having come to the conclusion, that she would have succeeded, even if your Majesty and the Queen had opposed it, she does not feel indebted to your Majesty in this matter."[102] It seems that the Spanish king had indeed urged his dying wife to name Elizabeth as her successor. The Venetian ambassador reported that

Philip had dispatched his confessor to Mary in order to persuade her to leave the crown to her half sister. But he found Mary "utterly averse to give Lady Elizabeth any hope of the succession." Instead she railed against her, full of "inveterate hatred" for all the wrongs she had committed. She did eventually relent, however, and sent a message to her husband expressing herself "muy contenta [much pleased]" with his suggestion.[103] But even though she had indicated her intention to make Elizabeth her heir by the time of their Hatfield and Richmond meetings that summer, she still refused formally to declare it. According to one report: "What disquiets her most is to see the eyes and hearts of the nation already fixed on this lady as successor to the Crown, from despair of descent from the Queen."[104]

As the Queen's condition worsened, Philip's ambassador at court urged him to return to England and be by his wife's side. But he clearly had no intention of doing so. In stark contrast to Mary, who had so mourned his absence that she had sent frequent letters and gifts to him, including some "game pasties" that she knew were his favorites,[105] and had lovingly pored over all his letters in her loneliness, Philip could not bring himself to pay his wife one final visit, even though he knew she was dying. Instead he sent a message to the Privy Council that "We are moved to send a person to England to attend certain business, visit her [the Queen], and excuse our absence."[106] His neglect was said to have hastened Mary's decline, and some even blamed her illness entirely upon it.[107] Mary's sole comfort now was her faith, which convinced her that she would soon enjoy the eternal peace of heaven. In her increasing delirium, she told the ladies who were gathered around her bedside that her dreams were filled with little children "like angels playing before her, singing pleasing notes."[108]

On October 28, she was conscious enough to add a codicil to her will, finally acknowledging that there would be no "fruit of her body," and confirming that the crown would go to the next heir by law. Even now, she could not bring herself to name Elizabeth in person: that last show of affection at Hatfield and Richmond had clearly been a sham. The old resentments against her sly younger sister, daughter of that "Great Whore" who had destroyed her mother, had risen to the fore again. She cried that Elizabeth "was neither her sister nor the daughter of the Queen's father, King Henry, nor would she hear of favoring her, as she was born of an infamous woman, who had so greatly outraged the Queen her mother, and herself."[109]

Only when faced with the most extreme pressure from her council did Mary finally relent. On November 8, Sir Thomas Cornwallis, controller of the Queen's household and secretary to the Privy Council, arrived at Hatfield to tell Elizabeth that her half sister had finally named her as heir. Shortly afterward, Mary dispatched her most trusted servant, Jane Dormer, to relay her final wishes to Elizabeth. These were to uphold the Roman Catholic faith "as the Queen has restored it," to be good to her servants, and to pay her debts.[110] Elizabeth's response was noncommittal. With the crown now within her grasp, where her elder sister was concerned, she no longer had any need to pretend.

On November 14, Feria reported that there was now "no hope of her [Majesty's] life; but on the contrary, each hour I think that they will come to inform me of her death, so rapidly does her condition deteriorate from one day to the next."[111] Three days later, between four and five o'clock in the morning, Queen Mary finally slipped from a life that had been marked by tragedy and heartache. A messenger was immediately dispatched to Hatfield with the news. Upon hearing that her half sister was dead and that she was now herself queen, Elizabeth proclaimed: "My Lordes, the law of nature moveth me to sorrowe for my sister: The burdaine that is fallen uppon me maketh me amazed."[112] Although some twenty-eight years later she claimed to have shed great tears for Mary, she is the only witness to record them.[113] Philip, meanwhile, remarked that he felt "reasonable regret" when he was told of his wife's death. If this was a poor reward for Mary's passionate adoration of him, it was at least a true reflection of his feelings toward her while she had lived.

Her late sister's reign had taught Elizabeth a number of lessons. One of the most important was the danger of placing one's trust—and love—in a man. She had seen Mary's suffering firsthand; had experienced Philip's open flirtation while her half sister pined for him, hu-

miliated by his increasingly obvious distaste and his prolonged absences. This confirmed the view Elizabeth had developed in childhood
that if a sovereign was to marry, then politics, not love, should govern
their actions. The catastrophic nature of Mary's experience might well
have put Elizabeth off marriage altogether; certainly it had intensified
her fear of childbearing.

Although she had shown flashes of the strength and majesty of her
father, Mary had on the whole been subservient to her husband. Even
when Philip was abroad, she had written to him almost daily, entreating him for guidance. In short, she had played the role more of a queen
consort than a queen regnant, losing the respect of her councillors and
her people, whose fears about the weakness of a female ruler had apparently been realized.

Mary had been naturally trusting, straightforward, and stubbornly
principled, while Elizabeth was cautious, reserved, and pragmatic. The
latter had seen how damaging Mary's inflexibility had proved to her
authority. As a result, she came to appreciate the need for compromise
and consultation, as well as for not openly committing to any single
policy. One of Elizabeth's most famous traits as a ruler would be her
ability to play her cards close to her chest at all times. She would drive
her ministers and ambassadors to distraction with her "answers answerless" and procrastinations, at one instant leading them to believe
she intended one thing, and the next another.

"Not being Mary or behaving like her proved a golden rule for Elizabeth."[114] Yet Mary had done much of the groundwork in reconciling
England to the concept of female monarchy. Admittedly, she had made
serious mistakes as she struggled to define her role, but it was she, not
Elizabeth, who had had to bear the brunt of all the prejudices against
a queen regnant, among both her councillors and her subjects. By the
time Elizabeth came to the throne, certain important precedents had
been set, including the 1554 act confirming that a "sole" queen should
rule as absolutely as a king. In his address given at Mary's funeral service, the Bishop of Winchester declared that she had been "a queen
and by the same title a king also. She was a syster to her that by the like
title and wryght is both king and quene at this present of the realm."[115]

The brevity of Mary's reign and her struggle to come to terms with her role meant that there was still uncertainty about what exactly a queen regnant should be. For Mary, the primary difficulty had been in trying to reconcile her role as queen with that as wife. The marriage settlement had confirmed her superiority over her consort, but the laws and social conventions governing marriage gave men the upper hand. Ever the traditionalist, Mary had struggled to accept that she should rule over her husband, and her personal adoration of him had complicated matters further. It had been much easier at her accession, when she had been able to set herself up as a virgin who was "married" to England. Although she had continued to use this imagery after her marriage to Philip, it had lacked impact. Upon her death, those searching for praise of the late queen had to hark back to the days of her virginity. George Cavendish remarked that the "virgin's life" had "liked thee best," even though she had subsequently "knit with a king."[116]

This would prove one of the most powerful lessons for Mary's successor. If Elizabeth was to triumph as a queen regnant, then she must truly be married to her country. In short, she must remain a virgin. She had long harbored this intention, but for different reasons. Now, having learned from her late sister's example, it became a linchpin of both her political strategy and her public image as England's new queen.

CHAPTER 7

The Queen's Hive

When Elizabeth was proclaimed queen in the city of London, there was great rejoicing. All across the capital, church bells were rung, and at night bonfires were lit, around which thousands of people gathered to drink and make merry, just as they had a little over five years ago to welcome Mary's reign. The new queen proceeded to order a lavish funeral for her late sister, entrusting the Marquess of Winchester with the arrangements, which in total cost some £7,763 (equivalent to more than £1.3 million or $2 million today). A magnificent procession bore the coffin, on top of which was a life-sized effigy of Mary, through the city of London.

But it was with the epitaph to Mary that Elizabeth's mask of respect began to slip. Upon reading it, she furiously decreed that it was too praiseworthy, and, worse, that it made no mention of herself. Upon her orders, the following lines were therefore added:

Marie now dead, Elizabeth lives, our just and lawful Queen
In whom her sister's virtues rare, abundantly are seen.

She also spared herself the expense of erecting a tomb above the unmarked vault in which Mary was interred, and during the course of her reign, pieces of stone from the alteration work within Westminster Abbey were piled on top of it.[1]

Just days after Mary's death, Elizabeth ordered her ladies to take a thorough inventory of the royal jewels to insure that there were none missing, and also sent a message to Philip II demanding that he hand over the jewels that he and his father, Charles V, had given to the late queen. Elizabeth also failed to honor any of the provisions of her half sister's will, or the three requests that Mary had made as she lay dying.

In government, too, Elizabeth made it clear that there would be no harking back to the past. Although she published an edict declaring her intention not to change anything "which had been ordained and established by the Queen her sister during her reign," it was soon clear that she would do nothing of the sort.[2] Less than a month later, the Count de Feria reported to Philip II: "The kingdom is entirely in the hands of young folks, heretics and traitors, and the Queen does not favour a single man whom Her Majesty, who is now in heaven, would have received and will take no one into her service who served her sister when she was Lady Mary. On her way from the Tower to her house where she now is, she saw the marquis of Northampton, who is ill with a quartan ague, at a window, and she stopped her palfrey and was for a long while asking him about his health in the most cordial way in the world. The only true reason for this was that he had been a great traitor to her sister." He concluded that Elizabeth was "as much set against her sister as she was previous to her death."[3]

As well as favoring those who had opposed Mary's regime, Elizabeth also promoted the former allies of her late mother. Matthew Parker, in whom Anne had confided her hopes for her daughter's future, was appointed Archbishop of Canterbury. Baron Norris of Rycote, son of Henry Norris, who had been executed on the charge of adultery with Anne, was given great honors by Elizabeth because, it was said, she was mindful that his father had "died in a noble cause and in justification of her mother's innocence."[4] Henry Carey, son of Anne Boleyn's sister and therefore a cousin of Elizabeth, was created Baron

Hunsdon immediately after her accession. She demonstrated a real affection for him throughout his life, calling him "my Harry," and signing her letters "Your loving kinswoman."[5]

The Howards, who were related to Anne Boleyn on her mother's side, occupied key posts in government throughout the reign. William Howard, Lord Effingham, was appointed Lord Admiral by Elizabeth. As her mother's first cousin, he was one of her nearest maternal kinsmen and became one of her most influential advisers. His credit was further enhanced by the fact that he married the Queen's other Boleyn cousin, Katherine Carey. Meanwhile, the son of George Boleyn, Anne's brother, was appointed dean of Lichfield. Various other Boleyn relatives were promoted during Elizabeth's reign, including the Sackvilles, Staffords, Fortescues, and Ashleys.

Now that she herself was queen, all of the suppressed outrage that Elizabeth had felt toward Mary during the latter's reign found full expression. Even as Mary lay dying, Feria observed that her half sister was "highly indignant about what has been done to her during the queen's lifetime."[6] Elizabeth seemed determined to right these wrongs by becoming a much greater queen than Mary had ever been. The years of fear and uncertainty had fostered within her a considerable strength of purpose, as well as a formidable authority. "She seems to me incomparably more feared than her sister and gives her orders and has her way as absolutely as her father did," observed Feria just weeks after her accession.[7] This still held true the following year, when the Venetian ambassador remarked that Elizabeth "insists upon far greater respect being shown to her than was exacted by the late Queen Mary."[8]

Elizabeth consciously set out to distance herself from her late sister. In the propaganda of her reign, she was the "clear and lovely sunshine" that dispersed the "stormy, tempestuous and blustering windy weather of Queen Mary."[9] True enough, by the time Elizabeth inherited the throne, England was racked by religious division, political uncertainty, economic hardship, and disastrous foreign wars that had resulted in the loss of Calais, her last stronghold in France. But this in itself presented the new queen with a distinct advantage. Surely anything now would be preferable to Mary's brief but catastrophic reign?

According to one contemporary chronicler, this was exactly what people thought. "And it was not without ground, that the nation concluded such great hope of being happily governed under this lady, both
in regard of her mild and serene beginnings: whereas the former
Queen's first steps into her government, was nothing but storm and
ruffle, violation of laws, terrors and threatenings, imprisonments and
executions."[10] Elizabeth made sure that throughout her reign, she
would always be compared favorably to her half sister. This served to
remind her subjects, at times of crisis, that no matter what they faced,
it was as nothing to the sufferings inflicted upon the English people by
her dogmatic, pro-Spanish predecessor.

Attractive, charismatic, and vivacious, Elizabeth had already won
the love of her new subjects. The Spanish ambassador remarked that
she had "many characteristics in which she resembled her mother."[11]
Sir Robert Naunton, who wrote an account of notables at Elizabeth's
court, believed that she had inherited all her best traits from Anne.
"Her mother was . . . as the French word hath it, more debonaire, and
affable, virtues which might well suit with majesty and which de-
scending as hereditary to the daughter did render her of a more sweet
temper, and endear her more to the love and liking of the people, who
gave her the name and fame of a most gracious and popular prince,
the atrocity of her father's nature being allayed in hers by her mother's
sweet inclination."[12] This account is a rare, almost exceptional, example of Elizabeth being compared favorably with her mother, as opposed to her father. It is telling that it was published after Elizabeth's
death, for Naunton would not have got away with such criticism of her
father in her lifetime.

But he did have a point. The magisterial arrogance that Elizabeth
had inherited from her father was tempered by the charm and
charisma that had been so much a part of Anne's character. It was
these qualities that won Elizabeth the love, as well as the respect, of
her people. She had as keen an instinct for public relations as her
mother had, and unlike Anne, she had the opportunity to exercise this
instinct to staggering effect. While Anne may have charmed a throng
of (largely male) courtiers and had a gift for political ostentation, the

fact that—in most people's eyes—she had usurped the rightful queen put her at a disadvantage from which she never had time to recover. Elizabeth had no such weakness. She succeeded a deeply unpopular queen whose religious fervor and pro-Spanish policies had caused widespread resentment and unrest. Against the aging, barren Mary, Elizabeth was a shining symbol of youth, hope, and prosperity. Little wonder that she ascended the throne on a wave of popular support and rejoicing. That she was able to sustain this beyond the temporary euphoria that almost always accompanies a new reign is at least in part a testament to the charisma and acumen that she had inherited from her mother.

Beneath the euphoria at her accession lay the deep-seated preju-dices against female rulers that had existed for centuries. Her sister's reign had strengthened these. In 1558 John Knoxe published *The First Blast of the Trumpet Against the Monstrous Regiment of Women,* in which he argued that it was abhorrent to nature for women to rule over men, and pointed to all the examples of disastrous female sovereignty, in-cluding Mary Tudor and Mary, Queen of Scots. "It is more then a mon-stre in nature that a Woman shall reigne and have empire above a Man," he argued, and decried "howe abominable, odious, and de-testable is all such usurped authoritie."[13] In a similar vein, Thomas Becon, a Norfolk clergyman, vehemently complained in a prayer to God: "Thou has set to rule over us a woman, whom nature hath formed to be in subjection unto man . . . Ah, Lord! To take away the empire from a man, and give it to a woman, seemeth to be an evident token of thine anger towards us Englishmen."[14]

The vast majority of Elizabeth's subjects firmly believed that the proper role of women in society was to be subservient to fathers, hus-bands, and brothers. They had neither the intelligence nor the strength of character to make their own way in the world. If they could barely manage a household, then how on earth could they rule over a king-dom? These sentiments were echoed by the leading Elizabethan writer Sir Thomas Smith, who claimed that women "can beare no rule," and should not "mingle with public affairs; for they were by nature weak and fearful, and easily forced into obedience and submission by men with their superior strength and courage."[15] Such prejudices went right to the heart of Elizabeth's new government. Even her closest adviser, William Cecil, was furious when one of the Queen's messengers discussed with her a dispatch for her ambassador in Paris, exclaiming that it was "too much for a woman's knowledge."

Whereas Mary had confirmed such prejudices during her own reign, Elizabeth set out to confound them. Although she shared her male subjects' views on the inferiority of women, she saw herself as an exception and was determined to stamp her authority upon all aspects of her court and government. But she knew that to win the respect of her ministers and subjects, she must first win their love. She therefore set about planning a series of carefully crafted public relations exercises.

The first and most important was her coronation. Elizabeth's inspiration was that staged by her mother some twenty-five years earlier. This would be the first demonstration of the elaborate symbolism and image cultivation for which Elizabeth, the Virgin Queen and Gloriana, would become so famous. Like Anne, she ordered the preparation of Latin verses that would be sung as her procession made its way to Westminster Abbey. Lavish scenery was constructed, much of it in the same classical style that Anne had ordered to line her own processional route. One of the vignettes included a representation of Anne as queen, which Elizabeth would have seen as she passed by. As a further tribute to her mother, she adopted the same manner of dress, with a heavily brocaded silk surcoat and a mantle of ermine. The final touch was to commission a special crown to be made that was lighter than the one that would be used in the coronation ceremony itself and therefore more comfortable to wear during the procession. This idea had been borrowed from Anne, who had done exactly the same thing for her own coronation.

Elizabeth had studied the account of her mother's coronation with meticulous care, judging by the similarity with her own ceremonials. But the influence extended far beyond that brief episode; indeed, the symbolism used by Anne Boleyn in her coronation pageantry would be adopted by her daughter throughout the latter's reign. One of the

most obvious examples was Anne's emblem, the white falcon. Elizabeth now adopted this as her own, and it became ubiquitous in her
palaces, adorning swords, fireplaces, virginals, and books.

But by far the most significant imagery borrowed by Elizabeth
from her mother's coronation was the notion of the queen as a Virgin
Mary on earth. As with so many of Anne's other ideas and influences,
she would make it her own, but better. The image of her as a divinely
appointed sovereign would be invaluable to her at a time when many
still harbored doubts about the legitimacy of her claim to the throne.
But her use of this symbolism would go far beyond that. Elizabeth did
not just mimic the Virgin Mary; she became her: untouched and un-
touchable, a divine presence here on earth. This, more than any other,
was the image that would win her the love and devotion of her subjects, and secure her a place in history.[16]

Watching all of this from the sidelines were Elizabeth's adoring old
servants, Kat Astley and Blanche Parry, both of whom had been accorded places of honor at the coronation. While the crowds that
thronged to see their new queen were presented with her public persona, Blanche and Kat witnessed her private one, as they clustered
around her in the curtained enclosure behind the altar of St. Edward's
Chapel in Westminster Abbey, which had been assigned as Elizabeth's
changing room. It was an immensely proud moment for the two
women who had stood by her in the greatest danger and had shared
her hopes for the future.

Elizabeth's lavish coronation was just part of her strategy to gain
popularity. "She lives a life of magnificence and festivity such as can
hardly be imagined, and occupies a great portion of her time with
balls, banquets, hunting and similar amusements with the utmost possible display," observed a Venetian envoy with some astonishment a
year after Elizabeth's accession.[17] The new queen had certainly made a
point of establishing her court as a spectacle of glorious, ostentatious
display; a theater of art, music, dancing, and lavish dress. The cultural
and political heart of England, it was the place to which all of the principal men of the realm flocked to pay reverence to their sovereign and
clamor for her favor. Elizabeth, like her father, had a natural gift for

public relations and fully appreciated how politically important it was to stage an obvious display of wealth and majesty at her court. "They are intent on amusing themselves and on dancing till after midnight," exclaimed one foreign visitor, who was shocked by such "levities and unusual licentiousness." Another described the May Day festivities at court, which had included "decorated barges, gunpowder, performers falling into the Thames . . . the Quen [Queen's] grace and her lordes and lades lokyng out of wyndows," and later remarked: "ther was grett chere tyll mydnyght."[18]

Elizabeth's banquets were always spectacular. Her reign saw the introduction of many exotic foods from the New World. These included rich spices such as cinnamon and ginger, as well as pineapples, chillies, potatoes, tomatoes, and chocolate. The food was prepared as much for visual effect as for taste, and there was a strong sense of theater throughout. A host of different colors, materials, and props were used to make and serve the food. Peacocks were reared for consumption, but their feathers were also used to decorate cooked foods. The Queen had a famously sweet tooth, and her cooks let their imaginations run wild when it came to preparing the confectionery for her banquets. On one occasion, an entire menagerie was sculpted in "sugar-worke," including camels, lions, frogs, snakes, and dolphins, along with more fantastical figures like mermaids and unicorns.[19]

Elizabeth would be offered tens, if not hundreds, of separate dishes during the course of one banquet. Her first course might comprise a choice of beef, mutton, veal, swan, goose, or capon, while the second provided lamb, heron, pheasant, chicken, pigeon, and lark. Baked fruits and custards would be served with each course, rather than as dessert, and the pièce de résistance would be the exquisitely crafted confectionery.

All of this was washed down with wine and ale—water rarely being clean enough to drink. Even breakfast would be accompanied by ale, which was brewed with malt and water, sometimes with flavors added such as mace, nutmeg, or sage. Wine was generally imported, although some fruit wines were produced in England. A form of cider referred to as "apple-wine" was also prepared, along with mead, an alcoholic drink sweetened with honey. Elizabeth was as sparing in her consumption of drink as she was of food, and never overindulged in either. The same could not be said of her courtiers and guests, who made the most of all the sumptuous fare on offer.

However decadent and carefree the court entertainments might seem to the casual observer, they were all carefully controlled by Elizabeth, who quickly established a strict etiquette and ceremony from which no courtier was allowed to stray. Ever watchful of her reputation as a young, unmarried queen, she was determined to insure that merriment would never descend into drunkenness, or flirtatiousness into sexual transgression. Thus a contemporary was able to recall: "The court of Queen Elizabeth was at once gay, decent, and superb."[20]

All of this was played out against the backdrop of some of the most magnificent palaces in Europe. The new queen's chief residence was Whitehall Palace in the heart of London, a vast maze of buildings that sprawled over twenty-three acres. The exteriors of the main buildings were decorated with elaborate paintwork, while principal rooms within were decked out with rich gold fabrics and carvings. The Queen's bedroom, which overlooked the river Thames, was the most sumptuous of all, complete with a gold ceiling and a bed "ingeniously composed of woods of different colours with quilts of sik, velvet, gold and silver embroidery."[21] Another of Elizabeth's London palaces was Greenwich, the place of her birth. Built on the banks of the Thames, it was conveniently accessed by barge, and when the Queen chose that method of transport, she ordered that perfumed oil be burnt to disguise the noisome smells of the river. The inside of the palace was adorned with the many expensive gifts that Elizabeth received during her reign, including a tablecloth made from peacock feathers.

Upriver, west of the capital, lay Richmond Palace, built by the Queen's grandfather, Henry VII. Viewed from the river, it was a fairytale palace with clusters of domed towers and turrets behind a high curtain wall, set within some of the most beautiful gardens in England, with sweet-smelling flowers and herbs, and orchards that yielded apples, pears, peaches, and damsons. Richmond had the luxury of a sophisticated plumbing system that supplied the residents with fresh spring water, and was also less drafty than the other palaces. Nevertheless, Elizabeth was slow to appreciate its virtues, and it was only later in her reign that she became a frequent visitor.

Further upriver was Hampton Court, "the most splendid and magnificent royal Palace of any that may be found in England—or, indeed, in any other kingdom," according to one foreign visitor.[22] With its eight hundred rooms, all lavishly decorated with rich hangings and furniture, Hampton Court was designed to inspire awe. The apartment in which Elizabeth sat in state was hung with tapestries garnished with gold, pearls, and precious stones, while her throne was studded with "very large diamonds, rubies, sapphires, and the like, that glitter among other precious stones and pearls as the sun among the stars."[23] Although the splendor of the palace undoubtedly impressed visitors, the Queen found it uncomfortable and unhealthy, and after falling dangerously ill there in 1562, she rarely visited.

By contrast, Elizabeth was extremely fond of Windsor Castle, where she tended to spend the summer months. Although the castle dated back to the time of William the Conqueror, it boasted a host of modern luxuries. Principal bathrooms all had running water, and the walls and ceilings were covered with mirrors, a priceless commodity. The Duke of Wurttemburg, who visited in 1592, marvelled at the many "costly things" on display, including "a genuine unicorn horn."[24] The castle's best attraction was the expansive Great Park, in which the Queen could indulge her passion for hunting. Later on in the reign, she also acquired Nonsuch in Surrey, a fantasy palace built by her father in the style of the great chateaux of the Loire.[25] Although the staterooms were magnificent, the palace was so much smaller than the others that tents had to be set up on the grounds to accommodate Elizabeth's vast entourage.

If the Elizabethan court was a carefully stage-managed production, then the Queen herself was the director. She was also the center of all the entertainments, intrigues, and flirtations. In the culture of chivalric love that permeated the court, she was—at least to the untrained eye— the object of all men's devotions. She was the unattainable mistress, at once both aloof and alluring. Although the vast majority of Elizabeth's flirtations were nothing more than playacting on both sides, she nevertheless demanded absolute fidelity, both emotional and political, from her male courtiers and would brook no rival for their affections. As one recent commentator has observed: "There could only be one queen-bee in the hive."[26]

In fusing her courtiers' personal desires with their political ambitions, the new queen created a highly volatile atmosphere in which men vied with each other for favor and advancement. "The principal note of her reign will be that she ruled much by faction and parties, which she herself both made, upheld, and weakened as her own great judgement advised," observed Sir Robert Naunton, who had firsthand experience of life at Elizabeth's court. "She was absolute and sovereign mistress of her grace and . . . those to whom she distributed her favours were never more than tenants at will and stood on no better ground than her princely pleasure and their own good behaviour."[27]

Although the new queen revelled in the entertainments on offer at her magnificent court, this did not distract her from the business of government. It was widely expected that one of her first priorities would be publicly to rehabilitate her mother. She immediately passed a statute in Parliament that provided legal certainty to her right to the throne, but this gave only a brief mention of Anne Boleyn and did not attempt to legitimize her marriage to Henry VIII. This was in stark contrast to the act passed by Mary, which had sought a complete rehabilitation of Catherine of Aragon. As such, it has been interpreted as a sign of antipathy or indifference by Elizabeth toward her late mother, but it could more accurately be seen as one of politics. Elizabeth knew that in promoting Anne's innocence, she would be casting blame upon her father, to whom she owed her position as queen.

But neither did she show any inclination to have her mother's remains reburied in a manner befitting her status. Even the future James I, who showed a callous lack of feeling toward his own mother, Mary, Queen of Scots, during her lifetime, ordered a lavish new tomb for her in Westminster Abbey. Elizabeth, meanwhile, seemed content to let her mother's bones remain in an old arrow chest, buried beneath the chapel of St. Peter ad Vincula at the Tower, that traditional resting

place of traitors. This may seem to show a cold indifference toward her
late mother, but even though she was now queen, Elizabeth was still
obliged to tread carefully where Anne was concerned. It would have
taken little to inflame the already discontented Catholics, both at
home and abroad, who resented this heretical queen and looked to
Mary, Queen of Scots, as the rightful heir. If Elizabeth had made such
a public show of loyalty to a queen whom most Catholics still viewed
as a usurper by having her remains reinterred, she would have been—
literally—digging up the past.

By contrast, it was with Henry VIII that Elizabeth publicly identified herself. "She prides herself on her father and glories in him," observed the Venetian ambassador, Michiel.[28] Her speeches would be
littered with references to him as she endeavored to overcome the perceived weakness of her sex by reminding her people that she was a chip
off the old block. She variously referred to herself as "my father's
daughter" or "the lion's cub." By reminding her subjects that she was
Henry VIII's daughter, she was also reiterating the legitimacy of her
claim to the throne.

For the most part, Elizabeth avoided making direct references to her
mother. The bitter experiences of her childhood had taught her this
pragmatism. She had seen how much trouble her half sister had brought
upon herself by her blindly principled defense of her own mother. She
had also seen Mary spend her reign trying to bring back the past: restoring her mother's legitimacy, reinstating Roman Catholicism—and all for
nothing. By the time of her death, she had been reviled by large swathes
of the population. Elizabeth had already suffered enough as a result of
her mother's disgrace, and no matter how fervently she might believe in
Anne's innocence, she was astute enough to realize that little good could
come of dedicating her life to proving it. The best compliment that she
could pay to Anne's memory would be to establish herself as the greatest queen that England had ever known.

This is not to say that when the opportunity presented itself, Elizabeth refrained from trying to restore her mother's reputation. A prime
example of this came in the summer of 1561, when news reached her
ears of a defamatory tract that had been published on the Continent,

condemning Anne as a heretical whore. Elizabeth's ambassador in France, Nicholas Throckmorton, reported to William Cecil that one Gabriel de Sacconay had "devised" and printed the work, "wherein he has spoken most irreverently of the Queen's mother." Anne Boleyn was denounced as a "Jezebel" and compared to the "heathen wives of Solomon" for persuading Henry VIII to turn his back on the "true" Church of Rome. Their "foul matrimony" was a result of lust, and Anne had met with just punishment for her wickedness. Cecil was aghast. If he told the Queen of it, a political storm would follow that would not only stir up trouble in England but also jeopardize her already fragile alliance with France. His procrastination had damaging consequences. Hundreds of copies of the book were printed, and it spread like wildfire across Paris and beyond. This forced Cecil's hand. In mid-September, more than a month after he had heard of it, he reluctantly told his royal mistress the truth.

Elizabeth's reaction was one of furious indignation. She wrote at once to Throckmorton, ordering him to go with all haste to the Queen Mother, Catherine de Medici (who as regent for her young son, Charles IX, was the real source of power at the French court), and demand that the "lewd" book be suppressed immediately. He duly did so, but although Catherine expressed her shock and disgust at the perpetrator of such a scandalous publication, she did not comply with Elizabeth's demand. Rather, she asked to see a copy of the book so that she "might cause it to be considered, and thereupon give order for the matter." She then consulted her son, Charles, who also demurred, while making a show of intending swift remedial action. He and his mother had every reason not to hurry with suppressing the book: Their support for Mary, Queen of Scots' claim to the English crown was well known, and it would do no harm to that cause if doubt were cast upon the parentage of her rival, Elizabeth.

Eventually an order was issued under Charles IX's name for de Sacconay to "alter the offensive passages" and sell no further copies in the meantime. But this did little to satisfy the English queen, who demanded that the book be destroyed completely. Furthermore, she wanted its author to be severely punished for his slanders. The French king and his mother continued to delay with fine words and promises, but Elizabeth pursued them doggedly, sending dispatch after dispatch to Throckmorton. In the second week of October, he was able to report that the king had at last issued a command for all of the books to be confiscated. But Elizabeth knew that the damage had already been done, and her resentment against Charles and Catherine continued to fester. Thus, when her ambassador suggested that she thank them for their pains, she completely ignored him. He continued to urge the necessity of doing so, pointing out that the Queen Mother was highly offended by Elizabeth's lack of courtesy. At length, in late November, she reluctantly sent a note of thanks via her ambassador in Spain—a studied discourtesy that made it clear that she had not forgiven the French for slandering her mother.[29]

There is also another, rather more personal piece of evidence that reveals Elizabeth's true feelings toward her mother. Elizabeth was well known for her love of expensive and elaborate jewelry, but one of her most cherished possessions in a collection that comprised thousands of items was a comparatively simple ring, fashioned from mother-of-pearl and embossed with tiny rubies and diamonds. It opened to reveal two portraits. One was of the Queen herself, shown in profile. The other was of a lady wearing a French hood and a dress with a low, square neckline. She stares directly ahead, and her features are remarkably similar to Elizabeth's, for she has the same high cheekbones and piercing dark eyes. The lady was Anne Boleyn.

By the time she became queen, Elizabeth was already surrounded by a tightly knit group of female attendants who were referred to as the "old flock of Hatfield."[30] Principal among them were Kat Astley and Blanche Parry, who had both served her for more than twenty years. One of Elizabeth's first actions upon becoming queen had been to bring Kat back to court and appoint her chief gentlewoman of the privy chamber. This was the most prestigious post in the royal household and gave Kat unrivalled access to her mistress. As well as attending her during the day, she would often sleep in the Queen's bedchamber at night. Her duties also involved overseeing all of the other ladies in the privy chamber, from the women and ladies of the privy chamber to the maids of honor.[31]

Meanwhile, Blanche was appointed second gentlewoman of the privy chamber. The accounts note that "Blaunche Apparey" was given seven yards of scarlet silk, fifteen yards of crimson velvet, one and a quarter yards of "gold yellow cloth," and three-quarters of a yard of "gold black" by the Queen so that she could be suitably attired for such a role.[32] As gentlewoman, Blanche received an annual stipend of £33. 6s. 8d.,[33] as well as board and lodging for herself and her servants (who grew in number as the reign progressed), horses and stabling, and various perks such as a guaranteed place in the royal carriages whenever the court was on the move. Knowing Blanche's fondness for literature, Elizabeth awarded her the additional responsibility of keeper of the Queen's books.

Two other ladies from Elizabeth's Hatfield household also enjoyed promotions upon her accession. Lady Elizabeth Fiennes de Clinton was appointed gentlewoman of the privy chamber. This was specified as being "withoute wages"—an indication of her highborn status, for the salaried members tended to be drawn from the lower ranks of the nobility. The same position went to Elizabeth St. Loe, better known as "Bess of Hardwick," who had also served Elizabeth at Hatfield and now basked in her reflected glory. At the age of thirty-one, she was one of the oldest members of the household, for the new queen liked to surround herself with young ladies in order to enliven the court after the staidness of her sister's reign. She was no less favored for that, however, and over the next two years, she won a great deal of respect, from both her royal mistress and the men and women of the court. Although by nature arrogant and self-serving, Bess had a good deal of political shrewdness, and succeeded in ingratiating herself with the new queen. Their relationship would become much stormier in the years to come.

One of the youngest women to find a place in Elizabeth's court was Lady Anne Russell. She was just ten years old when she was appointed a maid of honor in 1559. Anne was the daughter of Francis Russell, second Earl of Bedford, and may have been raised in Elizabeth's household before she became queen. Katherine Howard, daughter of Henry Carey, had certainly spent time with Elizabeth as a child, and she was now honored with the position of gentlewoman of the privy chamber, a privilege that was usually reserved for older married ladies.

The new queen also showed favor to ladies who were associated with her late stepmother. A former gentlewoman extraordinary of Katherine Parr's household, Mrs. Eglionby, was appointed mother of the maids, and Elizabeth Carew, who had been a member of Katherine's privy chamber, was given a prestigious position. Neither did Elizabeth overlook those who were associated with her closest male favorite, Robert Dudley. It was largely due to his influence that his sister, Mary, was appointed a gentlewoman "without wages" of the privy chamber in January 1559. This placed her at the heart of the royal court, and with her brothers Robert and Ambrose both in positions of considerable influence, she looked set to enjoy a glittering political career.

Above all, though, in selecting the women who would serve her as queen, Elizabeth was motivated by the desire to honor her late mother. She appointed a significant number of women from Anne Boleyn's side of the family, including her cousins, Philadelphia and Katherine Carey, daughter and sister respectively of Henry Carey. Katherine's daughter Mary later became another member of the household. The privy chamber lists at Elizabeth's accession and the years that follow contain regular mentions of the Careys, as well as the Knollyses, Howards, and various other Boleyn relatives.

Although Elizabeth's ladies formed a necessary element of her life at court, she was determined to restrict both their activities and their numbers. In contrast to the hundreds of male councillors, ambassadors, noblemen, and place-seekers who flocked to her court, there were only thirty or so women there at any one time. This was a marked decrease from both her half sister's and her father's reigns, when there had been upward of a hundred women at court, including wives, sisters, and daughters of courtiers, as well as the queens' attendants. Elizabeth, on the other hand, made it clear that the women who frequented her court were there either by necessity or sufferance. In short, if they were not members of her personal household, they were not welcome. Male courtiers were positively discouraged from bringing their wives to court, for this would destroy the myth of romantic enslavement to the Queen. Indeed, any woman who did accompany her husband would go hungry or sleepless, for the privilege of free board and lodging, which was afforded to most male courtiers, did not extend to them. Elizabeth's household itself was significantly reduced. The number of ladies and gentlewomen of the privy chamber and bedchamber was decreased from twenty to eleven, and there were now just six maids of honor. This was the lowest number of female attendants any queen had had for almost forty years.

This made the competition for places in the royal household fierce, for it was practically the only way that women could guarantee a presence at court. Indeed, if they were unmarried, there were hardly any other options available to them. At a time when virtually every other profession was an exclusively male preserve, the queen's household was one of the very few institutions in which women had a role to play. The fact that the new queen gave precedence to her mother's relatives and those who had served her before meant that most of the highborn ladies from across the kingdom who clamored for places went away disappointed.

The situation did not improve as the reign progressed. Elizabeth was keen to insure continuity in her household and encouraged long service. It was common for her ladies to stay in post for decades, and it was often death rather than dismissal or resignation that terminated their employment. Throughout the entire forty-five years of the reign, only twenty-eight women were appointed to salaried posts in the privy chamber, with the majority of these going to ladies of the same families. This made the privy chamber practically a closed shop.

Although small in number, Elizabeth's ladies were an indispensable part of her court, not just for practical reasons but also because they created a backdrop against which their sovereign mistress could be displayed to maximum effect. Everywhere the Queen went, she was flanked by an entourage of ladies who served to enhance her magnificence. One observer noted that in her coronation procession, she was accompanied by "a notable trayne of goodly and beawtifull Ladies, richly appointed."[34] On another court occasion early in the reign, one of those present enthused: "The rich attire, the ornaments, the beauty of the Ladyes, did add particular graces to the solemnity, and held the eyes and hearts of men dazeled between contentment and admiratione."[35] Immaculately dressed in the fashions dictated by Elizabeth, these ladies formed a decorative presence, pleasing to the eye—but not too pleasing. It was imperative that no woman should outshine the Queen; rather, they should emphasize her peerless beauty and magnificence. Thus, while Elizabeth appeared at court bedecked in lavish gowns of rich materials and vivid colors, her ladies were obliged to wear only black or white. No matter how attractive they might be in their own right, the plain uniformity of their dress would draw all eyes to the star of the show. To test the effect this created, the Queen once asked a visiting French nobleman what he thought of her ladies. He immediately protested that he was unable to "judge stars in the presence of the sun."[36] This was exactly the response that Elizabeth required, and it neatly defined the role she had created for the women at her court.

The structure of her household was also carefully controlled by the new queen. This was based around her public and private persona and was reflected by the layout of the rooms, which followed a similar pattern in all of the royal palaces. Beyond the gallery, great hall, and great chamber was the Queen's suite of rooms, and the further a courtier was able to progress into them, the more important he was deemed to be. First there was the presence chamber, to which most courtiers flocked in the hope of gaining an audience with the Queen. To get there, they would often have spent months writing letters, sending gifts, or offering bribes, but even then they were not guaranteed an audience. The presence chamber was filled by noblemen and other supplicants, who would regularly spend many hours waiting for Elizabeth to emerge from her more private rooms beyond—often entirely in vain.

Beyond this, there was the privy chamber, a day room for the sovereign, where she tended to take her meals. Only the most exalted members of court were admitted, such as Privy Councillors, ambassadors, or close favorites. Toward the end of Elizabeth's reign, her disgraced favorite, the Earl of Essex, lamented that although he had been given "access" to the presence chamber, he had been denied "near access"—in other words, admittance to the privy chamber, which he observed was given only to those whom the Queen "favours extraordinarily."[37] Yet there was still a more exclusive sanctum, and this was the Queen's bedchamber. It was exceptionally rare for any male courtier to gain admittance to this inner sanctum, for it was here that Elizabeth would seek refuge from the hustle and bustle of the court, the pressures of her council, or the solicitations of ambassadors and supplicants for her favor. It was an almost exclusively female domain, for the Queen would be served by a small number of her most trusted or highest-ranking ladies. They would see the private face—or "backstage persona"—that she kept hidden in the public rooms beyond.[38]

The bedchamber, privy chamber, and presence chamber would each be staffed by a select group of women who were assigned positions of varying status. After the chief and second gentlewomen of the privy chamber, there were the ladies of the bedchamber, followed by the ladies of the privy chamber, and, finally, the ladies of the presence chamber. In addition, there were the maids of honor—unmarried young ladies whose presence was largely decorative and who could move among the three chambers. They were under the supervision of the "mother" of the maids, although the chief gentlewoman would also insure that they did not step out of line. Lower down the scale, there were the chamberers, who undertook most of the menial duties within the household, such as cleaning, laundry, and mending. Finally, there were the women "extraordinary": unsalaried ladies who were kept in reserve for when the regular attendants were sick or otherwise absent from court, and only paid when their service was required.

Elizabeth was hardly ever alone. She herself admitted that she was "always surrounded by my Ladies of the Bedchamber and maids of honour."[39] At least one of them would sleep in the same room as her,

usually upon a truckle bed at the end of the Queen's own bed. Whatever she wished for, day or night, they would make it their business to
procure at any cost. Those who proved loyal and capable she would reward with gifts and friendship; those who vexed her would suffer punishments and reproofs that often far outweighed the crime.

Service to the Queen was certainly no sinecure. By the time she ascended the throne, Elizabeth had grown used to being attended by an
entourage of ladies and servants, and she had become an exacting mistress. She expected all of her women to be in constant attendance
upon her and to put her needs above any personal concerns. Illness,
unless it was very severe, was no excuse for absence; neither was marriage nor domestic matters. If any of her ladies fell pregnant, they were
expected to return to court as soon as possible after the birth, leaving
their offspring to the care of wet nurses and governesses. She took a
dim view of any requests for leave in order to attend to a sick child or
spouse, seeing this as a great inconvenience to her own needs.

The Queen's female attendants carried out a range of duties according to their position and status. The ladies of the presence chamber were required to attend only "when the queen's Majesty calleth for
them," which was usually when she granted audiences to ambassadors
or other prestigious guests.[40] These were always impressive occasions.
In May 1559, Elizabeth hosted a visit by the Duc de Montmorency at
Whitehall Palace and staged lavish entertainments in his honor. The
weather being fine, she directed that a "sumptuous feast" be prepared
in the garden of the palace. Her servants constructed a large outdoor
gallery, which was bedecked with gold and silver brocade, and an arti-
ficial door was made from roses and other flowers. This theme was
continued throughout. "The whole gallery was closed in with wreaths
of flowers and leaves of most beautiful designs, which gave a very
sweet odour and were marvellous to behold, having been prepared in
less than two evenings to keep them fresh," enthused Il Schifanoya, a
Venetian envoy. Meanwhile, two separate tables were prepared for the
Queen and the duc, together with a large table "54 paces in length" for
the other guests. In two corners of the gallery were large semicircular
cupboards "laden with most precious and costly drinking cups of gold

and of rock crystal and other jewels." When supper was ready to be served, trumpets sounded, and the Queen progressed through the assembled guests to take her seat at the top table. Huge joints of meat and many other rich dishes were served at the banquet, which was accompanied by "music of several sorts." Afterward, the tables were swiftly cleared, and there was dancing "till the eleventh hour of the night." Elizabeth then retired, which gave the guests their cue that they should do the same.[41]

Not all the Queen's ladies were fortunate enough to attend such gatherings. The maids of honor provided the necessary backdrop on less formal occasions and were otherwise to be found running errands for their royal mistress. For the ladies of the privy chamber and bedchamber, the duties were more closely defined: they would wash her, attend to her makeup and coiffure, choose her clothes and jewels and assist her in putting them on, serve her food and drink, and carry out any other task she saw fit to demand.

Although the chamberers carried out the more menial duties, such as cleaning the Queen's apartments, emptying her washbowls, and arranging her bed linen, service for the ladies of her household was still exacting. It also followed a strict routine. Elizabeth herself admitted: "I am no morning woman,"[42] and she would eschew the company of her male councillors and suitors until she had been dressed and adorned. But she had a habit of going for early morning walks in the gardens of her palaces still dressed in her night garments and always accompanied by a train of ladies. Once back in the seclusion of her bedchamber, the ceremony of her enrobing would begin. By the standards of the day, the Queen was unusually fastidious about her personal hygiene and would take regular baths in a specially made tub that would travel with her from palace to palace. On other occasions, she would be washed by her ladies with cloths soaked in water from pewter bowls. They would clean her teeth with a range of largely ineffective products, including a concoction of white wine and vinegar boiled up with honey, which would be rubbed on with fine cloths.

This task performed, it would take at least an hour to dress Elizabeth in the robes that she or her ladies had chosen for that day. This was not something she could have done alone, even if she had wished to. Each layer of clothing—from the farthingales and bodices to the outer garments, ruffs, and scented gloves—had to be carefully fastened into place with pins and laces. Often the gowns had to be sewn on each day and then the stitches carefully unpicked before she retired at night. As Elizabeth's gowns became ever more ostentatious, so the task of dressing her became increasingly complicated and time consuming. The same was true for attending to the Queen's hair. In the early years of her reign, she displayed her long auburn tresses to maximum effect, curled, pinned, and adorned with priceless pearls and other jewels, but later she took to wearing ever more elaborate wigs.

After the Queen's ladies had completed the painstaking ceremony of her dressing, which took two hours in all, they would have a range of other tasks. These included ordering and caring for their royal mistress's considerable wardrobe and jewels. They would mend and in some cases make her garments, take stock of those that were sent to her as gifts, and carefully catalogue both these and the many jewels in the Queen's collection. The more valuable the item, the higher the status of the lady who cared for it. During Elizabeth's courtship with the Duke of Alençon, her suitor made a great show of proving his "love and goodwill" by presenting her with "a most beautiful and precious diamond, of the value of 5,000 crowns." In return, "the Queen, on her part, having commanded her lady in waiting to bring her a small jewelled harquebus of a very great price, made Monsieur a present of it."[43] The women would also take charge of any other gifts received by their royal mistress. Some of these required a good deal of care, for as well as the various jewels and ornaments with which Elizabeth was routinely presented, she also received a pet dog, a monkey, and a parrot in a gilded cage.

Many of the delicacies that Elizabeth enjoyed at mealtimes were prepared and cooked by her ladies, including the sweet drinks and confections for which she had a weakness. The main meal of the day was served at noon, followed by supper at around five o'clock. The Queen usually took her meals in her private apartments, where she would be served by a select group of her ladies "with particular solemnity . . . and it is very seldom that any body, foreign or native, is admitted at that time."[44] Some would also attend her on the occasions when she dined in state, and would observe great ceremony in doing so. One visitor to court described how the ladies who had been appointed for the task would enter the presence chamber and "make three reverences, the one by the door, the next in the middle of the chamber, the third by the table."[45] They would then proceed to serve each dish with great solemnity, which, given the fact that Elizabeth regularly had twenty or more to choose from, took some considerable time.

It was essential that this unmarried queen should be accompanied by her ladies at all times in order to avoid any slur on her reputation. They would attend her as she progressed in state to chapel, dealt with correspondence or other official business, and gave audiences to ambassadors, councillors, and favorites. That one or more of them also slept in the same room as her at night was necessary for reasons of security, as well as propriety, for the Queen was under constant threat of assassination, particularly after 1570, when the Pope issued a bull of excommunication and encouraged her Catholic subjects to rise against her. Although her guards were exclusively male, the women who served her were invaluable for their constant presence and vigilance. They would taste each dish before it was served to the Queen to insure that it was not poisoned, test any perfume that was sent to her as a gift, and carry out nightly searches of her private apartments.

The new queen had long been plagued by a host of ailments, which would flare up at times of stress. She suffered headaches, stomach pains, aching limbs, breathlessness, and insomnia. Little wonder that the Spanish ambassador predicted early on in her reign that she was "not likely to have a long life."[46] Elizabeth was no easy patient. Like her father, she perceived illness as a sign of weakness and would rail against her ladies and physicians as they attempted to nurse her, insisting that there was nothing wrong. On one occasion, she ordered some water to be fetched from the Derbyshire town of Buxton, famous for its health-giving springs, so that she might bathe in it and thus ease a persistent pain in her leg. But when the water arrived, she flew into a rage and sent it away again, because by then, rumor had got out that she was unwell.[47] There were other occasions when the Queen was so ill that she had no choice but to retreat to her private chambers, attended only by her ladies. One time, she hid herself away for three days and was said to be "very unapt to be dealt with . . . being trobled with an exstreame cowld and defluxion into her eyes, so as she cannot indure to reade any thing."[48]

When Elizabeth was in good health, her ladies were expected to be accomplished in all things that would tend to her comfort or amusement. They would entertain her by reading aloud—often in several languages—playing cards or gossiping with her about the latest scandals and events at court. The ability to ride was an essential prerequisite for members of her household, for they would accompany her on the many hunting expeditions she undertook during her reign, sometimes riding at breakneck speed to keep up with her.

Elizabeth was an accomplished dancer and loved to show off her prowess by performing energetic routines such as the galliard or volta. As her ladies were often required to take part in court masques or other entertainments, the Queen would spend long hours with them rehearsing the complicated steps over and over again until they attained perfection. When the dances were performed, she would watch the ladies like a hawk, calling out sharp reproofs if they put a foot wrong. "She takes such pleasure in it [music] that when her Maids dance she follows the cadence with her head, hand and foot. She rebukes them if they do not dance to her liking, and without doubt she is a mistress of the art," remarked Monsieur de Maisse, a foreign visitor to court in the 1590s.[49]

When Elizabeth retired to her private apartments at the end of each long evening of court entertainments, she would be attended by her ladies. Away from the prying eyes of her courtiers, they would carefully undress her, take off her makeup, and unpin her hair. The separation of her public and private personas was as essential to her authority as queen as it was to her vanity as a woman, and none but her ladies was permitted to see both.

The women who served Elizabeth at court received only modest payment for the myriad duties that they had to perform. Indeed,

some—like the maids of honor and ladies of the presence chamber—
were rarely paid at all. The ladies of the privy chamber and bedchamber received an annual salary of around £33 (equivalent to around
£5,600 or $8,945 today), and this sum remained the same throughout
their mistress's long reign. Moreover, unlike the men at court, Elizabeth's ladies were unable to supplement their income with lucrative
appointments or monopolies. They were, though, allowed "bouge of
court," which included food, accommodation, lights, and fuel. They
were also given livery (uniform) to wear, which eased the pressure on
their meagre allowance. Their meals tended to comprise leftovers
from the Queen's table, which amounted to little when shared among
around twenty women. Most of them would take their repasts in the
great chamber, and they were often obliged to sit on the floor, because
their dresses were so wide—according to the fashion—that they took
up too much room on the benches.

The living quarters allocated to the Queen's ladies tended to be
cramped and uncomfortable. In stark contrast to the rich tapestries
and velvet-covered furniture that adorned their mistress's state and private apartments, their own were often lacking in even the most basic
of facilities. At Windsor Castle, for example, the maids of honors'
chamber was little more than a makeshift enclosure, which lacked a
ceiling and was partitioned by hastily assembled wooden boards. They
were eventually obliged to request that these boards be made higher
"for that the servants look over."[50] Few women were afforded the luxury of a fireplace in their chambers, so these rooms would become unbearably cold during the depths of winter.

Sanitation was also poor, and there were neither bathrooms nor
flushing toilets for the courtiers. As a result, after a few weeks' residence in any palace, the stench would become "evill and contagious."[51]
Elizabeth and her court were therefore obliged to make regular "removes" so that the palaces could be thoroughly cleaned and all the
human waste disposed of. Although these removes gave her ladies
some variety, they often entailed even worse accommodations than
they endured in the royal palaces. William Cecil's house at Theobalds
was particularly dreaded, for almost all of the ladies and chamberers

were obliged to share just two small rooms, only one of which had any heating. Privacy was a rare luxury for the women of Elizabeth's household, and as one contemporary noted, solitude "was very hard to do in that place."[52] Life at court was therefore a good deal harder for most of the Queen's ladies than that which they would have enjoyed at their own houses and estates.

Moreover, Elizabeth was every bit as strict as her late mother had been toward her own ladies. She demanded exceedingly high standards and would tolerate no deviation from them. If they failed in any of their duties, she would fly into a rage and deal them slaps or blows. When one poor lady was clumsy in serving her at table, Elizabeth stabbed her in the hand. A foreign visitor to court observed: "She is a haughty woman, falling easily into rebuke . . . She thinks highly of herself and has little regard for her servants and Council, being of opinion that she is far wiser than they; she mocks them and often cries out upon them."[53] Her notoriously fickle temper, which she also inherited from Anne Boleyn, kept both her ladies and courtiers at heel. "When she smiled, it was a pure sunshine, that every one did chuse to baske in, if they could," wrote her godson, Sir John Harington, "but anon came a storm from a sudden gathering of clouds, and the thunder fell in wondrous manner on all alike."[54]

The Queen's women were expected not just to tolerate her frequent bouts of temper but also to show her utmost devotion and reverence at all times. By nature possessive, Elizabeth could not abide any of her ladies to take a leave of absence. They were expected to attend her even if they were ill, and although in theory each had specific times of attendance, in practice they were never off duty and could be summoned day or night.

Although service to the Queen often involved considerable personal sacrifice, there was still a clamor for places. This suggests that there were material benefits to service at court. Her ladies could expect to receive gifts of clothing, jewelry, and other accessories, from both the Queen and those wishing to gain access to her. Shortly before her coronation, it was reported that the Earl of Arundel had given two thousand crowns' worth of jewels to the Queen's ladies in an attempt to buy himself the office of Lord Steward. Toward the end of the reign, the French ambassador's wife "gave among the Queen's maids French purses, fans, and masks, very bountifully."[55] Often, if Elizabeth did not care for a gift that had been presented to her by a suitor or supplicant, she would pass it on to one of her ladies.

Undoubtedly the greatest perk of all for women at court was the unrivalled access to the Queen that their service entailed. As one recent commentator has remarked: "Guaranteed access to the monarch was guaranteed power."[56] The vital importance of gaining a personal audience with the sovereign in order to plead a case or promote a suit had been proved time and again. Katherine Howard's fate might have been very different if she had succeeded in gaining the king's presence. Elizabeth herself had begged to be allowed to see her sister, Mary, when she had stood accused of involvement in the Wyatt rebellion. She had known as well as her contemporaries that few, if any, persons would risk pleading the cause of those in disgrace.

Yet from the beginning of her reign, Elizabeth made it clear that she did not wish her ladies to meddle in politics. The Count de Feria reported to Philip II: "A few days after the Queen's accession she made a speech to the women who were in her service commanding them never to speak to her on business affairs, and up to the present this has been carried out."[57] Writing later in the reign, Rowland Vaughan recalled, "none of these (near and dear ladies) durst intermeddle so far in matters of common wealth."[58] But this belies the true influence that these women enjoyed. Elizabeth may have banned her ladies from discussing "business affairs," but it was almost inevitable that she would talk to them about the latest developments at court. Much as she might wish to leave the world of politics and diplomacy behind as she entered her privy chamber, it was not as straightforward as that.

CHAPTER 8

The Virgin Queen

> And, in the end, this shall be for me sufficient, that a marble
> stone shall declare that a queen, having reigned such a time,
> lived and died a virgin.[1]

Elizabeth spoke these words to the first Parliament of her reign in
February 1559. Few members of the audience believed her. They probably reasoned that her statement was either the result of maidenly
modesty or statecraft, aimed to increase her value as a potential bride.

It was inconceivable that the Queen would not wish to marry.
Surely, as a "weak and feeble woman," she was desperate to find a husband who could take over the reins of government and enable her to
fulfil her more natural function of childbearing. This was certainly the
assumption made by large numbers of her subjects, including most—
if not all—of her Privy Council, who were convinced that the only
way to secure the new regime was if their royal mistress produced an

heir. Shortly after her accession, Philip II had told Elizabeth that she should marry him in order to "relieve her of those labours which are only fit for men."[2]

The social conventions of the day dictated that marriage was the desirable state for all women, not just queens. Those who remained single were derided as freaks of nature, and a contemporary song claimed that women who died as virgins "lead apes in hell."[3] As well as relying upon a husband for spiritual and emotional guidance, it was widely believed that every woman needed an outlet for her sexual urges, which would otherwise cause serious harm to her well-being. One contemporary authority even claimed that women who did not have sexual intercourse would be tormented by "unruly motions of tickling lust," and would also suffer from poor complexions and unstable minds caused by a "naughty vapour" to the brain. Elizabeth herself admitted: "There is a strong idea in the world that a woman cannot live unless she is married, or at all events that if she refrains from marriage she does so for some bad reason."[4]

Disregarding their royal mistress's apparent aversion toward marriage, her councillors proceeded to cast about for a suitable candidate. Among the first to be put forward was the Archduke Charles of Austria, a cousin of Philip II, who wished to see this heretical queen safely allied to a Catholic prince. Although Elizabeth had expressly forbidden her ladies to meddle in political matters, she soon came to appreciate how useful they could be in the various marriage negotiations that followed her accession. This included acting as intermediaries with the archduke's representatives, the Count de Feria and Bishop de Quadra, who were the Imperial and Spanish ambassadors, respectively.

De Quadra was soon able to report to his master that the Queen favored the Archduke Charles's suit because "her women all believe such to be the case."[5] Meanwhile, Baron Breumer, another envoy of the archduke, commissioned François Borth, a young man "on very friendly terms with the ladies of the Bedchamber," to report anything of interest that he gleaned from his conversations with them.[6]

Elizabeth secretly engineered all of these conversations. She instructed Lady Mary Sidney, sister of Robert Dudley, to give private assurances to the Count de Feria and de Quadra that if they pressed the archduke's suit to the Queen, they would be sure of success. "The lady would not speak herself, but urged that I should go, and said if I broached the matter of the match to the Queen now she was sure it would be speedily settled." This placed Mary in a rather awkward position, for she knew that if Elizabeth married a foreign prince, this would undermine the position of her brother, who—although he was already married—seemed to be doing everything possible to secure her as his own bride.

Noting the ambassadors' skepticism, Lady Sidney reassured them that "it is the custom of ladies here not to give their consent in such matters until they are teased into it," and added "that if this were not true, I might be sure she would not say such a thing as it might cost her her life and she was acting now with the Queen's consent." De Quadra was so convinced by this that he wrote immediately to Philip II, predicting confidently that the marriage would take place. He later noted that Lady Mary had conducted herself so "splendidly" that she deserved a rich reward from his master.[7]

Having, as she thought, reached a successful conclusion with the ambassadors, Mary reported their conversation to her mistress. "It seems the Queen answered her that it was all well," reported de Quadra. When he next saw Lady Sidney, she told him that she had been instructed to say no more on the matter for now, "and she was obliged to obey, although she was sorry for it, as she knew that if she might speak she could say something that would please me; but this must suffice." Lady Sidney was exceeding her commission, and she was soon made to look a fool because of it. Just a few days later, the Queen told de Quadra that "her answer was that she did not want to marry him [the Archduke] or anybody else . . . She says it is not fit for a queen and a maiden to summon anyone to marry her for her pleasure."[8]

De Quadra and the Count de Feria assumed that Lady Mary had deliberately misled them. In vain, she protested that she had faithfully relayed Elizabeth's messages, and not wishing to believe that she herself had been duped, she assured them "now more than ever that the

Queen is resolved on the marriage." However, when her brother, Lord Robert, reprimanded her for "carrying the affair further than he desired," she was greatly alarmed, realizing that he had been against the proposal all along. Horrified at the thought that her brother intended to forsake his own marriage vows and take Elizabeth for himself, she went at once to de Quadra and urged him to speak directly to her royal mistress. Highly affronted at being so deceived by the English queen, he accused Lady Mary of tricking him, but upon calmer reflection he realized that she had been fooled just as much as he. "I am obliged to complain of somebody in this matter," he wrote to Philip, "and have complained of Lady Sidney only, although in good truth she is no more to blame than I am." Humiliated and aggrieved, Lady Mary claimed that she had been exploited by both the Queen and her brother and bitterly regretted having agreed to be an intermediary in the matter. "She says she will make known to the Queen and everybody what has occurred if she is asked," reported de Quadra with some sympathy.[9]

This episode was enough to discourage Lady Mary from thenceforth getting involved in anything other than her routine duties at court. But there were many others willing to take her place—notably Kat Astley, who was desperate to see her former charge take a husband at last. By now, Kat enjoyed a level of favor unmatched by any woman and by few men. Although councillors such as William Cecil and Robert Dudley were undoubtedly the most influential when it came to affairs of state, Kat acted as the Queen's confidante on both private and public matters, and the fact that she was the person who shared most of Elizabeth's time gave her unparalleled importance at court. Everyone knew it. As one foreign ambassador observed, Mistress Astley "had such influence with the queen that she seemed, as it were, patroness of all England."[10]

Her position gave Kat a unique insight into Elizabeth's increasingly intimate relationship with Robert Dudley. The Queen had known him since childhood, and the two had become close during her years of uncertainty under Mary Tudor, when Robert had been one of the few men brave enough to pledge his allegiance. Theirs was a meeting of minds as well as hearts, and they rapidly developed a relationship that many said went beyond the niceties of court gallantry. In fact, the pair acted more like lovers, even in public, as they shared intimate conversations and spent many hours together hunting, dancing, and enjoying other court pastimes.

Elizabeth used her ladies to facilitate meetings with her favorite. Often this involved a great deal of subterfuge, such as in November 1561, when she disguised herself as the maid of Katherine Howard (later Countess of Nottingham) in order to enjoy the secret pleasure of watching Dudley shoot at Windsor.[11] Meanwhile, Lady Fiennes de Clinton helped to arrange for Elizabeth to dine with Sir Robert at his house. Philip II's envoy heard of this and reported: "The earl of Leicester[12] came from Greenwich to the earl of Pembroke's house on the 13th, the rumor being that he was going to his own home. The Queen went there the next day disguised to dine with them, accompanied by the Admiral [Lord Clinton] and his wife."[13]

But Robert Dudley was already married, and the Queen was risking her reputation in showing him such unbridled favor. This was even more serious than the Seymour affair, because her marriage was now one of the most critical matters of state. By making an advantageous match—preferably with a powerful foreign ally—she could secure her throne and, of course, produce heirs. But the more she cavorted with this married favorite, the less desirable a bride she became to other suitors.

Before long, Elizabeth and Dudley were causing a scandal not just in England but also in courts across Europe. Desperate to salvage her mistress's reputation and prevent her from throwing away everything for which they had fought so hard, Kat did what nobody else dared and confronted the Queen. One day in August 1559, she flung herself at Elizabeth's feet and passionately implored her to see reason and put an end to the "evil speaking" by marrying one of the many suitable contenders for her hand. She pleaded that if she did not do this, the country would be plunged into civil war, because Dudley had such powerful adversaries. Rather than see this come to pass, Kat said, she would have "strangled her majesty in the cradle."

A shocked hush descended among the ladies of the privy chamber who were witnessing this extraordinary scene. Nobody had ever dared to speak thus to the Queen: she had made it clear, even to the most senior members of her council, that her marriage was not a fitting subject for lesser mortals to meddle with. Furthermore, her temper was already gaining notoriety, and one of the surest means of inflaming it was to challenge her authority. Now here was Mistress Astley chiding Elizabeth as if she were her own daughter rather than the Queen of England. But to their surprise, Elizabeth responded graciously, thanking her old governess for her words, which she said were "outpourings of a good heart and true fidelity." She went on to assure Kat that she would consider marrying in order to dispel the rumors and set her subjects' minds at ease, but added that marriage was a weighty matter and that she had "no wish to change her state" at present.

Kat knew Elizabeth well enough to see that there was little substance in her promise, and that it came from the politician rather than the woman. She therefore renewed her entreaties to end the liaison for her own good and that of her country. The thought of giving up Lord Robert was too much for Elizabeth to bear, and suddenly throwing off the pretense, she cried that she had "so much sorrow and tribulation and so little joy" in the world that she would not deny herself this one happiness. Then, with a flash of temper, she added that if she wished to lead an immoral life, "she did not know of anyone who could forbid her."[14]

For all her defiance, Elizabeth had clearly been shaken by her old governess's words. Coming from one with whom she had experienced the danger of scandal firsthand, they would have had a powerful impact. Although she chose not to follow Kat's advice to marry, the evidence suggests that she did heed her words about Dudley. Thus, when, a little over a year later, his wife was found dead under apparently mysterious circumstances, Elizabeth did not seize the opportunity to marry him but instead distanced herself in order to avoid being implicated in any scandal. How much this was due to Kat's advice and how much to her own shrewdness is not clear, but she must certainly have recalled the warning that had so shocked her before. Furthermore, it

seems that Dudley himself blamed Mistress Astley for the Queen's refusal to marry him. Relations between the two became distinctly hostile, and in January 1561 he took petty revenge on her by accusing her
husband of some misdemeanor. John Astley was subsequently "committed to his chamber" and banished from court for six weeks. It was
no doubt thanks to Kat's influence with the Queen that he was then reinstated with full honors.[15]

Although Elizabeth had made it clear that the idea of marriage was
not appealing to her, Kat persisted in her efforts to persuade her to the
contrary. As well as having her mistress's welfare at heart, perhaps
there was an element of penance in her actions. The memory of her
own indiscretion and recklessness over the Seymour affair no doubt
provoked a fierce determination to safeguard Elizabeth's reputation
above all else. She therefore championed the various suitors for Elizabeth's hand in marriage with great enthusiasm.

Among them was the new king of Sweden, Eric XIV, who became
one of the strongest contenders in the early 1560s. His chancellor, Nils
Gyllenstierna, arrived in London in spring 1561 to prepare the ground
for a visit by the king, who had been rumored to be coming to England
for some time. Although she received Gyllenstierna with all due courtesy, Elizabeth would not actually commit to marrying his master, and
eventually, after a year of fine words and false promises, the chancellor
returned home disappointed.

Kat Astley was determined not to let the matter rest there.
Throughout the negotiations, she had been working behind the scenes
to try to further the match, assisted by her husband, John, who was apparently equally keen to see it accomplished. Kat had enlisted the services of one John Dymock, a message bearer with whom she had
formed an acquaintance "in the time of her troubles." She knew that
Dymock had been commissioned by William Cecil to accompany Eric
XIV's agent, Arnold Walwicke, to Sweden "to see how things stood."
She therefore resolved to make sure that Dymock carried a favorable
report of her mistress's intentions toward the Swedish king to counter
any rumors to the contrary. Dymock later reported that Mistress Astley had "solemnly declared that she thought that the Queen was free
of any man living, and that she would not have the Lord Robert."

Kat and her husband then cooked up a further scheme to convince Eric of the Queen's intention to marry him. Their inspiration came from John's position as keeper of the jewel house. Dymock had planned to take some jewels to Sweden in order to sell them to the royal family for the new king's coronation. He first asked John Astley if the Queen might like to buy some herself. John duly went to show the jewels to his royal mistress, who took a liking to one of them— a large ruby—but said that she could not afford to procure it. Seeing his chance, Astley suggested that Dymock should offer to sell it to the Swedish king so that he could send it back to Elizabeth as a token of his affection. The Queen smiled at the suggestion and joked that "if it should chance that they matched, it would be said that there were a liberal king and a niggardly princess matched." Undeterred, when shortly afterward John saw her playing with a ring on her finger, he suggested that she should let Dymock take it to Eric as a token of her affection. Elizabeth demurred, but Astley succeeded in persuading her to send some other, less personal gifts, including a pair of black velvet gloves, "a fair English mastiff," and a French translation of *Il Cortegiano*, a guide to what constitutes the perfect courtier. Eric was more generous with the gifts he sent in return, which included two large jewels and a portrait of himself. He also sent Mistress Astley two sable skins "lined with cloth of silver and perfumed" as an incentive for her to help his cause.[16]

Kat was determined not to let him down, and committed an act of such audacity that she risked losing her position. Together with an associate in the privy chamber, Dorothy Bradbelt, she wrote to the Swedish chancellor in secret, urging him to persuade his master to come to England. She said that if he did so, he would be assured of success, for the Queen had confided to her and Miss Bradbelt that she wished him to renew his addresses in person. In order to dispel any doubts that the chancellor might have of their sincerity, they claimed that they "understood somewhat more than the common report is."[17] The letter was swiftly intercepted by Cecil, who ordered an immediate inquiry.

The full extent of Kat's meddling was soon uncovered. Such a gross transgression could hardly be overlooked, even in the Queen's favorite attendant, who now once more faced imprisonment and dismissal. It was falsely reported by the Italian ambassador in September 1562 that some who "were formerly high in favour with the Queen, among them Mrs Asheley," had been thrown in the Tower. In fact, Kat was kept under house arrest, along with her coconspirator, Dorothy Bradbelt. Much to the astonishment of contemporaries at court and abroad, she was released within the month and restored to her former position.

The speed with which Elizabeth pardoned her chief gentlewoman could have been due simply to the esteem in which she held her. But a more likely explanation is that she herself had instigated the whole conspiracy. While she no doubt had little intention of going through with a marriage to the Swedish king, a crucial facet of her policy was to keep foreign suitors interested by giving them just enough hope of success without actually committing to them. She was no doubt aware that rumors of her relationship with Robert Dudley would have reached Eric XIV's court and thought that the most effective way of dealing with this would be for the woman known widely as her closest confidante to dismiss them for her. If Kat had really acted without the Queen's knowledge, there was more than enough evidence to condemn her. That she was restored to favor so quickly suggests that Elizabeth had used her to intervene on her behalf. Clearly Mistress Astley's duties extended well beyond the privy chamber.

As suitor after suitor was rejected by the English queen during these first few years of her reign, rumors began to circulate that there was more than just politics behind her reticence to marry. Her reproductive health had been an issue from the earliest days of her infancy. As a baby, she had been displayed "quite undressed" to the French ambassadors in order to prove that there were no impediments to her betrothal to Francis I's third son, Charles, Duc d'Angoulême. Now that she was queen, the issue of her fertility became a matter of even greater diplomatic interest, for it seemed that the entire security of her regime rested upon her ability to produce heirs. Moreover, none of her prospective suitors wished to commit himself until he had been reassured that she was able to bear children. It was in the interests of her adversaries to prove that she could not—after all, this heretical bastard could surely not survive for long unless she produced an heir.

Upon Elizabeth's accession, the Scottish ambassador, Sir James Melville, was asked to deliver a proposal to her from the Duke of Casimir, son of the Elector Palatine. He declined the commission on the following basis: "I had ground to conjecture that she would never marry because of that story one of the gentlewomen of her chamber told me . . . knowing herself incapable of children, she would never render herself subject to a man."[18] This theme was taken up by Philip II's envoy, Feria, who in April 1559 claimed: "If my spies do not lie, which I believe they do not, for a certain reason they have recently given me I understand she will not bear children."[19] His letter immediately sparked a rush of speculation, which became even more intense when, two years later, his successor, de Quadra, asserted: "It is the common opinion, confirmed by certain physicians, that this woman is unhealthy and it is believed that she will not bear children."[20] One of these physicians was Dr. Huick, who was said to have counselled the Queen that marriage and childbirth ought not to be attempted because of her "womanish infirmity." This was reported by her first biographer, William Camden, who added that there were some "hidden causes, which many times stucke in her minde, did very much terrifie her from marying."[21]

When in June 1559 Elizabeth was bled by her physicians, this was taken as proof that something was wrong with her natural functions. "Her Majesty was blooded from one foot and from one arm, but what her indisposition is, is not known," reported a Venetian agent. "Many persons say things I should not dare to write, but they say that on arriving at Greenwich she was as cheerful as ever was."[22] The theory was that the Queen had been bled in order to correct the imbalance in her body caused by her lack of periods. Elizabeth had suffered from irregular or absent periods for most of her life. Such a private matter was impossible to conceal, given the number of her close attendants and the fact that her clothes and sheets would have been washed by laundresses. Even if the latter were inclined to be discreet, it became common among foreign ambassadors to offer substantial bribes for information, which few could resist. Before long, the Queen's menstrual cycle had become a matter of international concern.

Elizabeth herself helped to fuel the rumors by dropping hints to her ladies that she was barren and declaring that she "hated the idea of marriage" for reasons that she would not divulge to a living soul.[23] As the pressure on her to take a husband intensified, a note of hysteria crept into her responses, and many began to suspect that she harbored a deep-seated fear of childbirth as a result of her physical defects. Nevertheless, she was aware of how critically important it was to at least appear to encourage the advances of her various suitors, so she also expressed enthusiasm on the subject at times. This maintained her position as one of the most sought-after brides in Europe for more than twenty years.

Foreign envoys and ambassadors therefore continued their discreet inquiries into the English queen's physical health. Philip II ordered one of his emissaries to bribe Elizabeth's laundress for details, and the woman reported back that her royal mistress was functioning normally as a woman. The fact that the Spanish king continued to view her as a potential bride suggests that he believed her. In 1566 the French ambassador, de la Forêt, quizzed one of the Queen's physicians in order to ascertain whether she would be a suitable wife for the young French king, Charles IX. The physician's reply was unequivocal: "Your King is seventeen, and the Queen is only thirty-two . . . If the King marries her, I will answer for her having ten children, and no one knows her temperament better than I do."[24]

Even as late as 1579, when Elizabeth was in her midforties—well past the usual age for childbearing in those times—the question of her fertility was still being discussed. Her suitor at that time was the youthful Duke of Alençon. An anonymous tract among the Venetian state papers claimed: "It is impossible to hope for posterity from a woman of the Queen's age, and of so poor and shattered a constitution as hers."[25] This may have prompted William Cecil (now Lord Burghley), who had long favored a French alliance, to investigate the matter himself. He closely interrogated his royal mistress's doctors, laundresses, and ladies-in-waiting, and recorded his findings in a private memorandum. "Considering the proportion of her body, having no impediment of smallness in stature, of largeness in body, nor no sickness nor lack of natural functions in those things that properly belong to the procreation of children, but contrariwise by judgement of physicians that know her estate in those things, and by the opinion of women, being more acquainted with Her Majesty's body," he concluded that there was a high probability "of her aptness to have children." Another paper on the subject also claimed that the Queen was "of the largyest and goodlyeste statuer of well-shaped women . . . and one whome in the syght of all men natuer can not amend her shape in eny parte to make her more lykely to conceyve & bere chyldrene withowte perell." Elizabeth herself proudly declared: "I am unimpaired in body."[26]

By the mid-1580s, when any hope of the Queen marrying and producing heirs had been abandoned, rumors again began to circulate that she had been infertile all along. Mary, Queen of Scots, claimed that her former guardian, Bess of Hardwick, had told her that Elizabeth was "not like other women," and that even if she had married, it could never have been consummated. To add weight to her claims, she added that an ulcer on the Queen's leg had dried up at the same time as her monthly periods had ceased. In the following decade, Elizabeth's godson, Sir John Harington, observed: "In mind, she hath ever had an aversion and (as many think) in body some indisposition to the act of marriage." This theme was taken up by the poet and playwright Ben Jonson. He claimed that Elizabeth "had a membrana on her, which made her incapable of man . . . At the comming over of Monsieur [the Duke of Anjou], ther was a French chirurgion who took in hand to cut it, yett fear stayed her, and his death."[27] Written during the reign of her successor, this account was almost certainly intended to discredit the late queen, but it has been quoted time and again in subsequent accounts, and its widespread currency has lent it undue credibility. Even some modern writers have speculated that Elizabeth either had an abnormally thick hymen or suffered from vaginismus, a condition that makes sexual penetration extremely painful.

The truth of the matter is that for every account of Elizabeth's infertility and fear of sex, there is at least one claiming that she regularly slept with her male courtiers and had several bastards by them. The earliest such rumor had appeared in 1549 in the aftermath of the Thomas Seymour scandal. As queen, there were many tales of her bearing Robert Dudley children, and in 1587, a young man going by the name of Arthur Dudley persuaded the king of Spain that he was their illegitimate offspring.[28] Meanwhile, a widow named Dionisia Deryck claimed that Elizabeth "hath already had as many children as I," although only two of them had survived into adulthood. Even Ben Jonson, who asserted that Elizabeth was "incapable of man," added that she had "tryed many."[29] Another contradictory account was by Sir James Melville. Having claimed that the English queen was infertile at the beginning of her reign, he went on to boast that he had tried to scare her off childbirth by relating how painful Mary, Queen of Scots' labor had been. If he had truly believed her to be barren, he would have had no need to do so.

It is interesting to strip away the rumors and counter-rumors and examine the evidence that exists about Elizabeth's physical state. The medical examinations that were carried out as part of the various marriage negotiations almost all confirmed that she was perfectly healthy and able to bear many children. These were corroborated by the laundresses' accounts. However, we know that Elizabeth displayed a number of symptoms that might suggest that she would have had difficulty conceiving or giving birth. For a start, there was the irregularity of her periods. "She has hardly ever the purgation proper to all women," observed the papal nuncio in France.[30] On its own, this does not prove that she was infertile. Indeed, its cause may have been the fact that she ate sparingly and was often described as being "very thin." She also sometimes appeared very pale—"the colour of a corpse," according to one account—which suggests that she could have been anemic.[31]

Nevertheless, Elizabeth's family history would not have given her a great deal of confidence about her ability to bear children. Her mother had suffered two, possibly three miscarriages, and had been unable to bear another healthy child after her firstborn. Her half sister, Mary, had endured the humiliation of two phantom pregnancies. Moreover, she, like Elizabeth, had suffered from frequent stomach pains and nausea, which had often laid her low for several days at a time. Although Elizabeth liked to boast of her physical vigor, she regularly experienced gastric attacks. "Her Majesty [was] suddenly sick in her stomach," reported William Cecil on one such occasion, "and as suddenly relieved by a vomit."[32]

Recent commentators have speculated that the Queen might have suffered from androgen insensitivity syndrome. This theory was first put forward by Michael Bloch, the Duchess of Windsor's biographer, who in the 1980s drew comparisons between Elizabeth and the Duchess, a known sufferer of the syndrome. Victims of this condition are born with male XY chromosomes but develop outwardly as female, owing to the body's failure to produce male sex hormones. Depending on the severity of the symptoms, the female reproductive organs can either be impaired or entirely absent, making sexual intercourse difficult or impossible. Women with this condition tend to be tall and lithe, with "strident personalities," thanks to the dominance of testosterone in their bodies. Elizabeth certainly fits this outward description: She was unusually tall for a woman, and was very slim and small breasted. That she had a forthright manner is beyond question. This chimes with sixteenth-century medical opinion, which claimed that "Such [women] as are robust and of a manly Constitution" were likely to be sterile.[33] And yet there is little evidence, apart from Jonson's dubious testimony, that she had any of the internal symptoms. Indeed, even the outward ones could have been the result of genetics rather than the syndrome. Her father had been very tall, and her mother had had a slight frame and small breasts. Interesting though the theory is, it is at best only speculative.[34]

Although she later succeeded in making a virtue of her virgin state, in the early years of her reign, Elizabeth was strongly criticized for refusing to marry. Without a legitimate heir born of her own body, how could she hope to hold on to her crown? In fact, there were sound political reasons why Elizabeth was reluctant to take a husband. In the sixteenth century, for the vast majority of women, marriage involved complete subordination to their husband's will. If he proved violent, abusive, or adulterous, they were expected to endure it. Furthermore, wives would relinquish any property they held and would have precious few legal rights over their husbands. As queen, Elizabeth stood to lose a great deal more than most women. Even though her half sister's marriage treaty had strictly defined the powers of her new husband, the reality had been rather different, and Mary herself had encouraged Philip to take the reins of government. But Elizabeth was far more independent than Mary had been, and with her formidable intellect and indomitable will, it would have been difficult for her to submit to the authority of any man. When she told Sir James Melville that she was resolved never to marry, he shrewdly replied: "Your Majesty thinks, if you were married, you would be but Queen of England; and now you are both King and Queen. I know your spirit cannot endure a commander." In a similar vein, Heironimo Lippomano, the Venetian ambassador in France, commented upon "the ambition which the Queen has by her nature to govern absolutely without any partner." As if to corroborate this, when provoked by her overbearing councillors, Elizabeth angrily declared: "I will have but one mistress here, and no master!"[35]

Yet Elizabeth's determination to remain single went beyond political reasoning. She occasionally showed flashes of profound unease when the subject of marriage was raised. In the mid-1560s, she told the French ambassador that she would leave herself entirely vulnerable if she took a husband, as he could "carry out some evil wish, if he had one." A German envoy was taken aback when she snapped that "she would rather go into a nunnery, or for that matter suffer death," than marry.[36] She went even further a few years later, fiercely declaring that she hated the idea of marriage every day more, for reasons that she could not divulge to a twin soul, if she had one, much less "to a living creature."[37]

The executions of her mother, Anne Boleyn, and stepmother, Katherine Howard, had brought home to Elizabeth in the most horrifying manner possible just how dangerous royal marriage could be. Perhaps she had equated it with violent death from that time forward; she had certainly grown to appreciate that love must never be allowed to come before affairs of state. In 1561, when conversing with an envoy from Scotland, Elizabeth admitted that certain events in her youth had made her afraid of marriage. In view of this, it is hardly surprising that she should exclaim: "So many doubts of marriage was in all hands that I stand [in] awe myself to enter into marriage, fearing the controversy."[38]

It is perhaps going too far to conclude that the events of Elizabeth's childhood had caused a deep-seated fear of love and sex, a frigidity that would make the idea of marriage seem abhorrent to her when she became queen. If this had been true, she would probably have shown a general antipathy toward men, and this was far from being the case. Elizabeth was one of the most notorious flirts in history. At times, her behavior toward some of her male courtiers was so outrageously suggestive that many believed she must be having affairs with them. She famously tickled Robert Dudley's neck during the ceremony to make him an earl. She even personalized her relationships with her councillors by giving them pet names: Cecil was "Sir Spirit," Dudley "Eyes," and Hatton "Lids." She delighted in being at the center of a game of courtly love that she herself had created, and loved to show herself off to her courtiers and foreign dignitaries as both a queen and a woman. Alluring and flirtatious one moment, she was cruel and aloof the next, which drove her adoring male courtiers to distraction. Edward Dyer warned Hatton, one of her greatest admirers: "First of all you must consider with whom you have to deale, & what wee be towards her, who though she does descend very much in her Sex as a woman, yet wee may not forgett her Place & nature of it as our Sovraigne."[39] This was exactly what Elizabeth had intended: She would flirt with her courtiers to her heart's content but would never let them forget her supremacy as queen. They flattered her with poetry and prose, wore her colors as they jousted, and fought each other for her favors. But they were never allowed to get too close.

Elizabeth made no secret of her fascination with the opposite sex and was clearly a woman of passion. However, it is unlikely that she ever allowed her relationships with her courtiers to go beyond flirtation. Highly sexual though she seemed, she was nevertheless too much mistress of her emotions to give way to such dangerous temptations.

Quite apart from the prospect of an unwanted pregnancy, it would have rendered her virtually worthless in the international marriage market if there was any proof that she was not chaste. Besides, as she was forever surrounded by her ladies, it would have been virtually impossible to conduct an affair in secret. She herself once pointed out: "I do not live in a corner. A thousand eyes see all I do."[40]

Although the new queen loved to flirt with her courtiers, this did not make her any more likely to marry. Rather, she presented herself as the "bride" of England and the "mother" of all her people. From the beginning of the reign, her speeches are littered with references to this metaphorical state. In 1559 she replied to the House of Commons' petition to marry by telling them to "reproach me so no more . . . that I have no children: for everyone of you, and as many as are English, are my children." Another time she declared: "I assure yow all that though after my death you may have many stepdames, yet shall you never have any, a more naturall mother, than I meane to be vnto yow all."[41]

Just as Elizabeth was "married" to her country, so she expected her ladies to be "married" to her service. Even though marriage did not preclude service in her household—indeed, most of those ladies who did take a husband with her permission resumed their duties shortly afterward—the Queen still viewed it as an irritating disruption to her established routine. She would regularly lecture the maids of honor and other unmarried ladies of her household on marriage, and would "much exhort all her women to remain in virgin state." This was an unpalatable subject to the giddy young girls whose heads were full of flirtation and "enticinge love," but their royal mistress persisted. "She did oft aske the ladies around her chamber, If they lovede to thinke of marriage? And the wise ones did conceal well their liking thereto; as knowing the Queene's judgment in this matter," observed her godson, Sir John Harington.[42]

One poor girl who gave a different answer forfeited her happiness for good. A daughter of Sir Robert Arundell, she had only recently arrived at court when her sovereign mistress repeated the oft-asked question to her and the other assembled ladies. Before anyone could warn her, the girl piped up that "she had thought muche about marriage, if her father did consent to the man she lovede." Breaking the shocked silence that followed, Elizabeth smilingly assured her: "You seeme honeste, i'faithe . . . I will sue for you to your father." The delighted girl waited excitedly for news, and it seemed that she would soon have her wish, for the Queen honored her promise and spoke to Sir Robert about the matter. He expressed surprise at the news, claiming that he had not known that his daughter "had liking to any man," but gladly gave "free consente to what was moste pleasinge to hir Highnesse." "Then I will do the reste," Elizabeth assured him. Soon afterward, she summoned the girl to her and told her that her father had given his written consent to her marriage. Ecstatic at this news, the girl exclaimed: "I shall be happie and please your Grace." "So thou shalte," replied the Queen, "but not to be a foole and marrye. I have his consente given to me, and I vow thou shalte never get it into thy possession: so, go to thy busynesse. I see thou art a bolde one, to owne thy foolishnesse so readilye."[43]

This episode reveals a rather vindictive side to Elizabeth's nature, and it would come to the fore time and again in her reaction to her ladies' love affairs. The fact that she was known to so fiercely disapprove of their marrying created a vicious circle in which her ladies were often too afraid to seek her permission and therefore married in secret, which in turn provoked even greater wrath when their actions were discovered. It was without doubt the surest way to lose the Queen's favor. But in expecting her ladies to conform, Elizabeth often gravely overestimated the strength of their commitment to her, as well as their self-discipline, which was rarely as great as her own.

Queen Mary had apologized on her deathbed for delaying the marriage of Jane Dormer, one of her favorite ladies, to the Count de Feria because she had not been able to bear parting with her. Her successor showed no such sentiment. Even if she knew that one of her ladies was genuinely in love, as had been the case with Mistress Dormer, this did little to alter her implacable opposition to their courtships. Driven by desperation, some of these ladies fell prey to temptation before the marriage had been sealed. Although Elizabeth was anxious to establish a strict moral standard at her court, the fact that so many young men and women were crowded together created an atmosphere charged with sexual tension. Moreover, the need for secrecy added a certain frisson to their courtships, as they snatched furtive encounters with their lovers in the many private alcoves and chambers of the royal palaces.

In a court filled with rumor and gossip, and in which privacy was a rare commodity, it was inevitable that these secret marriages and pregnancies would be discovered. As one court official noted during Henry VIII's reign, "There is nothing done or spoken but it is with speed known in the court."[44] Once their secret was out, the Queen's wrath was invariably formidable. In her fury, she would lash out against her ladies for defying her orders and would inflict severe punishments upon them. Some were thrown into the Tower or Fleet Prison (a notorious London prison built in 1197), while others were stripped of their titles and banished from court. Only very occasionally did Elizabeth forgive their transgression and allow them to continue in service.

The severity of the Queen's reaction was at least partly justified. She was in loco parentis to the young maids of honor, and as such had a responsibility to insure that they either married well or remained chaste. Taking an unsuitable husband or, worse, falling pregnant out of wedlock could spell financial disaster for the girl's family, whose reputation and estates often depended upon their offspring making profitable alliances. Moreover, any scandals involving the ladies in her household would reflect badly upon her and cause her suitors to doubt whether the moral standards at her court were really so unimpeachable. Her mother had been no less strict in controlling the activities of her ladies in order to safeguard her own reputation. It was said that Anne "wolde many tymes move them [her ladies] to modestye and chastertie, but in especiall to the maydons of honour, whom she wolde call before her in the prevy chamber, and before the mother of the maydes wold geove them a long charge of their behaviours."

Elizabeth also wished to keep politics out of her privy chamber, and was aware that if her ladies married influential men at court, they would inevitably be drawn into supporting their husbands' causes. When Mary Tudor was still on the throne, Lady Elizabeth Fiennes de

Clinton had lobbied Elizabeth to restore her husband to a position of power when she became queen. She received many more such requests after her accession by ladies within her household and was loath to grant any of them. Neither did she wish the words she had uttered to her ladies in private to leave the confines of her privy chamber. But if one of those ladies was married to a courtier, this was almost bound to happen. The fact that they had such unparalleled access to the queen was what made her ladies so attractive to the men at court.

Although she may have had good reason to punish her ladies for marrying, it is difficult to absolve Elizabeth completely of the charge of being "angry with any love," as one of her courtiers put it.[45] When Mary Shelton's secret marriage to an unnamed gentleman at court was discovered, the Queen was "liberall both with bloes and yevell words," and the beleaguered girl ended up with a broken finger. As one onlooker observed: "never woman bought hir husband more deare than she hath done."[46] Furthermore, Elizabeth sometimes refused permission for marriages that were, on paper at least, entirely suitable. For example, when in 1563 her cousin Katherine Carey was betrothed to the son of Lord Howard of Effingham—an extremely advantageous match for her—Elizabeth was so enraged that she immediately dismissed the couple from court. Later in the reign, she gave her consent for Elizabeth Russell to marry Lord Herbert but then deliberately created difficulties in order to delay the nuptials.

Little wonder that the ladies at court continued to follow the hazardous course of marrying in secret. But as two members of her household discovered, if those ladies were of royal blood, the cost of doing so could be much greater than simply the loss of their position.

CHAPTER 9

Cousins

From the moment she took the crown, Elizabeth had been faced with the threat of rival claimants. Given her mother's history and the subsequent confirmations of her own illegitimacy, this was inevitable. Her religion provided a further excuse for the Catholic powers both within England and across Europe to try to supplant this heretical queen with a candidate of their own. Ironically, given the prejudices against female rule that had been demonstrated during Mary's reign, all the leading contenders for the English throne were women. Descended from Henry VIII's sisters, they were also cousins of the new queen. Principal among them was Mary Stuart, daughter of Mary of Guise and James V of Scotland, who was the son of Henry VIII's elder sister, Margaret Tudor. James's sister, Lady Margaret Douglas, was another potential claimant. But Henry had excluded this branch of his family from inheriting the throne, so it was the descendants of his younger sister, Mary, who at first seemed to pose more of a threat.

Elizabeth had already witnessed an attempt to place a member of this latter branch upon the throne and thus usurp the rightful order of succession. That member had been Lady Jane Grey, the "Nine Days Queen." Her surviving sisters were Katherine and Mary.

The Grey sisters were the daughters of Henry Grey, Duke of Suffolk, but their royal blood came from their mother, Frances Brandon, who was the eldest daughter of Henry VIII's sister, Mary Tudor. Born around 1540, Katherine was eighteen years old when Elizabeth became queen. Her sister Mary was five years younger. Katherine and Mary's childhood had not been a happy one. Raised at Bradgate Hall in Leicestershire, they had been bullied and beaten by their ambitious parents, who saw them as little more than a commodity in their political schemes. Unlike their elder sister, Jane, neither Katherine nor Mary had sought solace in learning, and they had failed to impress Roger Ascham during his visit to Bradgate in 1550. But Katherine at least had beauty to recommend her, and was described by one historian as a "pretty featherbrain."[1]

Katherine's good looks and royal blood made her a desirable bride from a young age. As part of his plans to seize power after Edward VI's death, the Duke of Northumberland arranged her marriage to Henry Herbert, the eldest son of his ally, William Herbert, first Earl of Pembroke.[2] Meanwhile, the eight-year-old Mary was betrothed to her cousin Arthur, Lord Grey of Wilton, whose father was also a member of the Duke of Northumberland's faction at court.

On May 21, 1553, at the age of just thirteen, Katherine was married to Henry Herbert on the same day that her sister Jane married Guilford Dudley. The ceremony took place at the Duke of Northumberland's house and was "celebrated with great magnificence and feasting."[3] Katherine's marriage was not consummated[4]—a convenient fact that enabled Pembroke to have it dissolved when Northumberland's plot failed and Mary Tudor ascended the throne. Her future now looked bleak, as did that of her younger sister. Mary's betrothal had also been swiftly broken off in the aftermath of Lady Jane Grey's arrest and execution. Their father had escaped punishment for his part in the attempted coup but was subsequently executed for supporting the rebel

Thomas Wyatt. The Grey family name was now so sullied that it seemed unlikely that any man would wish to ally himself with either of the two girls.

Perhaps judging that the family had been punished enough for their part in Northumberland's attempted coup, Queen Mary showed Katherine and Mary considerable favor. Not only did she invite them to court, but she gave their mother, Frances, precedence with Lady Margaret Douglas at state occasions, ahead of Elizabeth, who, as heir to the throne, should have enjoyed this privilege. She seemed to have a soft spot for Katherine, who was given an esteemed position at her coronation in October 1553, along with a sumptuous new red velvet gown to wear. Katherine also attended Mary's marriage to Philip II the following year.

Throughout Mary's reign, Katherine and her sister were treated as princesses of the blood, and contemporary accounts record that "their trains were upheld by a gentlewoman" at all important court gatherings, a privilege accorded only to members of the royal family.[5] Their standing was further enhanced when Queen Mary appointed them ladies of the bedchamber, the most sought-after post in her household. By then, Katherine was so high in favor that it was even rumored that Mary intended to name her as her successor, although there is little evidence to substantiate this.

In 1554 Frances Grey married Adrian Stokes, a man half her age, and largely retired from court circles. Katherine was placed in the care of Anne, Duchess of Somerset, widow of the former Lord Protector, although she continued to reside at court. Among her companions in the Queen's bedchamber were the duchess's daughters, Margaret and Jane. It was with the latter that Katherine formed a particularly close attachment. Both girls lacked prudence and were more preoccupied with potential suitors than with their duties at court. Jane was delighted when her brother, Edward Seymour, Earl of Hertford, caught Katherine's eye.

Katherine and Seymour had first met at Hampton Court early in Mary's reign, and their friendship had developed during a visit to Hanworth in 1555. Lying some fifteen miles west of London, Hanworth

was the home of Edward and Jane's mother—and Katherine's patroness—the Duchess of Somerset. In August of that year, Jane, whose
health had always been delicate, fell ill and was excused from her duties at court in order to make a recuperative visit to her mother's estate, accompanied by her friend Katherine. They arrived to find
Edward Seymour, who was also paying a visit to his mother at Hanworth. Katherine was evidently delighted at the chance to renew their
acquaintance, and it was noted that they spent a great deal of time together. By the end of the visit, Katherine had fallen deeply in love with
the earl, and her feelings seemed to be reciprocated.

Edward Seymour was one of the most handsome men at court. Although not very tall, he had a slender frame and dark coloring, with an
aquiline profile and large, deep-set eyes. He was only about a year
older than Katherine and when they had first met, they had both been
in their midteens. Although several of his contemporaries criticized
him for being spoilt and conceited, Katherine was blind to his faults.
She called him "my good Ned" and "my sweet Lord," and was utterly
compliant to his will. Her judgment of men, as of many other things,
would prove fatally flawed. For his part, although Seymour was undoubtedly attracted by Katherine's physical beauty, he was perhaps
drawn more to her royal blood. Ambition ran in his family, and it is
possible that in Katherine he saw a path to the throne.

Their courtship continued after Katherine and Jane had returned to
their duties, and Edward became a regular visitor to court. But any
plans the couple may have had to marry were abruptly cut short in November 1558 when Queen Mary died. While she had always shown
kindness and favor toward Katherine and might well have assented to
the match, her half sister, Elizabeth, was of an altogether different
mind.

From the very beginning of her reign, Elizabeth made it clear that
she disliked the Grey sisters—Katherine in particular. The Spanish ambassador noted that "the Queen could not abide the sight of her" and
that she bore her "no goodwill."[6] Highly sensitive to questions about
her own legitimacy, Elizabeth also naturally distrusted any other persons of royal blood, particularly those whose place in the succession

had been confirmed by act of parliament. However justified her feelings toward the Greys, her judgment of the elder sister, Katherine, was clouded by the additional factor of jealousy. At eighteen, Katherine was seven years younger than the new queen. She had inherited the characteristic red hair and long nose of her Tudor relatives but also shared a good deal of her maternal grandmother's famous beauty. With her rosebud lips and delicate features, she was both a younger and a prettier version of Elizabeth, and the latter bitterly resented it.

Elizabeth immediately made it clear that she had no intention of upholding the favor that had been shown to Katherine and Mary Grey by her late sister. Although she could hardly banish them from court without drawing more attention to them than she wished, she could offer them less prestigious posts than they had enjoyed during Mary's reign. She therefore demoted them from ladies of the bedchamber to maids of honor, whose service was largely confined to the presence chamber. She also made it clear that she "does not wish her [Katherine] to succeed, in case of her death without heirs."[7] This infuriated Katherine, who had inherited a streak of the Tudor arrogance. The Count de Feria observed that she was "dissatisfied and offended" by the new queen and made no secret of the fact. "She has spoken very arrogant and unseemly words in the hearing of the Queen and others standing by," observed another courtier. Before long, it was widely reported abroad that the Spanish "take her [Katherine] to be of a discontented mind, as not regarded or esteemed by the Queen."[8]

Her rift with Elizabeth made Katherine Grey a natural ally of the Spanish, who were already conspiring against this heretical new queen. Although nominally a Protestant, Katherine's religious sympathies were flexible enough to convert to Catholicism if that would best serve her interests. From very early on in the reign, she therefore became the focus of Spanish plots to remove Elizabeth from the throne. Feria, with whom she had established a close friendship, groomed her for the task. "I try to keep lady Catherine very friendly," he reported to his master, "and she has promised me not to change her religion, nor to marry without my consent."[9] When Feria was recalled, Katherine made a point of becoming acquainted with the new Spanish ambassador, Bishop de Quadra. Before long, he was as fervent an advocate of her claim as his predecessor. In November 1559, he told Philip II of "the ruin which, as I think daily threatens the Queen. She would be succeeded by Lady Catherine, who would be very much more desirable than this one."[10]

De Quadra had been instrumental in the rather madcap plot earlier that year whereby Katherine would be "enticed away" to Spain on a ship that would lie in wait on the Thames, and would marry Philip II's degenerate son, Don Carlos. It is not clear whether Katherine herself knew of the plot, although the closeness of her relationship with Philip's ambassador suggests that she probably did. Even though it was all rather far-fetched, when the plot was discovered, it caused widespread fear "that if the Queen were to die your Majesty [Philip II] would get the kingdom into your family by means of Lady Catherine."[11]

The scheme had been prompted by an event involving another of Elizabeth's rivals. Shortly before she had ascended the throne, her cousin Mary Stuart had married François, Dauphin of France and son of King Henry II, in a splendid ceremony at Notre-Dame Cathedral in Paris. Although Mary was only fifteen years old, she had been betrothed to François for ten years, during which time she had lived at the French court. As well as having every expectation of becoming queen of France, Mary was already queen of Scotland. Upon hearing of her birth in December 1542, her father, James V of Scotland, famously lamented that his Stuart dynasty had "come with a lass [and] it will pass with a lass." He had died shortly afterward, leaving his kingdom to his only surviving child.[12]

The French were delighted with the young Queen of Scots when she arrived in 1548. From a very young age, her beauty was universally praised, along with her charm and grace. In 1553 the Cardinal of Lorraine wrote to tell Mary of Guise of her ten-year-old daughter's progress: "She has grown so much, and grows daily in height, goodness, beauty and virtue, that she has become the most perfect and accomplished person in all honest and virtuous things that it is possible to imagine . . . I can assure you that the King is so delighted with her that he passes much time talking with her, and for an hour together she amuses him with wise and witty conversation, as if she was a woman of twenty-five."[13] This was not mere flattery: Mary was an attractive girl—like her mother, unusually tall, with deep auburn hair that set off her pale skin to dramatic effect. She was also accomplished in the courtly arts of music, singing, dancing, embroidery, and riding. These she greatly preferred to the more academic elements of her education. She was certainly no intellectual like Elizabeth and was taught only the rudiments of languages. She could speak English but not write it, and her letters were almost always in French.

Like Elizabeth, Mary spent most of her childhood without her mother. Mary of Guise had stayed behind as regent of Scotland and was able to visit her daughter only once, in 1550. But this is where the parallel in their upbringing ends, for while Elizabeth learned the hard way—neglected by her father, declared a bastard, and often in danger of her life—Mary was the pampered princess, surrounded by flatterers and attendants who met her every need and taught her to accept queenship as a matter of right. In stark contrast to Elizabeth's childhood, although Mary's gave her short-term happiness and security, it did little to equip her for the monumental task that lay ahead.

The fact that Mary had been a queen almost from birth and had not—yet—had to fight for that exalted position made her rather arrogant. That she was likely to one day add France to her crowns gave her an even greater sense of invincibility. Moreover, she had long been taught to believe in her right to the English throne. She was the granddaughter of Margaret Tudor, and even though Henry VIII had excluded his elder sister's descendants from the succession, Mary, like her aunt Margaret Douglas, paid little heed to this. It was this fact, more than any other, that would set her on a collision course with Elizabeth.

In November 1558, Mary and her new husband received the news that Elizabeth was now queen of England. Viewed by many at the French court—not to mention in her own kingdom—as an illegitimate heretic, Elizabeth was also denounced as a wanton whore, tainted by her mother's adulterous liaisons and her own scandalous relationship with Thomas Seymour. Mary, on the other hand, had a flawless reputation as a newly married queen, and a Catholic one at that.

Encouraged by those hostile to Elizabeth, Mary wasted no time in asserting her claim to the English throne. The very day after Elizabeth's accession, she and her new husband began to style themselves king and queen of England and included the English royal arms in Mary's shield. The French king, Henry II, supported their claims, declaring that Elizabeth was an illegitimate usurper. At the wedding of his daughter Princess Claude early the following year, he ordered Mary's servants to wear the arms of England on their livery, quartered with her own.

This first act of open aggression on Mary's part dealt her relationship with Elizabeth a fatal blow. Insecure as she was about the validity of her own claim, the English queen needed little provocation where rival claimants to her throne were concerned. But Mary had provided her with enough ammunition for a lifetime's enmity. As Elizabeth's seventeenth-century biographer put it: "Hereupon Queene Elizabeth bare . . . secret grudge against her, which the subtill malice of men on both sides cherished, growing betwixt them, emulation, and new occasions daily arising, in such sort, that it could not be extinguished but by death."[14]

The threat posed by Mary increased still further in July 1559 when Henry II died unexpectedly and Mary became queen of France. Even though her Guise uncles assumed the power of regency, Mary's status and power had never been greater. She was now queen of two countries, and sandwiched between them was England. In the long power struggle that would ensue between the two women, Mary definitely had the upper hand.

Mary's aggressive stance prompted Elizabeth to change tack toward the Grey sisters. She promoted both women to posts in the bedchamber and went out of her way to show favor toward Katherine. "The Queen calls Lady Catherine her daughter, although the feeling between them can hardly be that of mother and child, but the Queen has thought best to put her in her chamber and makes much of her in order to keep her quiet. She even talks about formally adopting her."[15]

Such an obviously false display of affection did little to heal the rift between Elizabeth and her cousin. She had clearly appointed Katherine to her bedchamber for no other reason than to keep a closer eye on her. One of Cecil's agents reported that the Grey sisters were "straytely" looked to and their movements carefully watched by the Queen and her officials.[16] Nevertheless, rumors and plots continued to surround Katherine, giving her an ever more inflated sense of her own importance. By November 1559, the Spanish had hatched a plan to marry her to Archduke Charles, one of Elizabeth's own suitors. "The Archduke might be summoned to marry Lady Catherine to whom the kingdom falls if this woman dies," de Quadra told Philip.[17]

Meanwhile, Elizabeth and her ministers were spending long hours debating the vexed question of how to deal with the challenge posed by Mary Stuart. On the surface, they made friendly overtures, Elizabeth offering to send a portrait of herself to Mary, together with an assurance of her great affection toward her. The new queen of France responded in like manner, declaring that she would be delighted to receive the portrait and assuring her cousin that "her affection is fully reciprocated."[18]

But all the while, Elizabeth and her council were planning an altogether more aggressive solution to the threat posed by Mary. In early 1560 they sent troops to Scotland to support a Protestant anti-French uprising there. This decisive action led to the conclusion of the Treaty of Edinburgh in July, whereby the French forces agreed to evacuate Scotland and leave it to the government of a largely Protestant noble coalition. It also confirmed Elizabeth's sovereignty of England and stipulated that Mary would no longer lay claim to that kingdom. This news came hot on the heels of that received by Mary in late June, informing her of the death of her mother. Although she had seen precious little of her since moving to France, Mary was grief stricken and wept uncontrollably for several days. Worse was to come, for in December that year, her young husband, François, also died. At a stroke, her political position had been all but destroyed. She now had no direct link to the French dynasty, and even her authority in Scotland was under serious threat as a result of Elizabeth's wily tactics there.

If the English queen enjoyed a little respite from the threat posed by Mary Stuart, trouble was already brewing with her other main rival.

Although Lady Katherine Grey had shown herself willing to consider Philip II's candidates for her hand in marriage, she was already deeply committed to Edward Seymour. Their courtship soon came to the notice of William Cecil, who was rumored to favor the match because of its political implications. With the likelihood of Elizabeth marrying his archenemy, Robert Dudley, which he was sure would be her downfall, it was said that Cecil viewed Katherine Grey and Edward Seymour as alternative candidates for the throne. After all, both had royal blood: Katherine was the great-niece of Henry VIII and Edward was the nephew of the late queen Jane Seymour, mother of Edward VI. Their combined claim to the English crown was therefore formidable. As well as his political motives, Cecil also had personal ties with the Grey family, and in correspondence they always called him "cousin."[19]

Katherine's ever-ambitious mother, Frances, also approved of her daughter's courtship, no doubt grasping its political significance. When Katherine confided her feelings to her in the spring of 1559, Frances immediately offered to intercede with the Queen and Privy Council on her behalf. Aware that her own health was fading, she wrote to the Queen assuring her that the marriage "was th'onlie thinge that shee desired before her death, and shold be an occasion to her to die the more quietlie." However, in the first of what would be a series of misfortunes to befall Katherine and Seymour, Frances died before she had the chance to finish the letter.[20]

Although Edward had promised Katherine's stepfather, Adrian Stokes, that he would "meddle no further in the matter," he maintained the courtship, and "the love did continue, or rather increase," at least on Katherine's part. Aided by Edward's sister, Jane, they enjoyed a series of clandestine meetings. Jane would often accompany Katherine to the earl's lodgings on Cannon Row in the city of London, and when he visited court, she arranged for the couple to be alone together in the private "closet" that she had within the maidens' chamber.

It was largely thanks to Jane that the couple made the foolhardy decision to marry in secret. In the summer of 1560, when she and Katherine were in attendance upon Elizabeth at Hampton Court, she persuaded her friend that if Edward proposed, then she should accept.

She subsequently arranged for them to meet when the court returned to Westminster. Seymour told Katherine that "he had borne her good Will of longe tyme and that becawse she sholde not thincke that he intended to mocke her he was content if shee wold to marrie her." Katherine replied "that shee liked both him and his offer," and agreed to be his wife. There was no formal betrothal, apart from "kissing and embracing and joining their hands together," but Katherine was apparently content that Seymour was in earnest.[21]

If Seymour had any regrets, his sister insured that he could not renege on his promise. She hatched a plan for the couple to wed in secret on the very next occasion that the Queen left court. Because Elizabeth's ladies would usually be expected to accompany her on any such visit, Katherine and Jane had to think of an excuse to absent themselves. Accordingly, when in November 1560 Elizabeth announced her intention to go to Greenwich,[22] Katherine begged to be excused on account of a toothache.[23] The Queen readily agreed that both girls could stay at court, no doubt relieved at the prospect of being without her cousin's irksome company on this occasion.

Barely an hour after her departure, Katherine and Jane stole out of the maidens' chamber, scurried through the palace orchard, and descended the steps to the river. The tide was low enough for them to make the short walk to Seymour's house along the sands that bordered the Thames. Upon arrival at Cannon Row, Jane hastily led Katherine into her brother's house. Seymour was in his chamber, apparently unaware that they would be coming, because none of them had known exactly when the Queen would go to Greenwich. Being somewhat taken aback at their arriving "suddenlye upon him," he recovered himself enough to entertain Katherine while his sister went to find a priest. Their later testimonies suggest that Jane had literally dragged the first priest she saw off the street and either bribed or otherwise persuaded him to perform the ceremony.

Their vows were duly exchanged, and Seymour gave Katherine a ring made from four or five interlinking circles, inscribed with words of everlasting love.[24] After the speedy wedding had been concluded, Jane, apparently anxious to insure that at least some of the traditional formalities were observed, offered the couple "Comfects and other Ban-quetting meates and beare and Wyne." But Seymour and Katherine were far more interested in consummating their union, so, "perceiving them ready for bed," she discreetly withdrew. The contemporary ac-counts of the wedding describe what happened next in salacious detail. The newlyweds hastily "unarrayed themselves" and "went into naked bedd in the said Chamber where they weare so married." Sharing Sey-mour's urgent desire to consummate the marriage, Katherine ne-glected to take off the "coverchief" that she wore on her head. Once in bed, they enjoyed "Companie and Carnall Copulation . . . divers tymes," and "laie sometimes on th'oneside of the Bedd and sometymes on th'other." Eventually sated, they dressed and rejoined Jane, who had been waiting patiently downstairs. The tide having risen, the two women had to make their way back to court by boat, and Seymour waved them off at the water's edge, giving his new wife a kiss before she embarked.

Over the next few weeks, thanks again to Jane's intervention, the couple enjoyed regular meetings in secret. Still in that first insatiable rush of early marriage, they took extraordinary risks to satisfy their sexual desires. Seymour later admitted that he "laie with the saied Lady Catherine divers tymes in the Queene's howses both at West-minster and at Greenwich in the Chamber of the saied Lady Cather-ine." They snatched these hasty couplings during the day, when most of the other ladies of the Queen's privy chamber were about their business.

No matter how discreet the couple were, it was almost impossible to keep a secret for long in the crowded world of the Elizabethan court, where ambitious men and women were forever on the lookout for news or scandal that would bring down their adversaries. More-over, Katherine had her own attendants, and they certainly knew of her clandestine meetings with Seymour because they admitted him to her chamber and then discreetly "went always out when he came in."[25] Before long, rumors of their courtship had begun to circulate.

When Cecil heard of it, he warned Seymour that there was talk of "good Will" between him and Katherine Grey, but the earl denied it.[26]

Cecil also went to see Katherine. Along with the latter's companions in the privy chamber, Lady Fiennes de Clinton and the Marquess of Northampton, he "did seriouslie advertize her to beware the Companie and familiaritie with the saied Earle."[27] Katherine dismissed their concerns, denying that there was any "intimacy" between them. According to one contemporary source, she subsequently received a more unusual warning. It was said that Blanche Parry, who was friends with the Queen's astrologer John Dee, was fond of telling her fellow ladies' fortunes. When she read Lady Katherine's palm, she drew back in horror and exclaimed: "The lines say, madam, that if you ever marry without the Queen's consent in writing, you and your husband will be undone, and your fate worse than that of my Lady Jane [Grey]."[28] If this story is true, then Blanche had shown an extraordinary gift for fortune-telling. But it is perhaps more likely that her friend Cecil had told her of the courtship and asked her to warn Katherine, given that she had paid little heed to his own counsel.

Of course, it was far too late to warn the couple. Not only were they married, but Katherine was already with child. Whether Seymour realized this or whether the novelty of his marriage had already started to wear off is not certain, but within a few weeks of the wedding, he began to talk about going overseas. According to his own account, he asked his new wife several times if she was pregnant, assuring her that if she was, he would stay with her. She apparently told him that she was not certain, but that he should go ahead with the voyage anyway.

Taking Katherine at her word, around March 1561, Seymour set sail for France, ostensibly "for the sight of other countries and commonwealths whereby he desires to come to knowledge of things meet for his estate," but more likely to escape the storm that was gathering at court.[29] While he later attested that he wrote to Katherine at least twice during his absence, she complained that she never heard from him. Worse still, she learned that he had "sent divers other tokens to divers Ladies and Gentlewomen of the Court" during this time. Tormented by jealousy and terrified by the now certain knowledge that she was with child, Katherine wrote an urgent letter to her husband. She told him that she was pregnant and "therefore prayed him to retorne and declare how the matter stoode betweene them." But having some distance from the situation had finally given Seymour a sense of just how foolhardy a course they had followed in marrying without the Queen's consent, and aghast at hearing of this latest development, he resolved to stay in France until the whole affair had blown over. It was a naïve strategy.

Abandoned by her husband and terrified at the prospect of discovery, Katherine received another blow on March 20 when her only confidante, Jane Seymour, died. She had always been of a weak constitution, but given that she was just nineteen or twenty years old, her death came as a shock, especially to the Queen, with whom she had been "in grett faver," despite her flightiness. She had also been popular with the other ladies of the privy chamber, all of whom grieved her loss sorely. Elizabeth ordered a lavish funeral at Westminster Abbey, which was attended by two hundred mourners, including the ladies of her household and some of the highest officials at court, all of whom dressed in black.[30]

As well as losing a dear friend, Katherine had also lost the only witness to her wedding with Edward Seymour, for she had no idea where to find the priest who had performed the ceremony. In her desperation, her hopes alighted upon an unlikely savior: Henry Herbert, to whom she had been betrothed in 1553. She tried to persuade him to renew their attachment, somehow thinking that he would either not notice her pregnancy or accept the child as his own. It seems that he was initially prepared to consider her proposal, for he sent a gracious reply together with some tokens of his affection. But as the weeks went by, rumors of Katherine's condition began to circulate at court, and they eventually reached Herbert's ears. He immediately wrote her a letter filled with furious indignation, demanding that she return his letters and gifts and railing against her for using "the enticement of your whoredom . . . to entrap me with some poisoned bait under the colour of sugared friendship." He assured her: "I will not now begin with loss of honour to lead the rest of my life with a whore." To Katherine's horror, Herbert told her that her "deserts" were by now "openly known to all the world," and he threatened to tell the Queen.[31]

It seemed that he had made good his threat in June 1561, when, for no apparent reason, Elizabeth suddenly turned on Katherine, making it clear that she had greatly offended her. "Whatsoever is the cause I know not, but the Queen has entered into a great misliking with her," reported Sir Nicholas Throckmorton's agent at court.[32] Had Elizabeth guessed Katherine's secret? If she had, then she did not confront her lady-in-waiting at this stage. Perhaps she enjoyed watching Katherine's obvious torment as the weeks went by and her pregnancy became ever more difficult to conceal.

In late July 1561, when Katherine was in her seventh month, she was obliged to accompany her royal mistress and the rest of the court as they went on progress to East Anglia. The discomfort she suffered on the long and arduous journey can only be imagined. Perhaps it was this that convinced her that she could not carry on any longer, for just a few days after the court had arrived at Ipswich, she decided to confess everything to one of the other ladies attending the Queen.

The lady she chose was Elizabeth (Bess) St. Loe, whom she had known since the age of seven, when Bess had entered the Grey household as an attendant to Lady Frances. Although Bess had subsequently left the household to marry, she had kept in touch with the Greys and had always been a friend to Katherine. The latter therefore felt that she could trust her with the secret. On the night of August 8, after the Queen had retired, Katherine sought out Bess and in whispered tones confessed everything. Her confidante was so appalled when she heard of the girl's recklessness that she burst into tears, realizing that the marriage amounted to nothing less than treason. Worse still, Katherine had not just married without the Queen's consent; she had allied herself to a family that had long proved a troublesome threat to the Crown and was uncomfortably close in blood to the Tudor dynasty. Furious at being dragged into the scandal by association, Bess reprimanded Katherine for her foolishness, telling her "she was sorrie therefore becawse that shee had not made the Queene's Majestie pryvie thereunto."[33] She then told Katherine to go to bed while she thought upon the matter, and hurried off to join her husband.

The next morning, Katherine noticed little clusters of courtiers staring and apparently talking about her. The thought that "her beeing with Childe was knowen and espied out" and that her marriage was common knowledge filled her with horror.[34] In her panic, she cast about for an ally who could intercede with the Queen on her behalf. Robert Dudley was then high in favor, and he was also a kinsman of Katherine—albeit a distant one—because her late sister, Jane, had married his brother, Guilford. Anxiously waiting out the day, she stole from her bedchamber late that night and made her way to Dudley's quarters. Coming upon him on a sudden, she went down on her knees and gave a tearful confession of her misdemeanors, begging him to present the matter in a favorable light to Elizabeth and use his influence to win her forgiveness. But Dudley was as horrified by the tale as Bess had been. Alarmed lest the sight of a pregnant young woman in his bedchamber at so late an hour should set tongues wagging, he ordered Katherine back to her rooms.

The following day, Dudley went to the Queen and told her everything. Whether he tried to present it favorably is not recorded, but it would doubtless have done little good. Elizabeth was incandescent with rage, convinced that the whole affair was a plot to oust her from the throne. She immediately ordered Lady Katherine's arrest and that of Bess St. Loe, both of whom were to be "clapt" in the Tower and interrogated by her commissioners. "Our pleasure is, that ye shall, as by our Commandment, examyn the Lady Catharyne very streightly, how many hath bene pryvee to the love betwixt the Erle of Hertford and hir from the beginning; and let her certenly understand that she shall have no Manner of Favor, except [unless] she will shew the truth," Elizabeth ordered Sir Edward Warner, lieutenant of the Tower. Meanwhile, she directed him to put Bess "in awe of divers matters confessed by the Lady Catherine and so also deal with her, that she may confess to you all her knowledge in the same matters," adding: "It is certain that there hath been great practises and purposes and since the death of the Lady Jane she [Bess] hath been most privy." Clearly the Queen believed that there had been a widespread conspiracy. "Sondry Personagees have delt herin; and when it shall appeare more manifestly, it shall increase our Indignation ageynst hir, if she will forbeare to utter it."[35]

In vain Bess protested that she had known nothing of the affair until Katherine had confided in her. The Queen was adamant that one of her political guile and cunning must have been in on it from the start. And so Bess continued to languish in the Tower, pleading for her release. This was eventually granted after several months, but Elizabeth was not prepared to allow her back to court. She was sent home in disgrace to her estates in Derbyshire, where she remained for the next few years.

News of the affair spread like wildfire. On August 12, Cecil wrote to the Earl of Sussex: "The tenth of this [month] at Ipswich was a great mishap discovered. The Lady Catherine is certainly known to be big with child, as she saith by the Earl of Hertford, who is in France."[36] News had also reached Katherine's husband, who was aghast to learn that his worst fears had been realized. Fearful lest he escape, Elizabeth let it be known that she "intended no evil to him or Lady Catherine, but on her account desired to have him in England, in order that it might be decided by law that the Lady Catherine was his wife, whom he had married for his pleasure." But Seymour was not fooled. He held out for as long as he could before finally obeying the royal summons. Upon arrival in London that September, he was immediately "cast into the Tower," and it was reported that "his life is in peril, as also his wife's."[37]

The Queen's commissioners, led by the Archbishop of Canterbury, spent days "streightly" questioning the couple in an attempt to uncover the truth about the affair. They were also to find out whether the marriage had been lawful—or, indeed, whether it had taken place at all. Elizabeth made it clear that they should conclude it had not. This would be the simplest way of neutralizing the threat their union posed to her crown. In her distress, Katherine gave as detailed an account of her courtship and marriage as she could. Her husband did the same, and even though there was a high degree of correlation between the two, their interrogators seized upon the minor differences and concluded that their statements were "contradictory."[38] When confronted with this, Katherine broke down and pleaded that "shee was then in such troble of minde for feare of the Queene's Majesty's displeasure and for absence of her husband and her Imprisonment and beeing great with Child that she was not then soe well advised in her saied Answeares as shee hath sithence considered the same."[39]

Katherine's and Seymour's statements might have differed on points of detail, but they both insisted that the marriage had been legal. Even so, their verbal testimonies were not enough; they needed proof either in the form of documentary evidence or witnesses. With Jane Seymour dead, the only other witness to the wedding was the priest who had performed the ceremony, but as neither Katherine nor her husband knew his name or place of abode, he would be almost impossible to track down. Furthermore, even though they were able to describe certain aspects of his appearance and dress, they admitted that they probably would not recognize him if they saw him again.[40] Despite the lack of evidence to support the marriage, Elizabeth was still gravely troubled by the threat that she perceived to her crown. "The Queen is not without anxiety about it," reported the Spanish ambassador. The root of her suspicion was the fear that "some greater drift was in this."[41] It was rumored that the wedding had been "effected with the connivance and countenance of some of the nobles," among them Elizabeth's chief minister, William Cecil. However, her commissioners had found nothing to substantiate this.[42]

It was the Queen who would have the final say. Much as she despised Lady Katherine, she might have been more inclined to clemency had it not been for the event that took place on September 24. A little over a month after her arrest, Katherine gave birth to a son in the Tower. All Elizabeth's fears seemed to have been realized. It was bad enough that two rival claimants had been united in marriage. Now they had a male heir whose claim would be even stronger than their own because he combined the royal blood of both his parents. An Italian agent at court reported that the Queen was "particularly embittered" by the news, and was "already . . . bent on having the child declared a bastard by Parliament."[43] Determined to legitimize her baby son, Lady Katherine arranged for him to be baptised in secret "by a lady." He was christened Edward, after his father, and was later created Viscount Beauchamp. But unless Katherine could prove that her marriage to the boy's father had taken place, his illegitimacy would be assured.

The birth of Katherine's son could not have come at a worse time for Elizabeth. Just a month before, Mary Stuart had arrived back in Scotland. She had stayed in France for eight months after the death of her husband, unsure of what to do next. Her hesitation had allowed others to seize the initiative. Lady Margaret Douglas, Countess of Lennox, had welcomed the news of the young French king's death, for she spied a chance to realize her long-standing dynastic ambitions. His widow had a strong claim to the English throne, and Margaret resolved to ally their families in order to create an incontestable heir. She therefore hatched a plan to marry her eldest son, Henry, Lord Darnley, to his beautiful young cousin. With Darnley as king of Scotland, he and his formidable Lennox family could build up resistance to the English queen from among the Catholic nobles and their European allies. She duly dispatched her son to France on the pretext of offering his condolences to Mary.

When she heard of this, Elizabeth at once suspected the countess's motives. Margaret Douglas had long been her adversary. The fact that both Henry VIII and Mary Tudor had for a time considered making her heir to the throne gave her a lofty sense of self-importance that never diminished, despite her being subsequently declared illegitimate and passed over in favor of Elizabeth. Her desire to secure the crown for herself or her descendants was so strong that it became the driving force of her whole life. Although she and her husband had made a show of deference to Elizabeth upon her accession, it was clear that they bore her no goodwill, and they had soon retreated to their Yorkshire estates. The feeling was mutual. "Queen Elizabeth loved them not," remarked one contemporary, and she made it plain to everyone at court.[44] At Elizabeth's coronation, it had been expected that Margaret would bear her train—a privilege that was due to her as one of the highest-ranking ladies in the kingdom. But at the last minute, the new queen had replaced her with the Duchess of Norfolk. This was a studied insult, and Margaret had felt it keenly.

It seems that Margaret's son made quite an impression on Mary Stuart. Like his father, Henry was handsome, with a tall, athletic physique and fair hair. His looks were matched by his charming, courtly manners, and as well as being intellectually gifted, he was also a skilled musician. His proud mother was certain that a glittering future lay before him and that the surest means of achieving this was to marry him to the Queen of Scots. Not leaving it to chance, the countess moved to Settrington House on the East Yorkshire coast in order to facilitate communication with Mary. There she conspired with a host of secret agents from both sides of the channel, all united by the objective of securing the throne for Mary and Darnley. Margaret bombarded Mary with messages extolling Henry's virtues and insisting that he would make an ideal husband. She also persuaded the widowed queen of the political advantages that such a match would bring. "The Countess, to allure the Queen of Scots to her purpose, set forth her own title here, declaring what a goodly thing it were to have both the realms in one, meaning that her son should be King both of England and Scotland," reported a foreign agent. In order to secure Mary's loyalty, Margaret made herself as useful to her as possible, offering to "become an espial for her against the State here." She was as good as her word, for a short while later it was reported that "the Countess informs the Queen of Scots of all that passes."[45]

Margaret's efforts seemed to have paid off when, in August 1561, Mary decided to return to Scotland. Although she was still welcome at the French court, she no longer had a role there. The prospect of returning to her faction-ridden native country was hardly more appealing, but in the end it seemed the only possible choice. The easiest passage to Scotland was to cross the channel and then ride northward through England. Mary duly made peaceable overtures to the English queen, telling the latter's emissaries that she desired "amity" with their mistress, considering that "they were both in one isle, of one language, the nearest kinswomen that each other had, and both queens."[46] But when one of the emissaries urged her to prove this amity by ratifying the Treaty of Edinburgh—something that she had henceforth refused to do because it involved renouncing her claim to the English throne— Mary became evasive, claiming that she wished to discuss it with her council first. When she heard of this, Elizabeth retorted sharply: "We assure you your answer is no satisfaction," and refused to grant Mary the necessary passport.[47] Highly offended by such intransigence, the Scottish queen declared that she would make her way to Scotland without Elizabeth's help, adding that she was only sorry that she had asked for a favor she did not need.

Although her will had proved the greater, Elizabeth's reputation suffered by this hostile exchange. International opinion sided with Mary, who played up to her image as the victim of her cousin's petty vengeance. Elizabeth therefore tried to make amends by sending an envoy to congratulate Mary upon her safe arrival in Scotland. She assured her cousin that her dearest wish was "to unite in sure amity and live with you in the knot of friendship, as we are that of nature and blood."[48] The Scottish queen graciously accepted her goodwill and expressed her equal desire for friendship, claiming that she wished "to be a good friend and neighbour to the Queen of England."[49] Introducing a theme that she would repeat again and again over the years to come, Mary stressed the natural solidarity that she and Elizabeth should share as female rulers: "Yt is fetter for none to lyve in peace then for women: and for my parte, I praye you thynke that I desyer yt with all my harte." Thenceforth, she addressed her cousin as "sister" in all her letters, and urged her to write in her own hand in order to make their correspondence more personal.[50]

Mary's letters to Elizabeth have often been taken at face value. She is represented as genuinely well meaning and affectionate toward her cousin, in contrast to the English queen's suspicion and double-dealing. After all, she was still only eighteen, whereas Elizabeth was in her midtwenties and had a good deal more experience of political intrigue. One of Mary's own envoys admitted that "he finds no such maturity of judgement and ripeness of experience in high matters in his mistress, as in the Queen's Majesty [Elizabeth], in whom both nature and time have wrought much more than in many of greater years."[51] But Mary was not so innocent as all that, and it is likely that her effusive sentiments were aimed at achieving more than just sisterly affection. Thomas Randolph, the English ambassador to Scotland, certainly thought so: "Of this Queen's [Mary's] affection to the Queen's Majesty, either it is so great that never was greater to any, or it is the deepest dissembled, and the best covered that ever was. Whatsoever craft, falsehood or deceit there is in all the subtle brains of Scotland, is either fresh in this woman's memory, or she can fett [summon] it with a wet finger."[52]

Just a few days after her arrival in Scotland, Mary's true intentions were made clear. She dispatched William Maitland to secure an agreement from Elizabeth that she would name Mary as her heir. Maitland arrived at court in September 1561, and the initial signs were promising. He received a warm welcome from the English queen, who expressed great affection toward his mistress and claimed that she had forgiven her for laying claim to her throne three years earlier. She said that while this "had given me just cause to be most angry with her, yet could I never find [it] in my heart to hate her."[53] Mary evidently also had a number of supporters among Elizabeth's political elite, notably Sir Nicholas Throckmorton, the ambassador to France, who enthused: "The Queen of Scotland, her Majesty's cousin, doth carry herself so honestly, advisedly, and discreetly, as I cannot but fear her progress. Methinketh it were to be wished of all wise men and her Majesty's good subjects, that the one of these two queens of the Isle of Britain were transformed into the shape of a man, to make so happy a marriage as thereby there might an unity of the whole isle and their appendants." He also wrote to Elizabeth directly, urging her: "The best means that has been thought on for the quietness of the two Queens is . . . that Queen Elizabeth should for herself and her heirs peaceably enjoy the crown of England; and failing herself and her heirs, that the Queen of Scotland should be accepted next heir of England."[54]

When Maitland first presented the proposal to Elizabeth, she gave him an ambiguous answer. Although she acknowledged the strength of Mary's claim, she was careful not to say that it was the most valid among all of the various rivals to her throne, and instead declared: "When I am dead, they shall succeed that has most right. If the queen your sovereign be that person, I shall never hurt her; if another have better right, it were not reasonable to require me to do a manifest in-

jury." She went on to express concern about naming her successor because she had experienced firsthand how doing so had undermined
her sister's position as queen and soured their relationship. "Ye think
that this device of yours should make friendship betwixt us, and I fear
that rather it should produce the contrary effect. Think you that I
could love my winding-sheet? Princes cannot like their own children,
those that should succeed them . . . How then shall I, think you, like
my cousin, being once declared my heir apparent . . . And what danger
it were, she being a puissant princess and so near our neighbour, ye
may judge; so that in assuring her of the succession we might put our
present estate in doubt."[55]

Even though this was hardly encouraging, the fact that Elizabeth
had not rejected the proposal outright gave Maitland hope. Moreover,
the English queen continued to express great favor toward her cousin.
He therefore persisted with the negotiations, convinced that if Mary
would ratify the Treaty of Edinburgh, then in return Elizabeth would
promise her the throne. But the reality was more complex than that.
Mary might have had the strongest blood tie to the English crown
from among the various claimants, but she was in theory excluded by
the terms of Henry VIII's will and the act of succession. Her religion
provided a further barrier: England was still reeling from the turmoil
created by Mary Tudor's attempts to restore Roman Catholicism, and
many of Elizabeth's council were strongly opposed to appointing another of that faith as heir.

Equally, though, it would be a dangerous course to reject Mary's
proposal altogether. France might have withdrawn its troops from
Scotland, but the two countries remained on friendly terms, and there
was the added danger that the Spanish king would lend his support.
Furthermore, Mary was growing impatient with the ongoing negotiations and insisted: "We know how near we are descended of the blood
of England, and the devices that have been attempted to make us as it
were a stranger from it," adding: "We will deal frankly with you, and
wish that ye deal friendly with us."[56]

With the situation in danger of reaching an impasse, it was at
length agreed that the two queens should meet. Mary was delighted

with the idea, eager to see her "dearest sister" in the flesh. She declared that it was "the thynge that I have moste desyered ever since I was in hope therof . . . let God be my wytnes, I honor her in my harte, and love her as my dere and naturall syster."[57] She was confident that she could charm Elizabeth, just as she had almost everyone who came into her presence. In her excitement, she bombarded Randolph with questions about her English cousin: whether her "lyvelye face" matched that in the portrait that Elizabeth had sent her in France, what she ate, how regularly she took exercise, "and many more questions."

"I see my sovereign so transported with affection," reported Maitland after his return to Scotland, "that she respects nothing so she may meet with her cousin."[58] Meanwhile, Randolph described how Mary had slipped a letter from Elizabeth "into her boosame nexte unto her schyne [skin]," and told him: "Yf I could put it nerrer my hart, I wolde." She then gave him a ring containing a heart-shaped diamond, which she ordered him to convey to "my good syster" as a sign of her love.[59]

For her part, Elizabeth seemed just as keen to meet her cousin, although she was motivated more by curiosity than affection. She had heard much of the young Scottish queen's beauty and charm, and she wanted to see for herself what all the fuss was about. Like Mary, she also believed that she would stand a much greater chance of bending her to her will if they were to meet. Despite objections from certain members of her council, she ordered that the meeting should take place in August or September 1562 at York, which was roughly halfway between their two courts. However, the plans were disrupted by events in France in July that year, when the English sent troops to support the Huguenots (Protestants) in the civil war that had broken out, while Scotland remained neutral. This made it impolitic for the meeting between the two queens to go ahead, and it was therefore postponed.

When she heard of this, Mary was so upset that she took to her bed and wept bitterly for days. There is no record of Elizabeth's reaction, but it might be supposed that she was more philosophical: after all, she had less to lose from the delay. While Mary was anxious to strengthen her position in Scotland by securing the succession in England, Elizabeth had no wish to name her heir so early on in her reign. Moreover, her subsequent actions suggest that she had not been as keen to meet her cousin as she had pretended. Perhaps she was afraid of having her jealous assumptions about Mary's beauty and charm confirmed. Whatever the case, the postponement of this meeting would have a profound effect upon the relationship between the two queens, because it would turn out to be a permanent one.

Despite this setback, the delicate negotiations between Elizabeth and Mary continued. At the same time, another Tudor cousin was causing the Queen disquiet. Lady Margaret Douglas had been delighted when Mary had arrived safely back in Scotland in August 1561. "She sat down and gave God thanks, declaring to those [near]by how He had always preserved that Princess at all times."[60] Now that the Scottish queen was but a day or two's journey north rather than across the seas, her plans for her son's marriage seemed tantalizingly close to fruition.

The Lennoxes tried to shroud these plans in secrecy, adopting code names in their correspondence, such as "the Hawk" for Mary. They also instructed that all letters should be burnt as soon as they had been read. But Margaret was so sure of success that she could not stop herself from boasting to anyone who would listen. Time and again, she denounced Elizabeth as an illegitimate usurper, and claimed that she and her heirs were the rightful sovereigns of England. One gentleman who visited her at Settrington reported that he: "heard her say that Queen Elizabeth was a bastard, and that God would send her [Margaret] her right one day." She also had her jester poke fun at the Queen in his entertainments. These included a sketch about Elizabeth's supposed love affair with Robert Dudley, whom the jester portrayed as a pox-ridden traitor.[61]

Having been raised amidst the intrigues and power games of Scottish politics, Margaret should have known better. The new queen had sent her away from court at the beginning of the reign, but was no less suspicious of her for that. Indeed, she had arranged for spies to be sent to the Lennoxes' houses at Temple Newsam and Settrington. When she read her agents' dispatches, Elizabeth was said to be "greatly alarmed" at her cousin's audacity in conspiring against her so openly, and was furious at her insults and mockery.[62] By November 1561, she had amassed more than enough evidence to throw the countess and her husband into the Tower.

Upon hearing that the Lennoxes had been consigned to that fortress, one foreign envoy scoffed: "The prison will soon be full of the nearest relations of the Crown."[63] With Katherine Grey and her husband holed up in the same place, this was no great exaggeration. Margaret and her husband were subjected to close questioning, as were all of their servants. Naturally, the couple both attested their loyalty to the Crown, but thanks to their lack of discretion, there were too many witnesses to testify the opposite. Just as she had during the Katherine Grey controversy, Elizabeth feared that there was more to the plot than met the eye and that the Catholic powers of Europe must be involved. She was not alone in this belief. De Quadra's secretary reported that Philip II was planning to throw his weight behind Lady Margaret's cause, seeing it as an ideal way of restoring Catholicism to England. It was predicted that if he did support her, then "eight or ten of the nobility would rise in favor" of her.[64] As a result, now that the Lennoxes were under her control, Elizabeth was determined to keep them there.

When the interrogations were concluded, charges were drawn up against the couple. These condemned their treasonous relations with Mary, Queen of Scots, as well as Margaret's outspoken claims about her own rights to the throne. The wily queen used these charges to reinforce the fact that Henry VIII had barred any of his sister Margaret's children from inheriting the English throne. As the charges spelled out, "this makes both against Queen Mary [of Scots] and the Countess." Furthermore, it declared that "the Countess . . . can claim nothing, being a mere bastard, for the marriage of her mother with Archibald Earl of Angus was found null from the beginning, as appears by a sentence of divorce."[65] Elizabeth was clearly determined to destroy Lady Margaret's reputation and thereby neutralize the threat that she posed for good.

As the months dragged on, Elizabeth showed no sign that she

would release the Lennoxes. At length she transferred the countess to the custody of Sir Richard Sackville at Sheen but kept her husband in the Tower. In vain Margaret pleaded for his release, claiming that he suffered from an illness that made him unable to bear "solitariness."[66] When her pleas fell upon deaf ears, she exclaimed that it was "strange and grievous" to her that the Queen should be so stubborn. If she had shown greater humility or remorse, perhaps the countess would have been released sooner. As it was, she did herself few favors by continuing to insist upon her right to the throne. Furthermore, she was increasingly "obstinate" in the answers she gave to the members of the council who visited her.[67]

Meanwhile, the commissioners in the Lady Katherine Grey case had concluded their interrogations. Although Elizabeth still suspected a conspiracy, they found that "no one appears privy to the marriage, nor to the love, but maids, or women going for maidens."[68] But neither had they found any evidence of the marriage itself, which made it easy to concede to the Queen's private demand that it be declared invalid. At the end of January 1562, "A definitive Sentence were pronounced by the Archbishop [of Canterbury], that he [Seymour] had had undue and unlawful copulation with her [Katherine], and that for such their excesse, both he and she to be punished."[69]

No immediate action was taken against the couple, and instead they were left to languish in the grim fortress of the Tower. There was a good deal of sympathy for them among the people, who believed their marriage to be valid and regarded Elizabeth's treatment of them as unnecessarily cruel. The birth of their son increased this sympathy still further, not least because males in the line of succession were a precious commodity. One of William Cecil's agents warned him: "There be abroad, both in the city and in sundry other places in the realm, broad speeches of the case of the Lady Catherine and the Earl of Hertford. Some of ignorance make such talks thereof as liketh them, not letting [scrupling] to say that they be man and wife. And why should man and wife be let [prevented] from coming together?"[70] The couple were popular with many who remembered the tragic events of Edward VI's reign, which had seen the execution of Seymour's father, the Lord Protector, and Katherine's sister, Jane. Their royal blood also gave them a certain cachet, and by imprisoning them in the Tower, Elizabeth merely increased their popularity.

It was not just her ordinary subjects who sympathized with the couple's plight; they had powerful supporters at court. Elizabeth's reign was still in its infancy, and she was the third monarch to rule in the space of just over a decade. Moreover, she showed little inclination to marry, so the succession was at the forefront of her councillors' minds. Opinion was divided as to which of the various rival claimants to the throne ought to succeed if this new queen's reign was to prove as short lived as those of her siblings. Robert Dudley led the camp for Mary, Queen of Scots, and William Cecil for Katherine Grey. Their claims were openly discussed, and it seems that Mary, at least, appreciated that she had a rival. In a letter to her, Sir Nicholas Throckmorton, the English ambassador in France, once referred to "your competitor . . . Lady Katherine."[71] Even though Mary was closer in blood to the throne, Katherine had more followers. "She has many supporters among them [the Council]," reported an Italian envoy in October 1561, "it is said even Cecil, the Queen's first secretary, who governs all the state."[72] Even though her religious sympathies were ambiguous, Katherine had at least been raised a Protestant, which made her more acceptable to many in government than the Catholic Mary.

The Queen would hear of neither candidate and hated the question of the succession to be raised. But the matter suddenly became pressing in October 1562. Elizabeth and her entourage were then in residence at Hampton Court. On the night of October 10, the Queen complained of feeling unwell and decided to ease her aching body by taking a bath. The next morning, her symptoms had worsened and she had a high fever. As her condition continued to deteriorate, her physicians confirmed everyone's worst fears: She had smallpox. This was one of the most deadly diseases in the sixteenth century, and there was no known cure. Survival depended more upon chance than the ministrations of physicians. As well as being deadly, smallpox was also highly contagious, so most of Elizabeth's ladies shrank from the task of attending her. Lady Mary Sidney alone insisted upon staying by her side, closeted in the stifling, sickness-filled confines of the royal bed-chamber.

Meanwhile, with Elizabeth's reign but four years old and the men-ace of Spain and France growing ever greater, England looked set for a war of succession. Few of her councillors expected her to survive, and as she lay in feverish delirium at Hampton Court, they called an urgent meeting to agree upon a successor. "There was great excitement that day in the palace," reported de Quadra, "and if her improvement not come soon some hidden thoughts would have become manifest. The Council discussed the succession twice . . . Some wished King Henry's will to be followed and Lady Catherine declared heiress."[73]

As the Queen's life hung in the balance, a group of her councillors met again, this time at the Earl of Arundel's house. The debate over whose claim was most valid was complicated by more self-interested motives, as the powerful nobles saw the chance to ally their blood to the throne. De Quadra noted: "The question of the succession was dis-cussed, and I understand they favoured Lady Catherine, who is sup-ported by the Duke [of Norfolk], perhaps with the idea that one of his little daughters may in time be married to Lady Catherine's son." Robert Dudley and the Earl of Pembroke, on the other hand, were said to be "much against" Katherine.[74] The debate raged on until the early hours, and still no resolution was reached.

In the meantime, Elizabeth had started to show signs of recovery, and against all the odds, she survived one of the most virulent diseases of the time. Neither was she permanently scarred with the unsightly pockmarks that it so often left behind. As the Queen was recovering, Lady Sidney grew increasingly sick, and it was soon all too obvious that she too had contracted the disease. As her fever worsened, it seemed that she would pay the ultimate price for her devout service. But, like her royal mistress, she survived. Sadly, though, she was not as fortunate in escaping disfigurement. "I lefte her a full faire Ladye in myne eye at least the fayerest," lamented her husband, "and when I re-torned I found her as fowle a ladie as the smale pox could make her, which she did take by contynuall attendance of her majesties most pre-cious person (sicke of the same disease) the skarres of which (to her resolute discomforte) ever syns hath don and doth remayne in her face."[75]

But the Queen was too preoccupied with affairs of state to give much thought to the sacrifice her lady-in-waiting had made. When she heard that her councillors had been meeting to discuss her successor, she "wept with rage."[76] She was still furious two months later, when, having resumed her duties, she railed at them for what she saw as unforgivable disloyalty. "She was extremely angry with them," reported de Quadra, "and told them that the marks they saw on her face were not wrinkles, but pits of small-pox, and that although she might be old God could send her children as He did to Saint Elizabeth."[77]

In the event, it was not to Elizabeth that God sent a child, but to her despised rival, Katherine Grey. Sir Edward Warner, lieutenant of the Tower, had taken pity on the lady, who was often in tears at being separated from her husband. A few months after the birth of her son, Sir Edward therefore began to allow the couple secret conjugal visits. By July 1562, Katherine was pregnant once more. The lieutenant was no doubt aghast at so undeniable a proof that he had disobeyed his orders, but this pregnancy, like the first, was hushed up, and it was only when the child was born in early February that the secret leaked out. "The x day of Feybruary was browth a-bed within Towre with a sune my lade Katheryn Harfford, wyff to the yerle of Harrford," noted Henry Machyn, a London tailor, in his diary. The couple had obviously established quite a cozy fraternity within their prison, for he added: "the god-fathers were ij [two] warders of the Towre."[78]

The Queen's fury knew no bounds. The Spanish ambassador observed that she was "the colour of a corpse" upon hearing the news, utterly appalled that her orders should be flouted with such monumental disrespect. Her rage was not lessened by the fact that this second child was also a boy. Now the woman whom many believed was the rightful heir to her throne had two male heirs, while Elizabeth herself remained unmarried and childless.

While she was conscious enough of popular opinion not to dare condemn Katherine and Hertford for treason, the Queen resolved to do everything in her power to humiliate them. Seymour was hauled before the Court of Star Chamber, which primarily dealt with political and treason cases, and accused of compounding his original offense of "deflowering a virgin of the blood royal in the Queen's house" by having "ravished her a second time." After being found guilty, he was fined the staggering sum of £15,000—more than £2.50 million or $4 million in today's money.[79]

Much worse was to come, at least for Katherine. The kindly Sir Edward Warner was peremptorily dismissed from his post, which put an abrupt end to the snatched meetings that Katherine and her husband had enjoyed.[80] She wrote to Seymour lamenting "my great hard fate to miss the viewing of so good a one [husband]." She added: "I . . . long to be merry with you as you do with me. I say no more but be you merry as I was heavy when you the third time came to the door and it was locked." Katherine went on to thank her husband for sending her a message, together with some money and a book, "which is no small jewel to me. I can very well read it, for as soon as I had it, I read it over even with my heart as well as with my eyes."[81]

Elizabeth was not satisfied that her cousin would not use her wiles to persuade another lieutenant to take pity on her, so she resolved to separate her from Seymour for good. In August 1563, the couple were removed from the Tower on the pretense of protecting them from the plague—an act of "compassion" on the part of the Queen, as her council's instructions claimed. The earl was sent to the house of his mother, there to be kept under close surveillance. Katherine was placed in the custody of her uncle, Lord John Grey, at his house in Essex. He received detailed instructions for her imprisonment, "which hir Majesty meaneth she shuld understand of yowr Lordship and observe, as some Part of hir Punishment; and therin hir Majesty meaneth herin to trye hir Disposition how she will obey that which she shall have in Comandment." Lord Grey was to insure that his niece would have no "conference" with anyone apart from his household staff and would "use hir self there in yowr Houss with no other Demeanor, than as though she were in the Towre."[82] Katherine wrote at once to Cecil, begging him to intercede with the Queen, whose forgiveness "with upstretched hands and down bent knees from the bottom of my heart most humbly I crave."[83]

But as far as Elizabeth was concerned, it was a case of out of sight, out of mind. Far more important to her were the actions of her Scottish cousin. She was growing increasingly resentful of the many effusive reports she received about Mary's beauty and allure. When one foreign diplomat remarked that she was reputed to be "very lovely," the English Queen snapped that she herself was "superior to the Queen of Scotland."[84] The fact that Mary was nine years younger than her did not help, particularly as Elizabeth faced almost daily reminders of her advancing age as a potential bride. Whereas before, she had been the most desirable bride in Europe, now her younger, fairer cousin was attracting all the attention. One contemporary observer would later claim: "the Queen of England never liked her, but was always and for a long time jealous of her beauty, which far surpassed her own."[85]

The idea that Elizabeth's view of Mary was colored by personal jealousy as well as political concerns is supported by the famous account of her meeting with her cousin's ambassador, Sir James Melville, in 1564. Setting aside the political matters that Sir James had been sent to discuss, Elizabeth quizzed him upon every aspect of Mary's personal appearance and accomplishments—from what color her hair was to how well she played the lute. Realizing that he would have to choose his answers extremely carefully in order to avoid incurring the English queen's notorious jealousy of her cousin, Melville kept his answers as equivocal as possible. But Elizabeth would not be satisfied, and demanded that he provide an assessment of the two women's comparative merits. "She desired to know of me, what colour of hair was reputed best; and which of the two was fairest. I answered, The fairness of them both was not their worst faults. But she was earnest with me to declare which of them I judged fairest. I said, She was the fairest Queen in England, and mine the fairest Queen in Scotland. Yet she appeared earnest. I answered, They were both the fairest ladies in their countries; that her Majesty was whiter, but my Queen was very lovely." Still Elizabeth persisted, asking next who was the taller—the answer to which question she must have been confident of, given that she was renowned for her height. When the beleaguered Sir James admitted that the Scottish queen was taller, Elizabeth snapped,

"Then . . . she is too high; for I myself am neither too high nor too low."

Determined to find something in which she surpassed her cousin, the English queen turned to Mary's accomplishments, which she was sure would be inferior to her own. She learned that her cousin was fond of hunting, reading history, and occasionally playing the lute and virginals. Seizing upon the latter, Elizabeth asked Melville if his mistress played well, to which he replied: "reasonably for a Queen." Later that day, Sir James was invited by Lord Hunsdon to accompany him to a quiet gallery, where, he said, he might listen to the Queen playing music without her knowledge. The ambassador duly did so and patiently listened "a pretty space" while Elizabeth, with her back to the door, played the virginals "exceedingly well." Suddenly she stopped playing and swung around. Feigning surprise at his presence, she gave him a playful slap and chided him for so intruding upon her privacy, "alleging she used not to play before men." The true motive behind this farcical episode was soon revealed, however, when the Queen asked—no doubt as casually as she could—whether Mary or she played best. "In that I found myself obliged to give her the praise," recalled Melville. Before he was permitted to leave her court, Elizabeth made sure he had seen her dance, and inquired of him "whether she or my Queen danced best." Again Sir James was forced to admit that Mary was her inferior in this respect.[86]

Satisfied that she had gained the upper hand, Elizabeth was apparently inclined to feel more affectionate toward her cousin. She assured Melville that she had a great desire to meet his royal mistress, but as that could not soon be brought to pass, she would console herself with looking at Mary's picture. Summoning him to her bedchamber, she opened a little cabinet in which were a number of miniatures wrapped in paper with the names of the sitter inscribed on each in the Queen's handwriting. Taking out Mary's picture, she made a show of kissing it very reverently. However, when Melville suggested that she should send her cousin the "fair ruby, as great as a tennis ball" that he had spotted among her jewels, Elizabeth quickly refused, and instead sent her a much smaller diamond. Determined to leave Sir James with a favorable impression, she expressed her earnest wish to meet her cousin. Taking her at her word, Melville offered to convey her to Scotland in secret, disguised as a page, while excusing her absence from court with the tale that she was keeping to her bed due to sickness. "She appeared to like that kind of language," he recalled, "only answered it with a sigh, saying, 'Alas! if I might do it thus.' "[87]

The Scottish ambassador was not fooled by this display of regret, nor by the many protestations of "sisterly love" that Elizabeth had made toward Mary. Upon his return to Scotland, he at once sought an audience with his royal mistress and relayed everything that had been said. With greater shrewdness than she showed on other occasions, Mary asked him "whether I thought that Queen meant truly toward her inwardly in her heart, as she appeared to do outwardly in her speech." Melville declared that in his opinion, "there was neither plain dealing, nor upright meaning; but great dissimulation, emulation and fear, lest her [Mary's] princely qualities should over soon chase her from the Kingdom."[88]

Elizabeth had clearly failed to disguise her true feelings toward Mary by her excessive compliments and declarations of love. Although such overblown sentiments were typical of the courtly language of the day, the fact that they were not backed up by action revealed them to be false. Her behavior during Melville's visit had been sparked as much by Elizabeth the woman as by Elizabeth the queen. The personal rivalry between the two women dangerously intensified their political rivalry. Ironic, then, that Mary should try to play on their sex as a means to unite them in some kind of sisterly bond: two female sovereigns trying to make their way in a world dominated by men. Far from uniting them, however, their sex proved a source of discord—at least on Elizabeth's part. "It is certain that two women will not agree very long together," observed a Spanish envoy.[89] Subsequent events were to prove him right.

One of the principal causes of discord between them was Mary's search for a new husband. This dominated her relations with Elizabeth throughout the early to mid-1560s. In contrast to Elizabeth, the Queen of Scots viewed the position of unmarried female ruler as something undesirable and, therefore, temporary. The English ambassador in Scotland attributed Mary's increasingly frequent bouts of melancholy to her unsatisfied longing for a husband.[90] The sooner she could find a man to share the burden of monarchy with her, the better. The choice that she made was naturally of intense interest to her cousin south of the border. If she married a prince of Spain or France, then England would find herself in a potentially dangerous position, threatened with invasion by Catholic potentates just across her northern border, rather than over the seas as they were now. It was therefore vital that Elizabeth have some say over the husband her cousin should take. To secure this, she used Mary's desire for the English succession as a bargaining tool. Even though in reality Elizabeth had little intention of naming the Scottish queen her heir, it was enough to convince Mary that she must consult her cousin on the question of her marriage.

Various candidates were put forward, including the Archduke Charles of Austria, whom the English queen strongly disapproved of due to the fact that he was still in theory one of her own suitors. He was also "in the Emperor's lineage," and as such constituted a grave threat to Elizabeth if he were to marry her Scottish cousin. She found Philip II's son Don Carlos equally unacceptable, and also the young French king, Charles IX. Randolph predicted that if Mary persisted in favoring such candidates, it "must needs bring a manifest danger to the private amity betwixt the Queens, an apparent occasion to dissolve the concord that is presently between the two nations, and an interruption of such a course as otherwise might be taken to further and advance such right or title as [Mary] might have to succeed [Elizabeth] in the crown of England."[91] Instead Elizabeth tried to steer her cousin toward an English husband, "with whom her Majesty would more readily and more easily declare, that she inclines that failing of children of her own body, you might succeed to her crown."[92] Ambrose Dudley, Earl of Warwick, was suggested, along with the Duke of Norfolk. Then, in March 1564, Elizabeth put forward a quite extraordinary candidate: Robert Dudley, Earl of Leicester—her own great favorite.

Quite what the English queen was thinking in suggesting such a match for Mary has bemused both contemporaries and historians.

Surely there could be no man in the world whom she would less desire to see married to her beautiful Scottish cousin, the woman of whom she was already intensely jealous? And yet she had suggested that the three of them live together at the English court, in what one historian describes as "a virtual ménage à trois."[93] Although Elizabeth may have taken the pragmatic view that it would be better to have a man loyal to herself on the Scottish throne than a Catholic potentate, given her personal antipathy toward Mary, it seems more likely that her promotion of Robert Dudley as a suitable husband was an act of petty revenge. She knew that when news had reached Scotland of the death of Dudley's wife under suspicious circumstances four years earlier, Mary had quipped that the Queen of England was about to marry her "horsekeeper,"[94] who had killed his wife in order to make way for her. Elizabeth had indeed seemed seriously to contemplate marrying Dudley, but she had abandoned such thoughts by the time she suggested him as a husband for Mary. Dudley, she said, was the man "whom she would have herself married, had she ever minded to take a husband. But being determined to end her life in virginity, she wished that the Queen her sister might marry him."[95] To put forward one of her own castoff suitors was therefore little short of an insult, particularly as he was well down the pecking order in the nobility of England. It was a ploy that Elizabeth had every confidence would be thrown out by her cousin.

To her horror, the plan seemed to have backfired. True, Mary had been suitably insulted by her cousin's proposal initially and had sent a pointed response that Elizabeth should look to her own marriage first. "She [Mary] wonders at the occasion of my sovereign's stay [from marriage], as much for her years, the wise counsel about her, sought of so many as she has been, and may be when she will, matched with the greatest for herself," reported Randolph. Mary claimed that "for herself, the remembrance of her late husband is yet so fresh, she cannot think of any other," and added a final sideswipe at her cousin by saying: "Her years are not so many but she may abyde."[96] This reminder of the difference in their ages would have touched a raw nerve with the English queen, who was increasingly sensitive about her advancing years.

Mary had a further tactic up her sleeve, and after a while, she decided to play Elizabeth at her own game by making a show of seriously considering Dudley as a husband. Sir Robert himself was apparently delighted by the idea, and wrote a number of "discreet and wise letters" to further his suit. The Queen of Scots duly professed to have "so good liking of him" that she was prepared to accept him as a consort. When Elizabeth heard of this, she was thrown into a panic. She had borne many sacrifices in her endeavor to become a great queen, but this was too much. Her horror at the prospect that Dudley might actually marry her cousin proved that the whole scheme had been a sham.

Elizabeth need not have worried. The suitor whom Mary favored was not Dudley but Darnley. The latter's mother had been working ceaselessly to bring this about. Lady Margaret Douglas had been present during Melville's visit to court in 1564, having been released with her husband early in the previous year. Elizabeth was gracious in her acceptance of their humble apology and showed the couple a great deal of courtesy. "My Lord and Lady Lennox are continual courtiers, and made much of," reported one observer.[97] The Queen invited them to accompany her on progress, and they were joined by their eldest son. Lord Darnley was apparently high in favor with the Queen, who affectionately referred to him as "yonder long lad." He carried the sword before her on official occasions and entertained her with music and compliments. "My Lord Darnley . . . is also a daily waiter and playeth very often at the lute before the Queen, wherein it should seem she taketh pleasure, as indeed he plays very well," an onlooker remarked.[98]

But this show of favor was fooling nobody—least of all the countess. She was under no illusions about Elizabeth's opinion of her, and, for her own part, she had far from given up the scheme to marry Darnley to the Scottish queen. Her confidence was boosted when, in the autumn of 1564, she learned that her husband had succeeded in reclaiming his estates in Scotland, which had been forfeited in 1545 when he had tried to advance Henry VIII's cause there. With Lennox back in favor at the Scottish court, the way was now clear for Mary to take his son as her second husband.

A modern reproduction of a copy made in 1600, from a lost original
from 1559, of Elizabeth I at her coronation.

Elizabeth dancing with Robert Dudley,
Earl of Leicester (ca. 1580), watched by her ladies.

Elizabeth receiving Dutch emissaries (ca. 1585). The lady
dressed in black is believed to be Blanche Parry.

Lady Katherine Grey and her eldest son, Edward (ca. 1561–63). Both of her sons were born while she and her husband were imprisoned in the Tower.

Lady Mary Grey, whom the Spanish ambassador described as "little, crookbacked, and very ugly."

Mary, Queen of Scots, Elizabeth's deadliest rival.

Margaret Douglas, Countess of Lennox
(ca. 1560–65), Elizabeth's ambitious
and much imprisoned cousin.

Arbella Stuart (ca. 1589), the last of
the women to lay claim to
Elizabeth's throne.

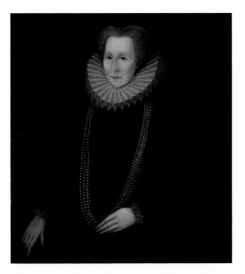

Elizabeth Talbot, Countess of Shrewsbury,
known as "Bess of Hardwick" (ca. 1590).
She was often at odds with her
royal mistress.

Elizabeth FitzGerald, Countess of Lincoln, known as "Fair Geraldine" (ca. 1575). She was one of Elizabeth's earliest friends.

Anne Dudley, Countess of Warwick (ca. 1569), Elizabeth's long-standing friend and attendant.

Lady Mary Sidney (ca. 1550–55), sister of Lord Robert Dudley and long-suffering attendant of Elizabeth I.

Helena Snakenborg, Marchioness of Northampton (ca. 1569), one of Elizabeth's favorite ladies at court.

Lettice Knollys (ca. 1577), whom Elizabeth referred to as "that She-Wolf" after she secretly married the Earl of Leicester.

Elizabeth Vernon (ca. 1600), who caused a scandal when she fell pregnant by, and secretly married, the Earl of Southampton.

Bess Throckmorton (ca. 1603), who fell from favor after marrying Sir Walter Ralegh in secret.

Elizabeth I in old age, with time and death in waiting (ca. 1620).

Elizabeth I's funeral procession (1603).

Lady Margaret was encouraged in her plans by Melville's visit to court in 1564. Although masquerading as an official diplomatic mission, he had secret instructions to confer privately with the Countess of Lennox about the proposed match between her son and his royal mistress. He found Margaret all graciousness and charm. "She was a very wise and discreet matron," he recalled in his memoirs, and went on to relate that she had sent "many good advices" to Mary, along with various tokens, including "a ring with a fair diamond," for she was "still in good hope" that the marriage would come to pass.[99]

Buoyed by her secret conference with Melville, Lady Margaret petitioned the Queen to allow Lord Darnley to go to Scotland, ostensibly to visit his father and take possession of his newly regained estates. Elizabeth knew full well what the real intention of such a visit would be and refrained from giving her cousin an answer while she considered it. It seems that she did briefly entertain the idea of Darnley marrying the Queen of Scots: At least he was a known quantity, and one whom Elizabeth could control through his mother. She also hoped "that he [Darnley] being a handsome lusty youth, should rather prevail, being present, than Leicester who was absent."[100] She duly gave him leave to go to Scotland in February 1565.

No sooner had Darnley embarked, however, than the Queen seemed to have second thoughts. Influenced by her councillors, who cautioned her about the danger of an alliance that would only serve to bolster Mary's claim to the English throne, she may also have heard that her archenemy King Philip of Spain had pledged his support to the marriage. In early June, he had written to his envoy at Elizabeth's court: "The bridegroom and his parents being good Catholics and our affectionate servitors; and considering the Queen's good claims to the crown of England, to which Darnley also pretends, we have arrived at the conclusion that the marriage is one that is favourable to our interests and should be forwarded and supported to the full extent of our power."

In panic, Elizabeth sent word expressly forbidding Darnley from marrying the Scottish queen, claiming that it would be "unmeet, unprofitable, and perilous to the sincere amity between Queens and their realms."[101] She also ordered both him and his father to return to England at once. When this went unanswered, she had Lady Margaret placed under house arrest at Whitehall, "where she may be kept from giving or receiving intelligence."[102] Then, in late July, came the news that Mary and Darnley were married. Hearing that her son was now king of Scots was as joyful to Margaret as it was abhorrent to her cousin. Elizabeth had never seriously entertained the idea of their union, and now the implications of it filled her with horror. In her rage, she was determined to take revenge upon the woman she knew to be the architect of it all. The Countess of Lennox was thrown into the Tower once more, accused of having "deceitfully asked leave for her son to go to Scotland," which had made the Queen "justly indignant." Elizabeth then sent word to Margaret's husband and son, reminding them of "the hard case of the Lady Margaret, now in the Tower, whose wellbeing must depend upon their behaviour there."[103]

Elizabeth also directed her fury at Mary herself. As Melville observed, all the "inward griefs and grudges" that the issue of the Scottish queen's suitors had caused between them now "bursted forth."[104] Throwing off the pretense of submissively acceding to her older cousin's guidance on the matter of her marriage, Mary defiantly declared "that she thanked the Queen of England for her offers, but that, it being a particularity of ladies to choose their husbands when it is in their power, she had chosen the said Earl [Darnley], and could not now prefer another."[105] She also took the precaution of asking the Pope for aid, "in case the Queen of England should be minded to do her an evil turn by way of showing her dissatisfaction, which is extreme, at her marriage with the Earl of Lennox."[106]

Mary's act of defiance was enough to destroy the fragile—and arguably false—harmony that existed between her and Elizabeth. "Whereupon ensued so great a coldness, that they left off for a considerable time from writing each to other, as they had formerly done by post . . . for in their hearts from that time forth there was nothing but jealousies and suspicions," remarked one contemporary at the Scottish court.[107] It also shifted the power balance between the two women. Prior to the marriage, Mary had played the part of a deferential supplicant for Elizabeth's favor, eager to submit to the will and guidance of her older, wiser cousin. Now she had seized the initiative and married Darnley against the English queen's express command. No longer the meek novice queen, she caused the English ambassador to exclaim: "a wylfuller woman, and one more wedded unto her owne opinion, withowte order, reason or dyscretion, I never dyd knowe or hearde of."[108] At a stroke, Mary had put herself on a far more equal footing with Elizabeth.

As well as feeling piqued at being so deceived by her cousin, the English queen was also genuinely alarmed by her choice of husband. Lord Darnley's claim to the English throne was a strong one—indeed, the fact that he had been born in England made his claim better than that of Mary, who was technically an alien according to English common law. The uniting of their claims posed a considerable threat to Elizabeth's interests and boosted the ranks of her Catholic dissenters. But the marriage also generated significant opposition in Scotland, particularly among the Protestant lords, such as the Earl of Moray and the Hamiltons, traditional enemies of the Lennox family. Heartened by the English queen's reaction to Mary's marriage, this powerful enclave rebelled against it. Although they failed to win enough support, they remained troublesome opponents to Mary's regime, and her hold on the crown began to look increasingly tenuous.

Elizabeth was still reeling from the news of Mary's marriage to Darnley when another scandal involving one of her cousins broke at court. Lady Mary Grey had disregarded everything she should have learned from her elder sister's example by marrying in secret. She was an unlikely romantic heroine. Having been born with a crooked spine, as an adult she grew to only around four feet tall. The Spanish ambassador described her as "little, crookbacked, and very ugly."[109] Mary was now twenty years old, and during her twelve years of service, first to Queen Mary and then to Elizabeth, she had kept a low profile. What happened in 1565 therefore came as a complete shock, not just to Elizabeth but to the entire court. Little Mary Grey, who had been the butt of so many jibes, suddenly found herself at the center of the biggest controversy the court had seen since her elder sister's disgrace.

Thomas Keyes was a steady, unremarkable member of the Queen's staff who would have warranted no more notice in the contemporary records than the entry of his wages in the account books had it not been for his part in the scandal. Each of the royal palaces in London had a river gate from which the Queen would embark for her excursions on the Thames. As her sergeant porter, Keyes's job was to supervise the gate at Westminster. He lived on-site, having comfortable apartments above the river gate, and it was here that this convivial and popular man often gave parties for the other servants at court.

The nature of Keyes's duties meant that he saw the Queen and her entourage on a regular basis. Exactly when he and Lady Mary Grey became more closely acquainted is not known; she evidently shared her sister's aptitude for clandestine liaisons. Besides, it was no doubt a good deal easier for the couple to keep their relationship a secret because of the simple fact that nobody would ever have expected it. To the casual observer, Mary Grey and Thomas Keyes were singularly ill suited. While Mary was of royal blood, Keyes was a minor gentleman from Kent—possibly related to the Knollys family but otherwise of a much lower status than the object of his affections. He was also a widower with six or seven children, which did not make him any more obvious a suitor. But the greatest difference between the couple was in their physical appearance. As well as being twice Mary's age, Thomas was very nearly twice her height. At six feet eight inches tall—an impressive height even now—he was a giant of a man for the time, said to be the tallest in London, if not the entire kingdom. He also had a girth to match his height, while Lady Mary was petite in all respects. It was one of the most dramatic examples in history of opposites attracting.

Lady Mary knew full well that unless she took matters into her own hands, she would be highly unlikely ever to marry. Even if the Queen had once been prepared to consider an alliance for her, Katherine's disgrace would have changed her mind for good. Moreover, unlike her elder sister, Mary had at least chosen a husband without any political pretensions whatsoever. On his own, Keyes posed no threat to the Crown, and Mary might have thought this would be enough to win the Queen's forgiveness—even approval—for the match. But she, like Katherine, had gravely miscalculated.

On August 10 or 12, 1565, after the wedding at court of his kinsman

Henry Knollys, some of the household guests adjourned to Thomas Keyes's apartments to carry on the celebrations. Later that evening, after the guests had departed, Mary and Keyes seized their chance and were married in those same apartments. It has been suggested that Keyes was drunk; certainly both he and Mary must have been out of their senses to take such a risk. But Mary had at least learned enough from her elder sister's example to insure that there were several witnesses, including her new brother-in-law, Edward Keyes.

Perhaps it was this fact that caused the secret to break much quicker than it had with Katherine Grey's marriage. A little over a week later, it had reached Elizabeth's ears. "Here is an unhappy chance and monstrous," reported William Cecil, "the Sergeant Porter, being the biggest gentleman in the Court, hath married secretly the Lady Mary Grey, the least of all the Court."[110] It was the talk of London that Mary should have so "forgotten herself in her hasty concluding of marriage."[111] But while most saw it as a huge joke, Elizabeth was consumed with rage. "They say the Queen is very much annoyed and grieved thereat," reported the Spanish ambassador, while Cecil noted ominously: "The offence is very great."[112] Far from seeing it as a harmless, even comic union, she was convinced that this was yet another plot by a Grey sister to take the throne. She railed that she would have no bastard of Keyes succeeding her.

The Queen's sergeant porter was thrown into Fleet prison, while Mary was committed to the custody of Mr. William Hawtrey at his Buckinghamshire house, Chequers.[113] All the witnesses of their "unequal match" were arrested and interrogated. Elizabeth put pressure on Archbishop Grindal to declare the marriage unlawful, but he refused, much to her annoyance. At least she could prevent the newlyweds from producing any children—she had learned her lesson from the fiasco with Katherine Grey. She therefore insured that they would be kept forever separate.

Elizabeth seemed to have a lot of prisoners to deal with that summer. The Countess of Lennox was not taking kindly to being incarcerated

in the Tower a second time. She bombarded the Queen and her council with furious letters, insisting that she had "not deserved" such treatment and demanding to be released. Elizabeth was said to be "greatly
offended" by one such letter, and this no doubt strengthened her resolve to keep Margaret where she was.[114] Besides, her cousin was the
only bait she had to entice the Earl of Lennox and Darnley back to
England. Far from contemplating Lady Margaret's release, she had all
her properties confiscated and searched. Her acquaintances were also
kept under close surveillance, and it was found that in her absence they
had been acting as informants to Mary, Queen of Scots.

On August 12, 1565, just days after the countess had been incarcerated in the Tower, Mary, Queen of Scots, urged Elizabeth to set "hir
moder in law the Lady Margaret" at liberty, considering that she was
"alsua [also] sa [so] tender of blude to hir majestie." She was still insisting upon this the following February, when she appealed to her
English cousin's better nature by claiming that she had heard that Lady
Margaret's health was suffering from her "strict imprisonment." Mary
also questioned "whether she merits punishment for desiring the welfare of her child"—not an argument that was likely to win favor with
the Queen.[115]

It is to Mary's credit that she continued to plead for her mother-in-
law's release, even though her marriage to Lord Darnley had already
started to unravel. Arrogant, feckless, and vain, he was a singularly unsuitable choice as consort. He was crowned by Mary shortly after their
marriage, and although he was still only a king consort, this newfound
status gave him an even more inflated sense of his own importance.
Rather than making conciliatory moves toward Moray and the Hamiltons, he seemed set upon doing everything he could to inflame their
hatred of him still further. Yet initially, at least, Mary remained blind to
his faults. Randolph reported: "This Queen in her love is so transported, and he grown so proud that to all honest men he is intolerable."[116] Mary had been little more than a child during her first
marriage; now, a fully grown and very desirable woman, she succumbed to her sexual impulses with reckless abandon, refusing to
heed the trouble her husband was causing. To her, he was "the lustiest

and best proportioned long man that she had ever seen."[117] By contrast, a growing number of her subjects could see him for the worthless wretch that he was. Even Sir James Melville, who was usually effusive in his praise of Mary, decried: "No woman of spirit would make choice of such a man, who more resembled a woman than a man. For he was handsome, beardless and lady-faced."[118]

But even Mary, passionate, romantic, and in love though she was, could not long ignore her new husband's increasingly volatile behavior. When he turned his cruelty and aggression upon her, she realized with horror what a mistake she had made in marrying him. Within months of the wedding, Elizabeth's ambassador in Scotland reported: "I know now for certain that this Queen repenteth her marriage, that she hateth Darnley and all his kin."[119] But it was too late. Marriage was a virtually unbreakable bond for women in the sixteenth century. There was also a more compelling reason for Mary to stay with her husband: by October 1565, she knew herself to be pregnant.

How Elizabeth received the news of this is not recorded in the surviving sources. She would certainly have been all too conscious that without any heirs from her own body, this child would have a strong claim to her throne. Darnley appeared to have no such awareness, and showed an ever-greater contempt for his pregnant wife. This reached a horrifying climax when, in March 1566, he executed a plot to assassinate her Italian secretary, David Rizzio, whose influence with the Queen of Scots he bitterly resented. As Mary was conversing with Rizzio at Holyrood Palace on March 7, she was aghast when Darnley and a group of armed men suddenly burst into the chamber. Before she knew how to act, she was seized by her husband and forced to look on in horror as he ordered his men to assassinate her secretary. Rizzio was dragged screaming into the adjoining room and stabbed to death. Mary was now a prisoner of her own husband.

Reacting with a courage and resourcefulness that won her the admiration of many—including her cousin Elizabeth—the heavily pregnant Queen of Scots staged a daring midnight escape to Dunbar after four days of captivity. There, with the help of the Earl of Bothwell, she assembled an army and marched upon Edinburgh, immediately seizing back power. Most of the rebels fled to England, while Darnley remained in deep disgrace with his subjects. Mary now loathed him with an intensity that far exceeded her earlier love.

At first, Elizabeth seemed disposed to believe the stories relayed by the plotters who had fled to England. Mary wrote a furious letter, marvelling that "she credits the false speaking of her unworthy subjects, whom she will hereafter know never deserved her favor or assistance to their mischievous enterprises." She went on to condemn their murder of her "most special servant" and their treasonable actions toward her, and demanded to know if her cousin was minded "to support them against her as she boasts, for she is not so disprovided but that other Princes will help her to defend her realm."[120] Perhaps this latter consideration prompted Elizabeth to change her stance, for her attitude rapidly underwent a complete transformation. She took to wearing her cousin's miniature portrait on a chain about her waist, clearly visible to everyone at court. She also proclaimed: "Had I been in Queen Mary's place, I would have taken my husband's dagger and stabbed him with it."[121] But this apparent display of female solidarity did not last, and before long, relations between the two women were as testy as they had been before.

In April 1566, Mary took up residence at Edinburgh Castle in order to await the birth of her child, an event that took place some two months later. After a long and difficult labor, she was delivered of a son, Prince James. The news that her greatest rival had given birth to a male heir was said to have thrown Elizabeth into a deep depression. According to Melville, the Queen was dancing and "in great mirth" after supper when Cecil, who had received a note from the Scottish ambassador, discreetly whispered its contents to her. At once "all her mirth was laid aside for that night; all present marvelling whence proceeded such a change; for the Queen did sit down, putting her hand under her cheek, bursting out to some of her ladies, that the Queen of Scots was mother of a fair son, while she was but a barren stock." The following day, Sir James went in person to Greenwich with his brother, Robert. In his memoirs, he noted that they "were met by some friends who told us how sorrowful her Majesty was at my news; but that she had been advised to show a glad and cheerful countenance; which she did, in her best apparel, saying, that the joyful news of the Queen her sister's delivery of a fair son . . . had recovered her out of a heavy sickness which she had lain under for fifteen days."[122]

Melville's account is not corroborated by any other source. Indeed, Philip II's envoy reported that "the Queen seemed very glad of the birth of the infant."[123] Given her notorious jealousy of her cousin, however, it is likely that for all her declarations to the contrary, she was bitterly aggrieved by the news that Mary had borne a son. She had good reason to be so politically, for the arrival of a male heir created a dangerous new dimension to England's relations with Scotland. But her resentment may have gone deeper than that. At the age of thirty-three, and with no apparent prospect of marrying or begetting an heir herself, Elizabeth was intensely jealous of her younger, more beautiful cousin, who was already on her second husband and had proven her fertility by becoming pregnant almost immediately after the marriage. She had given in to her desires and been rewarded for it. Meanwhile, Elizabeth had fought a long battle with her own feelings, eventually conceding—with great reluctance—that she must sacrifice her love for Robert Dudley to the interests of her crown. And for what gain? Little but ongoing hostility from both within and outside her kingdom, together with the ever-present threat of invasion by a Catholic potentate. Small wonder that she felt so bitter toward Mary.

The Scottish queen's insensitivity did not help matters. Even before the birth, she had taken every opportunity to gloat about her own fertility. In April 1566, she had sent a letter to Elizabeth, excusing her poor handwriting, which she claimed was due to her being "in her seventh month."[124] In fact, Mary's handwriting was always rather untidy, because she had none of Elizabeth's reverence for the written word and tended to scrawl her letters in haste. To use the rather dubious excuse of pregnancy was therefore more likely intended to provoke her cousin's jealousy. How much greater was Mary's pride when she gave birth to a male heir. She had fulfilled her duty as a woman and a queen. In the eyes of many of her contemporaries, Elizabeth had failed on both counts.

As a gesture of courtesy, Mary invited her cousin to be godmother to the new prince. Elizabeth made a show of being flattered, and sent the Queen of Scots a lavish gold font for the christening. Mary was apparently so enraptured upon receiving it that she declared that if she were to die, she wished her cousin to raise the boy. But Elizabeth's true feelings were soon revealed, when Melville suggested that acting as James's godmother would provide her with an ideal excuse to visit Mary in person—something that she had long claimed was her dearest wish. "Whereat she smiled, saying she wished that her estate and affairs might permit her."[125] Realizing that she had no intention of doing so, Melville urged her to appoint either William Cecil or Robert Dudley, her two closest advisers, to make the journey on her behalf. But again Elizabeth demurred, refusing to give him a direct answer. Eventually the Earl of Bedford was appointed for the task. This may have been a studied insult on the part of his royal mistress, for he was one of the staunchest Protestants at court and therefore rather unsuited to attend the christening of a Catholic prince.[126]

The birth of a male heir to the Queen of Scots intensified the pressure on Elizabeth to name her successor. By now, both her council and Parliament were fiercely divided between the supporters of Mary Stuart and Katherine Grey. "Although the Scotch Queen has a large party in the House of Lords, it is thought that Catherine would have nearly all the members of the lower Chamber on her side," reported the Spanish ambassador, de Silva. "It seems, therefore, that everything tends to disturbance."[127]

Mary was unable to push home her advantage, for she was beset with the increasingly volatile situation in her own country. The problem of what to do with her troublesome husband was becoming ever more acute. Hatred of Lord Darnley among the people, the nobles, and above all Mary herself had reached fever pitch. There was talk of putting him on trial for the murder of Rizzio and the imprisonment of the Queen, and also of using his fiendish practices as an excuse to secure a divorce from his wife. Mary sought the counsel of various Scottish lords, including Moray, Argyll, and Maitland, as well as the man with whom she was forming an increasingly close attachment: James

Hepburn, Earl of Bothwell. It was later claimed that among the solutions they discussed was to murder Lord Darnley. Unless new evidence comes to light, this can never be proved beyond doubt. However, subsequent events suggest that Mary was at least aware of, if not an active proponent for, plots to do away with her estranged husband for good.

However, in January 1567, she seemed to contemplate a reconciliation with Darnley. Upon receiving news that he was gravely ill at his father's house in Glasgow, she went to be with him.[128] Once there, she managed to persuade him to come home to Edinburgh, where he could convalesce in greater comfort, adding the promise that she would resume marital relations with him if he acceded. Her husband agreed to accompany her back to Edinburgh, and took up residence at the house of Kirk o'Field. It would prove a fateful move.

In the early hours of February 10, 1567, the citizens of Edinburgh were awoken by an almighty explosion. In the confusion that followed, it was discovered that Kirk o'Field had been blown up by a huge quantity of gunpowder. Although there were remarkably few casualties, two bodies were subsequently found on the grounds of the house. They were those of Lord Darnley and his servant. Neither, however, had been killed by the blast; they had been strangled or suffocated. This shocking news spread like wildfire across Scotland and throughout the courts of Europe. It was widely expected that Mary would hunt down her husband's killers and bring them to swift and brutal justice. But the Scottish lords had no such expectations. For them, Darnley's death was justice enough for all the wrongs he had committed.

As Mary procrastinated, suspicions about her involvement in the plot began to be voiced. For once, her English cousin showed unequivocal support and declared that she "saw no cause to conceive an ill opinion of her good sister of Scotland."[129] She then wrote an impassioned letter to her cousin, expressing her horror at the news and assuring her of her support. "My ears have been so deafened and my understanding so grieved and my heart so affrighted to hear the dreadful news of the abominable murder of your husband and my killed cousin that I can scarcely yet have the wits to write about it," she began. "I cannot dissemble that I am more sorrowful for you than for him."

Elizabeth went on to give such sound advice that there can be little doubt her concern was genuine. Assuring Mary that "you have no wiser counselors than myself," she entreated her: "O madame, I would not do the office of faithful cousin or affectionate friend if I studied rather to please your ears than employed myself in preserving your honour. However, I will not at all dissemble what most people are talking about: which is that you will look through your fingers at the revenging of this deed, and that you do not take measures that touch those who have done as you wished, as if the thing had been entrusted in a way that the murderers felt assurance in doing it." Elizabeth had evidently also heard the rumors about Mary's friendship with Bothwell, for she continued: "I exhort you, I counsel you, and I beseech you to take this thing so much to heart that you will not fear to touch even him whom you have nearest to you if the thing touches him . . . Praying the Creator to give you the grace to recognise this traitor and protect yourself from him as from the ministers of Satan." Fearing lest her cousin might suspect her motives, she assured her: "I wish as much good [to you] as my heart is able to imagine or as you were able a short while ago to wish," and ended by expressing "heartfelt recommendations to you, very dear sister."[130] This extraordinary letter was the first demonstration of genuine support by Elizabeth for her great rival. It had apparently taken this extreme crisis in Mary's life to inspire a true sense of solidarity with a fellow female sovereign. But it would prove all too short lived. The shocking events that unfolded soon afterward would be enough to make Elizabeth resort to her former hostility.

Kept under strict surveillance in the Tower and deprived of letters and messages, it was some time before the Countess of Lennox learned of the horrific news from Scotland. Just a few months earlier, she had been greatly cheered by the announcement of her grandson's birth. Margaret's ambitions for the crown of England now seemed closer than ever. Then came the news that brought her world crashing down around her. Her beloved son Henry had been murdered. Taking pity on her cousin, Elizabeth had asked William Cecil's wife and Margaret's old friend Lady Mary Howard to break the news to her. No matter how gently they might have done so, the full horror of it threw Margaret into a fit of grief so extreme that a physician had to be summoned. Even then she "could not by any means be kept from such passion of mind as the horribleness of the fact did require."[131]

The Queen ordered that the beleaguered countess be released from the Tower and conveyed to Sackville House, the site of her former imprisonment, to be looked after by Lord Dacre and Lady Sackville. Before long, grief had turned to fury, and Margaret railed against her daughter-in-law, accusing her of Darnley's murder. To have her main rival so discredited played right into Elizabeth's hands. She now showed great favor to the Lennoxes, restoring to them the income from their estates and granting them the (somewhat dilapidated) royal palace of Coldharbour in London as a residence. More important, she promised to help them to avenge Darnley's death. "She seems very sorry for their troubles," observed the Spanish envoy.[132]

There would be no such clemency for the Queen's other cousin, Lady Mary Grey. She had languished for two years at Chequers, sending frequent pleas that the Queen might forgive her. At the beginning of 1566, she wrote to William Cecil in her neat, childlike handwriting, apologizing "for trublynge you so oftenn withe my rude letters," but went on to bemoan "what a greffe the quenes maiestes desplessur is to me, which makes me to wyshe deathe rather thenn to be in thes greatte messery." She therefore begged him "to be a contenewall meane for me to her maieste, to gett me her maiestes favor agayen; trustynge if I myghte ons obtayne it never to forgo it, whill I lyve."[133]

In August 1567, Mary was sent to live in the custody of her stepgrandmother, Katherine Willoughby, Dowager Duchess of Suffolk. Although she had been kind to Mary in the past, this formidable matriarch was hardly pleased to see her now, especially as she had not been told of her arrival in advance. She was also incensed at the girl for bringing the Grey family into disrepute again. But the longer Mary was with her, the more the duchess pitied this poor creature, who cried for most of the day and began to refuse food. "All she hath eaten now these two days is not so much as a chicken's leg," the Duchess told Cecil, adding: "I fear me she will die of her grief."[134]

But Elizabeth had more pressing matters to attend to, for events in

Scotland were unfolding at a bewildering pace. In late April 1567, Mary went to visit her infant son at Stirling Castle. On her return a few days later, she and her entourage were intercepted by the Earl of Bothwell and a large troop of horsemen. Bothwell, who had already determined to have Mary as his wife, duly abducted her and took her captive to Dunbar. Quite what took place there has been the subject of intense debate ever since. Although some believe that Mary was already in love with Bothwell by this time, the evidence suggests that he raped her and that, as Melville observed: "the Queen could not but marry him, seeing he had ravished her and lain with her against her will."[135] Bothwell swiftly divorced his existing wife, and on May 6 he brought Mary back to Edinburgh, where she declared her intention to marry him. The ceremony took place on May 15 at the Palace of Holyroodhouse.

Mary's marriage to Bothwell was a telling indication of her deeply conventional views of queenship. She believed that the only way for a woman to rule effectively was by submitting to the direction of male advisers. Sir Nicholas Throckmorton once remarked that Mary had no confidence in her own intellectual ability but was content "to be ruled by good counsel and wise men."[136] While Elizabeth used her skill in polemic to face down opposition from her councillors, Mary would sit in council meetings quietly sewing as her advisers debated the issues at hand. Emotions aside, one of her main motivations in seeking a husband was to have someone to share the burden of rule. She justified her latest marriage on the basis that the unrest within her country "cannot be contained in order unless our authority be assisted and set forth by the fortification of a man."[137] The man she had chosen was far from being an equal partner: domineering, aggressive, and deeply scornful of women, he had forced her to submit meekly to his will.

The fact that Mary had married Bothwell, one of the prime suspects in the murder of Lord Darnley, spelled disaster for her rule in Scotland. The earl soon alienated those members of the political establishment who had promised to support the marriage, and they now joined a large confederacy of opponents to Mary's regime, which succeeded in occupying Edinburgh and taking over the Privy Council. Mary was forced to flee with her new husband to Borthwick Castle

and then Dunbar, where they were able to amass an army of supporters. On June 15, they confronted the army of the confederate lords at
Carberry Hill. As the battle wore on, the ranks of Mary's supporters
steadily dwindled, and she was eventually forced to surrender and face
imprisonment again. Bothwell, meanwhile, fled into exile.

Holed up in the island fortress of Lochleven, Mary faced the worst
crisis of her young life. Ousted from her capital—and, she feared, her
throne—she waited anxiously to hear what her fate would be. There
was talk of enforced abdication, exile, life imprisonment, trial for murder, even execution. Meanwhile, she was denounced as "the most notorious whore in all the world."[138]

Learning of the dramatic events that were taking place almost daily
in Scotland, Elizabeth felt a mixture of shock, bewilderment, and satisfaction. Suddenly, the fortunes of the two women had been reversed.
Not so long ago, she herself had been the heretical whore, daughter of
the concubine and lover of Robert Dudley—and no doubt many more
besides. By contrast, Mary had been the flawless, model princess,
chaste in her young widowhood and devout in her religion. Now Elizabeth had the moral high ground, and she was determined to enjoy it.
Adopting the style of a sternly virtuous matriarch, she expressed her
horror at her cousin's actions, telling her ambassador in Scotland that
she had a "great misliking of that Queen's doing, which now she doth
so much detest that she is ashamed of her."[139] She also wrote again to
Mary, admonishing her for not following her advice. "Madame, to be
plain with you, our grief hath not been small that in this your marriage
so slender consideration hath been had . . . for how could a worse
choice be made for your honour than in such haste to marry such a
subject, who besides other, and notorious lacks, public fame hath
charged with the murder of your late husband, beside the touching of
yourself also in some part, though we trust that in that behalf falsely.
And with what peril have you married him that hath another lawful
wife alive."[140]

Elizabeth realized, however, that she could not simply berate her
cousin and sit back to enjoy the smug satisfaction of being proved right.
Mary's predicament had thrown her own position into jeopardy, for if

the anointed queen of the neighboring kingdom could be so easily overthrown by her subjects, what message would this send to her own people—especially those recalcitrant Catholics who had opposed her rule? She was therefore obliged to offer Mary support as a fellow sovereign. That she did so out of self-interest rather than concern for her cousin's welfare, there can be little doubt. Nevertheless, her promise to Mary seemed genuine enough: "We assure you that whatsoever we can imagine meet for your honour and safety that shall lie in our power, we will perform the same that it shall well appear you have a good neighbour, a dear sister, and a faithful friend, and so shall you undoubtedly always find and prove us to be indeed towards you."[141]

The English queen was as good as her word. She alone stood out in her support for Mary among all the potentates of Europe. Even Mary's natural ally, France, proved unwilling to lend assistance, and her former mother-in-law, Catherine de Medici, was coldly disapproving. Elizabeth, on the other hand, defied opposition from among her own council in order to advocate Mary's rights as an anointed queen. She expressed her outrage that her cousin's subjects should dare to treat her with so little respect, and called for her immediate release. She also dispatched Sir Nicholas Throckmorton to Scotland with orders to demand Mary's restoration. But at the same time as championing her cousin's cause, Elizabeth also purchased some of her jewels, which were being sold off by the recalcitrant lords. These included a beautiful rope of pearls, which the English queen proudly showed off to her courtiers.

Apparently abandoned by all her former allies, Mary remained in captivity. On July 24, she was presented with the deeds of abdication and told that she must sign or face death. Desperately sick, having recently miscarried Bothwell's twins, she miserably accepted that she had little choice but to give up her throne. It was agreed that her son would be made king and that a regency council would rule during his minority. Meanwhile, Mary remained a prisoner at Lochleven. It was said by some that she owed her life to Elizabeth, whose lobbying had done just enough to warn off the rebellious lords from committing murder. If this was so, the events that followed some twenty years later would make it deeply ironic.

For once, it seemed that things were moving in Elizabeth's favor as far as her troublesome cousins were concerned. In January 1568, she received news that Lady Katherine Grey was seriously ill. Katherine had been moved to several different houses after leaving the Tower in 1563 and was now at Cockfield Hall in Yoxford, a bleak and remote estate in the Suffolk countryside with few amenities or comforts. Pining for her husband and despairing of ever being allowed her freedom, her health had begun to decline. From the very beginning of her separation from Seymour, she had lapsed into a deep melancholy, refusing to eat. Her first custodian, Lord John Grey, had been shocked by her wan appearance and had entreated her to take some food, but she had refused, saying: "Alas, Uncle, what a life is this to me, thus to live in the Queen's displeasure. But for my Lord and children I would I were buried."[142]

As the years of her imprisonment dragged on, Katherine continued to eat little, and in her weakened state she fell dangerously ill. Elizabeth was persuaded to send her physician, Dr. Symondes, to attend her, and when he arrived at Cockfield Hall, he realized at once that she was dying. Among the various household members who gathered around Katherine's bed early in the morning of January 27, 1568, was a gentleman who recorded her final hours. According to his account, she seemed to welcome death, glad to be free at last from the pain and sorrow of her existence. When one attendant tried to comfort her by saying that she would live for many years, she replied: "No, no. No lyfe in this worlde; but in the worlde to come I hope to lyve for ever. For here ys nothinge but care and myserye." She then turned to her keeper, Sir Owen Hopton, and joyfully exclaimed: "Even now [I am] goynge towards God as faste as I canne." Eager though she was for death, Katherine was still anxious to protect her children from the Queen's wrath and therefore begged her forgiveness for any wrongs done, professing herself to be a loyal subject. She urged Hopton to present her plea to Elizabeth "even from the mowthe of a dead woman" and ask that she "will be good unto my children and not impute my fawte unto them whom I wholy give unto her Maiestie. For in my lyfe they had fewe frends, and fewer they shall have when I am dede except her Maiestie." She went on to plead the same clemency for her husband, "for I knowe this my deathe wilbe heavy newe[s] unto him, that her Highnes wolde be so gracyous as to send him libertie to glad his sorrowful hart withal."

After delivering this touching speech, Katherine asked for the box containing her wedding ring, and entreated Sir Owen to deliver it to Seymour as a mark of her affection and to urge him to be "a lovinge and a naturall father unto my children." Then, looking down at her hands and seeing her nails were already turning purple, she cried: "Look you, here he ys come." A few moments later, "with a cherefull countenance," she said: "Welcome, Death," and, offering her soul up to Jesus, breathed her last.[143] She was just twenty-eight years old.

If this account was written with the intention of exciting Elizabeth's compassion, it did not succeed. While she made a show of regret upon hearing of Katherine's death, her overriding emotion was one of relief. "The Queen expressed sorrow to me at her death," remarked de Silva, "but it is not believed that she feels it, as she was afraid of her."[144]

Katherine had maintained the validity of her marriage right up until her death, and in her will she left the "necessary declarations" to prove it. Elizabeth would never countenance the idea of overturning her commission's verdict, thereby placing Katherine's sons in the order of succession. Even so, the boys remained a focus of attention for those who had favored her mother's claim, and their names would continue to plague the Queen during the years to come. Hounded as she had been by Elizabeth in life, Katherine may have derived some satisfaction from knowing that after her death, her sons would be a constant source of anxiety to the Queen. She would have been more satisfied still if she had known that her marriage to Seymour would eventually be found to be valid and her sons legitimate.[145] Raised by their father, who was released after Katherine's death, both boys were taught to honor her memory, and the elder evidently passed this on to his own son. Almost a century after his grandmother's death, he ordered that her bones be removed from the humble churchyard of Yoxford and reinterred with those of her husband in a magnificent baroque monument in Salisbury Cathedral.

With Katherine out of the way, Elizabeth once more turned her attention to Scotland, where her cousin Mary was beginning to recover her health. At the beginning of May 1568, she was strong enough to orchestrate a daring escape from the castle, assisted by a powerful force of supporters, including the Earl of Argyll, who had defected from the new regime. Together they headed for the stronghold of Dumbarton, and a battle ensued at Langside near Glasgow on May 13. Once again Mary's forces were easily defeated, and together with a small band of men, she fled south to Dumfries. Realizing that to turn back would almost certainly mean death, the beleaguered Queen of Scots made the fateful decision to go to England and throw herself upon the mercy of her cousin.

CHAPTER 10

Faithful Servants

The sudden arrival of Mary, Queen of Scots, in England threw Elizabeth into a quandary. Now, more than ever, she needed the support of her companions at court. And yet the closest of these had died some three years before. Kat Astley had been Elizabeth's chief councillor for almost thirty years and had shared all her most intimate secrets. She was the only woman who dared to speak plainly to the Queen, and even Cecil could not match her level of frankness. That she could thus address her royal mistress was due to the fact that Elizabeth knew she loved her more than any other, and time and again she had proved her unflinching loyalty. Her death therefore came as an appalling shock to the Queen—even more so because it was so unexpected.

In early July 1565, Elizabeth's former governess fell ill, but her life was not thought to be in any great danger. She died a few days later, on July 18. "Her Majesty went to see her the day before and I am told she is greatly grieved," reported the Spanish ambassador. When his master, Philip II, received this letter, he added his own comment to it, writing in the margin: "What a heretic she was!"[1] News of her death soon spread throughout the kingdom and overseas. For most, their immediate concern was who else to cultivate as an intermediary now that Elizabeth's closest attendant was dead. Hugh Fitzwilliam, an English agent at the French court, lamented: "now that Mistress Ashley is gone he has no friend about her to make his moan to."[2]

For the Queen, Kat Astley's death went far beyond mere inconvenience. She had lost the woman who had known her better than any other and who had been her most loyal friend, adviser, and confidante. She had stepped into the void created by the death of Elizabeth's mother and the neglect of her father, and had shown her more love, devotion, and loyalty than any other person living. Only death had been able to break their otherwise impregnable emotional bond— a bond that had been as strong as that between mother and daughter. Elizabeth was inconsolable in her grief and kept to her chamber for days, suspending all official business. If she had recalled her words to Kat when defending her desire to spend so much time with Robert Dudley, she might have reflected how much more appropriate they were now. Suddenly her life seemed filled with "so much sorrow and tribulation and so little joy."

Elizabeth lost another of her childhood companions four years later. Her cousin Katherine Knollys had been an important member of her entourage ever since the accession. As well as attending the Queen's person, Katherine had also been given special responsibility for the various gifts with which her royal mistress was presented. These were often unusual as well as expensive, notably the "chained monkey" that Elizabeth was once given as a New Year's present.[3] Throughout her service, Katherine was said to be "In favour with our noble queen, above the common sort."[4] As a sign of this esteem, the Queen appointed a number of Katherine's children to places in the household, including her precocious and attractive daughter Lettice. Meanwhile, her eldest son's wedding in 1565 was marked by a splendid court tournament.

But all of this had come at a price. It was as if Elizabeth felt that she had fulfilled her part of the bargain by giving Katherine a much-coveted place in her household and honoring her family. Now she looked to her cousin to prove how faithfully she could serve her in return. She made it clear that she expected Lady Knollys to put her needs above all others, including those of her many children. So demanding was her royal mistress that Katherine would "often weep for unkindness." As well as having to subordinate her children's welfare to that of the Queen, she also had to endure long absences from her husband, the courtier Francis Knollys, whom Elizabeth dispatched on various military and diplomatic missions at home and abroad. These included taking on the guardianship of Mary, Queen of Scots, upon her arrival in England.

Worn down by years of unstintingly selfless attendance upon the Queen, in late 1568 Katherine fell gravely ill at Hampton Court, where the court had taken up residence for the festive season that year. Hearing of this, Sir Francis repeatedly begged to be allowed to visit her and complained bitterly at Elizabeth's "ungrateful denial of my coming to the court." Neither was he appeased by a letter from William Cecil assuring him that his wife was "well amended." This merely prompted him to write to Katherine expressing an intense desire for them to retire from service and live "a country poor life."[5]

This wish, albeit whimsical, would never be fulfilled. Katherine's health deteriorated so rapidly that her royal mistress became greatly alarmed. "She was very often visited by her Majestie's owne comfortable presence," remarked Nicholas White, a protégé of William Cecil. Although Lady Knollys was no doubt consoled by this sign of Elizabeth's affection, it was for her husband that she pined. Without his comforting presence, she slipped further into decline and died on January 15, 1569, aged just forty-six. The Queen was thrown into such "passions of grief . . . for the deathe of her kinswoman and goode servant," that she "fell for a while from a prince wanting nothing in this world to private mourning, in which solitary estate being forgetfull of her owne helthe, she tooke colde, wherwith she was much troubled."[6]

Upon a visit to Elizabeth shortly afterward, Nicholas White noted that she was still beset by grief and would hardly be drawn to any other subject than that of her beloved servant and kinswoman. "From this she returned back agayne to talk of my Lady Knollys. And after many speeches past to and fro of that gentilwoman, I, perceyving her to harpe much upon her departure, sayd, that the long absence of her husband . . . together with the fervency of her fever, did greatly further her end, wanting nothing els that either art of man's helpe could devise for her recovery, lying in a prince's court nere her person." This rather untactful remark was followed by another: "Although her Grace were not culpable of this accident [Katherine's death], yet she was the cause without which their being asunder had not hapned." The Queen replied disconsolately that she was "very sory for her deathe."[7]

In her grief, Elizabeth was able to turn to her old servant Blanche Parry, who had replaced Kat Astley as chief gentlewoman of her privy chamber. After the loss of two of the Queen's closest attendants, Blanche represented a precious link with the past. She had given her royal mistress more than thirty years of unfailing loyalty and diligent service. None of the Queen's other ladies so richly deserved this promotion. Blanche was quick to grasp the full implications of her new role, for it made her one of the most powerful ladies at court. She now had the right of admittance to every room of the Queen's most private suite. She was with her mistress constantly, from when she rose in the morning, to her mealtimes and leisure hours during the day and her evening entertainments. Her access was unrivalled: She often spent hours alone with the Queen when no other soul was admitted—not even her most trusted advisers. With Blanche, Elizabeth could discuss matters of business, exchange gossip about courtiers, and share her most private feelings. She knew from long experience that her oldest gentlewoman would always prove discreet, trustworthy, and honest— an ideal confidante.

As well as the private moments with the Queen, Blanche was often also in attendance when her mistress received ambassadors or ministers and was able to listen in when important matters of state were being discussed. She was, as her biographer has claimed, "at the apex of the power structure of the Elizabethan Court."[8] The rest of the court was well aware of the closeness of their relationship, and

Blanche's influence there grew considerably as a result. Her nephew, Rowland Vaughan, although embittered against her, admitted that the court was "under the command of Mistress Blanche Parry."[9]

Even at the height of his power, Robert Dudley relied upon Mistress Blanche to intervene with Elizabeth on his behalf. In March 1566, he annoyed his royal mistress by absenting himself from court because of her prevarication over a grant of land. Blanche warned his agent, John Dudley, that the Queen "much marvelled she had not heard from you [Robert] since last Monday." She had tried to excuse Sir Robert's behavior, assuring her mistress that he would return to court with all haste. But she urged John to insure that his master wrote to the Queen before coming back into her presence; otherwise he might encounter a frosty reception. John pressed Sir Robert accordingly, assuring him that with their other ally among Elizabeth's ladies, Dorothy Bradbelt, being absent, "our best friend in the Privy Chamber is Mrs Blanche."[10]

Blanche's selfless devotion served as a benchmark against which all of Elizabeth's future servants would be assessed, including the two new additions who were recruited at around this time. The first was Mary Radcliffe, who entered the Queen's service as a maid of honor in 1564 at the age of fourteen. The story goes that her father, Sir Humphrey Radcliffe of Elstow, Bedfordshire, came to court at the end of 1560, and upon being presented to the Queen, offered his daughter to her as a New Year's gift when she came of age. Although this was an audacious circumvention of the traditional rules of appointment, Elizabeth was amused and agreed to accept her "gift." She did not regret her decision, for Mary soon proved to be an asset to her household, carrying out her duties with humble assiduity.

The Queen's favor toward Mistress Radcliffe soon became apparent. She paid her a stipend of £40 per year—a very generous sum, given that maids of honor were usually funded entirely by their families. As well as appreciating her conscientiousness, the Queen was impressed by Mary's apparent resolve to remain a virgin so that nothing would distract her from her duties. This caused a great deal of disappointment among the gentlemen of the court, for Mary was something of a beauty and had many admirers. One of them, John

Farnham, described her as being "as comely as ever" in a letter to his friend Roger Manners. She also had an impeccable sense of style. Sir Robert Sidney's agent at court, Rowland Whyte, wrote admiringly to his master: "Yesterday did Mrs Ratcliffe weare a whyt satten gown, all embroddered richly cutt upon good cloth of silver."[11]

Mary Radcliffe would go on to serve Elizabeth for the rest of her reign. The same was true of Helena Snakenborg, who was the only foreigner among the Queen's ladies. She was born into an ancient Swedish baronial family, and her father, Ulf Henriksson, was one of the late king of Sweden's most trusted supporters. In 1564, at the age of fifteen, Helena was appointed a maid of honor to Princess Cecilia, sister of the new king, Eric XIV, who had recently been a suitor to Queen Elizabeth. Later that year, Cecilia proposed to make a voyage to England. A staunch Anglophile, she claimed that she wished to make the acquaintance of the English queen, but it was rumored that she planned to revive the suit of her brother, Eric, to marry her.

The princess and her suite of ladies, which included the young Helena, duly embarked for England but were obliged to take such a round-about route in order to avoid hostile countries that it took them almost a year to reach their destination. During their journey, on which Cecilia was accompanied by her husband, she became pregnant, which must have made the passage even more uncomfortable.

Their arrival in London in September 1565 caused a great deal of excitement at court. De Silva, the Spanish ambassador, reported it to his master, Philip II: "On the 11th instant [of September] the king of Sweden's sister entered London at 2 o'clock in the afternoon. She is very far advanced in pregnancy, and was dressed in a black velvet robe with a mantle of black cloth of silver, and wore on her head a golden crown . . . She had with her six Ladies dressed in crimson taffety with mantles of the same."[12] The Queen gave them a lavish reception, which Helena Snakenborg described in a letter to her mother. "There came so many to visit us that there was no end to it. All wished us a hearty welcome to England."[13] Being so near her time, her mistress was conveyed at once to the place assigned for her confinement. Meanwhile, Helena and the other maids were besieged by a host of ladies and gentlemen who were dispatched by Elizabeth at regular intervals to inquire after Cecilia's health. The Queen herself paid a visit on September 14, and the following day, the princess gave birth to a son.[14]

Cecilia did not remain long in confinement after the birth, and she and her entourage became regular visitors to the court, as Elizabeth staged lavish receptions and entertainments in their honor. Although people were fascinated to see the Swedish princess and her new baby, the real star of the show was Helena. At sixteen years of age, she was blossoming into a beauty, with large brown eyes, naturally curly red hair, and a flawless pink and white complexion. Among her many admirers was Katherine Parr's brother, William, Marquess of Northampton. Some thirty-five years her senior, he was besotted from the moment he first saw her in Cecilia's apartments. Helena described their meeting in a letter to her mother: "Amongst the gentleman was a courtier who always came with the earliest arrivals and left amongst the last. When my gracious Lady had been 'churched' after the baby was born, the Marquess of Northampton (for that was the courtier's name) talked to my gracious lady about me."[15]

The marquess wasted no time in pressing his suit. He showered Helena with extravagant gifts, including richly embroidered clothes and priceless jewels. Parr was an experienced suitor, with charming, courtly manners. William Camden described him as: "A man very well practised in the pleasanter sort of studies, as Musicke, Love-toyes, and other Courtly dalliances."[16] It seems that Helena was beguiled by this cultured and affable older man, and she took all his protestations of love seriously, even though more cynical observers thought that he was merely toying with her. After all, she had little but her beauty to recommend her, for she brought with her no dowry. However, Parr's love proved to be true. A divorcee of fifty-two with no children, he was eager to find a companion with whom to spend his later years, and he was captivated by this alluring young Swedish girl. After just a few weeks' courtship, he proposed. Helena wrote in great excitement to her mother: "I answered him to this effect, that I was a simple maid, what would his Lordship want with one not equal to him in rank? His Lordship said he did not seek for riches: if only God would give him happiness of my loving heart, he would ask no other wealth. Even if I brought nothing but my shift and gave him happiness it would be a gift from God."[17]

But the love affair seemed doomed when Helena's mistress announced her intention to return to Sweden with all haste. Cecilia had lived an extraordinarily lavish lifestyle since her arrival in England, running up debts in excess of £3,500—more than £600,000 ($960,000) in today's money. Before the Queen would allow her to leave, she was obliged to sell whatever she could to raise the funds, including her own dresses. Her husband, meanwhile, had been arrested at Rochester after trying to flee the country.

Elizabeth knew all about the courtship between one of her leading noblemen and the pretty young Swedish girl. Although Helena could so easily have sparked the Queen's notorious jealousy, Elizabeth seemed as fascinated by her as everyone else. Moreover, she did not object to William Parr's attempts to secure her as his bride; indeed, she seemed just as eager to keep Helena in England as he was. In the end, it was she who came to the lovers' rescue by offering Helena a place in her household. The young girl was overjoyed. "I can never thank God Almighty enough for the joy he has given me in a foreign land," she wrote to her mother.[18] Princess Cecilia left for Sweden shortly afterward, in April 1566, declaring that she was "glad enough to get out of this country."[19]

Elizabeth had taken a big risk in interfering in the dispute among Princess Cecilia, William Parr, and Helena Snakenborg. By insisting that the latter remain in England, she was flouting the accepted protocol of the day, whereby a sovereign had authority over her own ladies—not other people's. Given that the king of Sweden was still a contender for Elizabeth's hand in marriage, she would not have risked causing a diplomatic incident so lightly. That she should take the risk at all seems to have been entirely due to her own attachment to Helena. In a very short space of time, the latter had endeared herself to the Queen. Although a great beauty, Helena was also disarmingly modest and had a gentle, kind disposition. She was also completely enthralled by Elizabeth and tried to emulate her in dress and manner. She even copied her signature, underlining the *H* with the same elaborate flourish that Elizabeth used for her *E*. Imitation was truly the best form of flattery in this case, for the Queen was charmed by Helena's admiring attentions.

It is an indication of how much Elizabeth liked her new protégée that she also flouted the conventions for appointing ladies to her service. Helena Snakenborg had no family connections in England, and any diplomatic worth she might have had was offset by the fact that she had caused great offense to the Swedish princess by remaining in the country. The Queen appointed her a maid of honor, and as this post was usually unsalaried, awarded her a host of additional privileges, including a comfortable apartment, a body of servants, and a horse. She also asked Blanche Parry to make sure that Helena had everything she needed.

Helena proved as conscientious in her duties as the Queen had anticipated, and before long she was an indispensable member of the household. In recognition, Elizabeth later promoted her to gentlewoman of the privy chamber. This placed her at the very heart of the royal court, for she was now with the Queen constantly and controlled access to her by the throngs of courtiers desperate for an audience. She would often accompany her royal mistress when the latter gave such audiences or went on excursions outside the palace, and was one of the most beautiful ornaments of the court. Edmund Spenser extolled her virtues in a poem he wrote later in the reign, casting her as "Mansilia" and the Queen as "Cynthia." It is clear that by this time, Helena had become one of the most prominent members of Elizabeth's entourage:

> No lesse praisworthie is Mansilia,
> Best knowne by bearing up great Cynthia's traine . . .
> She is the paterne of true womanhood,
> And onely mirrhor of feminitie:
> Worthie next after Cynthia to tread,
> As she is next her in nobilitie.[20]

But while Helena's career at court had flourished, her relationship with William Parr had not run so smoothly. Although he had divorced his first wife, Lady Anne Bourchier, in 1551, he did not feel able to marry again while she was still alive, because questions had been raised about the validity of that divorce. Helena seemed to accept this with sanguinity. It no doubt helped that her suitor kept her extremely well provided for. She told her mother that he had appointed ten of his own servants to wait upon her and give her "everything I can fancy," adding: "I cannot imagine I shall ever want for anything however beautiful or expensive that his Lordship can buy without his getting it at once for his Elin."[21]

The couple were rewarded for their patience in 1571, when Anne Bourchier died. They finally married that May after a courtship that had lasted almost six years. By that time, Helena was so high in favor with the Queen that Elizabeth offered the couple her own closet at Whitehall Palace for the ceremony and further honored them by attending it herself. The fifty-seven-year-old groom and his twenty-two-year-old wife went on to enjoy a happy but all too brief marriage, dividing their time between the marquess's houses in Guildford, Surrey, and Stanstead Hall, Essex, so that Helena could still attend court regularly. At the end of September 1571, they accompanied Robert Dudley, Earl of Leicester, to Warwick, but while there, the marquess fell gravely ill. He had been plagued by gout for some time, and he now took a turn for the worse. A lodging was found for him in Warwick, and Helena nursed him faithfully for several weeks. He never recovered, dying on October 28 after being married to his beloved Helena for just five months. Elizabeth bore the whole charge of the funeral and oversaw all the arrangements, even down to providing material from her own wardrobe for the mourning garments.

Parr had left his young widow so well provided for that she could have retired to a life of comfort. But Helena was anxious to continue her service to the Queen, and therefore returned to court as soon as her late husband's affairs had been settled. Elizabeth was overjoyed to have her back, particularly as there would now be no distractions to her service as a result of domestic affairs. Helena was equally glad to be with her royal mistress once more, and before long it was as if she had never been away.

Rather less favor was shown toward Lady Mary Sidney. Despite the enormous personal sacrifice she had made by nursing Elizabeth through smallpox and being permanently disfigured by it herself, Elizabeth was slow to appreciate her virtues. Yet Mary had all the qualities that Elizabeth looked for in the ladies of her household. At heart selfless and obliging, she had a patient, uncomplaining disposition and was eager to serve her royal mistress in whatever capacity she desired.

No matter how conscientious Lady Mary was in her attendance, she was hampered by the Queen's increasingly volatile relationship with her brother, Robert Dudley. The 1570s proved a difficult time for Lady Mary at court, thanks to the various scandals in which Lord Robert was involved. Regardless of how faithful, loyal, and devoted she proved herself to be, her royal mistress seemed to scorn her efforts and find her more of an irritation than a comfort. When Lady Mary appeared at court wearing a gown made from some velvet that her husband had sent her from Ireland, the Queen immediately demanded that he send her some of the same. There was a note of hysteria in Mary's letter to her steward in July 1573, urging him to procure the material at any cost. She insisted: "You may not slake [slack] the care hearof for she will tack it ill and it is now in the wourst tyem for my lord for divers consyderacions to dislyek her for souche a trifle."[22]

The Queen gave Lady Mary scant reward for her services, and the latter soon found herself in financial difficulties. As an unsalaried member of the privy chamber, she was dependent upon annual stipends from her husband, Sir Henry Sidney. But he too was sinking further into debt, thanks to the huge sums he had had to lay out in Ireland, and now in his Welsh estates. When he was offered a barony in 1572, he could scarcely afford the expenses that went with it, and his wife therefore begged William Cecil, Lord Burghley, that he might be excused from the honor unless it was accompanied by an increase in his estate.[23]

Although unsalaried, Lady Mary was entitled to free board and lodgings at court, but even here there were complications. After a brief absence from court, she found that her accustomed rooms (which, thanks to her brother's intervention, were spacious and comfortable) had been given to somebody else, and instead she had been allocated a rather shabby apartment that lacked heat and basic furnishings. After putting up with these uncomfortable conditions for more than two years, she eventually resolved to write to the Lord Chamberlain, Thomas Radcliffe, third Earl of Sussex, who was also her husband's brother-in-law. "Her Majesty hath commanded me to come to the courght and my chamber is very cold and my owne hangings very scant and nothynge warme," she complained. Her health had been irreparably damaged by the smallpox, and her living conditions at court had aggravated her various ailments. She told Sussex that her "great extreamyty of syknes" meant that she dare not "aventure to lye in so cold a lodginge without some forther healpe." She might have reflected just how much things had changed: In her brother's heyday, she had not only had luxurious rooms herself but also had been able to procure others for her friends and family.[24] Now she was forced to humbly beg for "3 or 4 lyned peacis of hangings" to keep out the drafts, and assured Sussex that as soon as the weather turned warmer, "they shalbe safely delivered agayne. And I shall think my sealfe most bownde unto yow yt your pleasure be to shew me this favour."[25]

Lady Sidney's pleas fell upon deaf ears. Eventually, worn down by such petty humiliations as much as by ill health, she left for her Welsh estates in the summer of 1579. Despite the often spiteful treatment her royal mistress had inflicted upon her, she had served her faithfully to the end. Her departure apparently prompted no acknowledgment on the part of the Queen, and it might have slipped from the records altogether had it not been for a remark made by Don Bernardino de Mendoza, the Spanish ambassador, on July 6 that a sister of the Earl of Leicester "of whom the queen was very fond and to whom she had given rooms at court" had retired to her own home.[26]

The changes in Elizabeth's household during the 1560s and early 1570s were distracting enough on their own, but they served merely as a backdrop to the issue that had come to dominate her life. Mary Stuart's arrival in England threw her relationship with Elizabeth into sharp relief.

As queens of neighboring countries, their rivalry had been intense enough; now that they were in the same kingdom, that rivalry assumed a different intensity altogether. Elizabeth's bewitching cousin, the woman who had by turn fascinated and repelled her, was suddenly tantalizingly close. There was no longer any obstacle to their meeting. But rather than bringing them closer together—emotionally as well as geographically—the Queen of Scots' fateful decision to flee south of the border would set the two cousins on a collision course that could end only in death.

On May 17, 1568, Elizabeth received news that her cousin had arrived at Workington and had been lodged at nearby Carlisle Castle. She immediately wrote to congratulate her upon her escape and instructed the bearer to convey certain messages relating to Mary's "estate and honour." Apparently not being able to resist spelling out some of these messages herself, she chided her cousin that if she had had "as much regard to her honour as she had respect for an unhappy villain, every one would have condoled with her misfortunes, as to speak plainly not many have."[27]

Despite this inauspicious beginning, Mary apparently had no doubt of her cousin's support. Indeed, she had given little thought to other options as she had lain exhausted at Dumfries. Yet these did exist. She could have remained there for long enough to muster the considerable number of supporters who still existed in Scotland. Or she could have fled to France, where, despite the recent disapproval of her actions, she would have been assured of at least a safe exile. Nevertheless, it was Elizabeth's outspoken support that convinced Mary that she would be the surest means of restoring her to her throne. As Melville later observed: "She never rested till she was in England, thinking herself sure of refuge there, in respect of the fair promises formerly made to her by the Queen of England."[28]

In reality, Mary's arrival in England had thrown Elizabeth into a dilemma. It had been comparatively easy to pledge her support when her cousin had been in her own kingdom, but now that she was under the English queen's jurisdiction, it was an entirely different matter. All the powers of Europe, not to mention the newly formed Scottish council, would expect her to take decisive action one way or the other: either to support Mary and send an army to Scotland to fight for her restoration, or to side with the regent Moray, who had long proved a good friend to England. Such a situation was not at all to Elizabeth's taste. She much preferred to manage diplomatic affairs from a distance, maintaining England's fragile alliances with the kind of "fair promises" that she had given Mary. Now the latter's presence in her own country had effectively forced her hand.

While she debated with her council over what to do, Elizabeth ordered that Mary should be kept under close guard at Carlisle. It was not made clear whether this was for her protection or imprisonment; indeed, Elizabeth herself did not yet know the answer to that. In the meantime, there were more immediate practical matters to see to. Upon her arrival in England, Mary had written to her cousin, pleading: "I have nothing in the world, but what I had on my person when I made my escape."[29] She therefore urgently needed fresh garments. With very bad grace, Elizabeth agreed to provide them, and selected some of the shabbiest gowns from her own wardrobe. When an embarrassed Sir Francis Knollys arrived at Carlisle with these, he first claimed that the clothes had been chosen for "lightness of carriage," but upon noting Mary's frosty silence, he changed his story and pretended that the Queen had misunderstood her request, assuming that the clothes were needed for her maidservants. Elizabeth clearly resented the dilemma in which her cousin had placed her and was determined to make her grateful for any charity—no matter how meagre—that she showed toward her. Mary's failure to write a letter of thanks is entirely understandable, but Elizabeth had the audacity to berate her for ingratitude and demanded to know how her gifts had been received. A hapless Knollys admitted: "her silence argues rather scornful than grateful acceptance."[30]

Before a decision had been reached over what to do with this "daughter of debate," as Elizabeth termed her, the Queen was anxious to assure the French king that she would do everything in her power to further Mary's cause. She instructed her ambassador in France, Sir Henry Norris, to "let the French King know she has provided all things for Mary's safety and means speedily to proceed in consideration how she may reduce her honourably in concord with her subjects."[31] Although subsequent events would prove the falseness of her words, they were enough to convince many at court that she was genuinely resolved to support her cousin. Even the ever-critical Spanish envoy, de Silva, believed that Elizabeth was doing everything she could to persuade her council to restore Mary. He told Philip II: "It is said that this Queen took the part of the queen of Scotland, but her views did not prevail as a majority of the Council was of a different opinion." He went on to summarize the dilemma that Elizabeth faced: "If this Queen has her way now, they will be obliged to treat the Queen of Scots as a sovereign . . . If they keep her as if in prison, it will probably scandalise all neighbouring princes, and if she remain free and able to communicate with her friends, great suspicions will be aroused."[32]

But not everyone was so anxious to see Mary restored. When her mother-in-law heard of her flight to England, she and her husband hurried at once to court. Dressed in deepest mourning, they went down on their knees before Elizabeth and begged her to avenge their son's killer. The countess's face was said to be "all swelled and stained with tears," and her wailing was so loud and so prolonged that the Queen lost patience and dismissed the couple from her presence, telling them that she would not condemn Mary unheard. She was now in the midst of a delicate diplomatic situation that had wide-reaching ramifications, both in her kingdom and across the Continent. She would not be bullied into acting precipitately by a cousin whom she had never liked. Besides, Mary had already complained that Lady Margaret had been allowed to voice her allegations to the Queen, whereas she had been denied the chance to defend herself in person.

As the weeks dragged on, Mary grew increasingly impatient with her cousin's apparent inaction. She had expected to be immediately invited to court, where the two queens could discuss the best way to restore her to her throne. In her impatience, she wrote an indignant letter to Elizabeth, complaining of her failure to meet her. The English queen wrote an equally indignant reply, chastising Mary for her impatience and urging: "Pray do not give me occasion to think that your promises are but wind." The fact that it was Elizabeth's promises, not Mary's, that were the issue was conveniently ignored. She ended on a more conciliatory note: "I assure you I will do nothing to hurt you, but rather honour and aid you," and signed the letter "Your good sister and cousin Elizabeth."[33] Not content with her response, Mary proceeded to bombard her cousin with a series of long and impassioned letters, all pleading with her to honor her promise of support.

Mary was not the only one who began to doubt the English queen's motives and those of her council. "They have their signs and counter-signs, and whilst they publicly unite and do one thing, they secretly order another; and as this Scotchman says, the Queen of England uses towards his mistress fair words and foul deeds," reported de Silva in July.[34] Eventually, pressured into action, Elizabeth convened a conference to inquire into Mary's grievances and those of her chief adversary, the Earl of Moray, in order to resolve their differences. The plan was that Mary would then be restored to her throne, although her powers would be strictly limited by the Scottish Council. If Elizabeth thought this would appease her cousin, she was soon proved wrong. Mary was highly affronted at having to prove the justice of her cause, having been sure that Elizabeth would defend her unquestioningly as a fellow sovereign. Nevertheless, she had little choice but to go through with the conference.

This was convened at York in October 1568. While she had no doubt hoped that she would be able to present her case in person and, better still, finally meet her English cousin, Mary was forbidden to attend and instead had to content herself with sending commissioners to represent her. Moray, however, was present, which was an ominous sign for Mary. After a month of presentations and debate, the conference was moved to Westminster, where the Queen could play a more active part in proceedings. However, by now it had ceased to be an assessment of the relative merits of each side and had become an investigation into Mary's guilt. Desperate to convince Elizabeth of this, Moray produced a series of correspondence that has since gained notoriety as the "Casket Letters." These purported to prove that Mary had had an adulterous affair with Bothwell while she was still married to Lord Darnley, and that she and her lover had plotted the latter's death. The letters have since been proved almost beyond doubt to be fakes, but at the time they were sufficient to condemn Mary in the eyes of most of Elizabeth's council and to destroy any hope of a compromise with Moray.

But what of Elizabeth? Did she believe her cousin guilty of the despicable crimes that were outlined in those scandalous letters? Her immediate reaction suggests that she did not. Terminating the conference on the basis that nothing had been sufficiently proved on either side, she gave Moray leave to return to Scotland. However, her actions toward Mary—both then and over the ensuing years—indicate that whether or not she truly believed in her guilt, it suited her to act as if she did. Any pretense that Mary was being kept under armed guard for her own protection was abandoned. When she was moved to the insalubrious Tutbury Castle in Staffordshire, it was as Elizabeth's prisoner.

The constable of Tutbury, George Talbot, Earl of Shrewsbury, was entrusted with the heavy responsibility of taking charge of the captive Queen of Scots. Shrewsbury was the fourth husband of the indomitable Bess of Hardwick, who had had a difficult relationship with the English queen ever since becoming embroiled in the Katherine Grey scandal. But as much as she might distrust her, the Queen still respected Bess for her strength of character and her determination to carve out a role for herself in a world dominated by men—just as she herself was having to do. Geography also helped to restore Bess to favor. The court was too small to contain the overbearing personalities of these two matriarchs, but now that Bess was many miles away in Staffordshire, the tension between them dissipated. As far as the Queen was concerned, Bess's qualities were best viewed from a distance.

Bess soon justified the Queen's trust, for she became just as much Mary's keeper as her husband was. She even placed her own spy in the Scottish queen's household "to give her intelligence of all things," and reported everything back to her royal mistress, who praised her for this "manner of service."[35]

Mary was outraged when she realized that, far from restoring her to her throne, her cousin was instead imprisoning her like a common criminal. So confident had she been in Elizabeth's "fair promises" that she had vowed to be back in Scotland at the head of an army within three months of her flight to England. Now here she was, some eight months later, a captive of the English queen, with apparently little prospect of ever returning to Scotland. In her fury, she protested that her imprisonment was entirely unlawful: she was a queen in her own right, and Elizabeth had no jurisdiction over her.

Mary was, of course, quite correct, and her cousin knew it. But Elizabeth herself was held captive by the political pressures that now surrounded Mary. Moray would brook no schemes for her restoration in Scotland. If she was sent across the channel, there was a danger that the French would seize the opportunity to intervene in Scottish affairs, which would almost certainly be to England's detriment. And if she remained in England, she would provide a figurehead for the growing number of plots against Elizabeth's regime. Little wonder that the ever-perceptive Lord Burghley told his royal mistress: "The Queen of Scots is, and always shall be, a dangerous person to your estate."[36]

While Elizabeth wrestled with the issue of what to do with Mary, the months dragged on. In a letter to Pope Pius V, Mary woefully referred to "the Queen of England, in whose power I am."[37] Meanwhile, the Venetian ambassador remarked: "The Queen of Scotland is now an object of compassion."[38] But Mary was far from being the hapless victim that she liked to portray herself. In fact, she was ruthless in her desire to reclaim not just her Scottish throne but also a promise of the English succession. Even Sir Francis Knollys, who had been somewhat beguiled by her during his visit to Carlisle in May 1568, admitted: "The thyng that most she thirsteth after is victorye," adding that she did not care how she attained it.[39] The fact that she was already drumming up support from among Elizabeth's enemies proves the point. Therefore, while in theory the English queen held all the cards and, with Mary as her captive, had the power to decide her fate, the reality was far more complex. At times, Elizabeth seemed almost intimidated by the beguiling Scottish cousin whom she had never met. Mary's unquestionable legitimacy made her feel insecure about Elizabeth's own claim to the throne, and her image as a Catholic martyr, a mother separated from her son, and a heroine in need of rescue understandably caused her English cousin many misgivings. Far from relishing her role as jailer, Elizabeth felt imprisoned herself. "I am not free, but a captive," she lamented in despair. On another occasion, she exclaimed to a visiting ambassador: "I am just as anxious to see Mary Stuart out of England as she can be to go!"[40]

The almost mythical status that Mary had begun to assume in her cousin's mind contributed to the latter's reluctance to meet her. Perhaps Elizabeth feared that she would fall under Mary's spell, as so many others had before her. She once told the French ambassador: "There seems to be something sublime in the words and bearing of the Queen of Scots that constrains even her enemies to speak well of her."[41] Mary herself seemed to believe that this was the reason why Elizabeth hesitated to meet her, and referred to it in one of her many letters begging for an audience: "Good sister, change your mind . . . Alas! do not as the serpent that stops his hearing, for I am no enchanter, but your sister and natural cousin."[42] Others believed that Elizabeth was reluctant to bring Mary to court because she was aware of how many supporters she had there. "It is not believed that Queen Elizabeth, for her own reasons, will allow the Queen of Scotland to come to the Court, suspecting probably some ulterior motive, inasmuch as the Queen of Scotland is nearest in blood to the Crown of England, and beloved by some of the chief personages, although they dare not say so openly; and this is thought to be the cause why no interview will take place."[43]

As Mary continued to besiege her cousin with increasingly insistent letters that she should arrange for them to meet, Elizabeth lost her temper. "In your letter I note a heap of confused, troubled thoughts, earnestly and curiously uttered to express your great fear and to require of me comfort," she began, "that if I had not consideration that the same did proceed from a troubled mind, I might rather take occasion to be offended with you than to relent to your desires." Chiding her for writing a letter "so full of passions," she went on to remind Mary that if she had only heeded her advice, she would not be in such a predicament: "I will overpass your hard accidents that followed for lack of following my counsels . . . I, finding your calamities so great as you w[ere] at the pit's brink to have miserably lost your life, did not only entreat for your life but so threatened such as were irritated against you that (I only may say it) even I was the principal cause [to] save your life."[44]

Far from being chastised by her cousin's letter, Mary seemed to step up her involvement in the various plots to place her on the English throne, or at the very least incite Elizabeth to name her as heir. "This Queen sees that all the people in the country are turning their eyes to the Queen of Scotland, and there is now no concealment about it. She is looked upon generally as the successor," observed Philip II's envoy as early as July 1569.[45] Later that year, the first serious threat to Elizabeth's throne came with the rising of the staunchly Catholic earls of Northumberland and Westmorland. This was prompted by the arrest of their ally at court, the Duke of Norfolk, whom Mary had been plotting to marry. Although the rebellion was crushed, it had prompted the Pope to act against this heretical English queen. In 1570 he issued a bull of excommunication, declaring Elizabeth a usurper and absolving her subjects from allegiance. It also sanctioned any rebellion against her and even encouraged assassination attempts. This intensified the threat posed by Mary, who increasingly identified herself with the Roman Catholic faith in order to become a figurehead for opposition to her cousin. By the following year, it was said that she was "favoured by all the Catholic party in England."[46]

Although she involved herself in ever more plots and conspiracies, Mary also hedged her bets by professing her loyalty toward Elizabeth, evidently still hoping that the Queen would one day restore her to her Scottish throne. Elizabeth gave her words little credence, especially when fresh evidence of Mary's intrigues came to light on such a regular basis. In 1571 the most serious plot yet was discovered. It had begun the year before, prompted by the Earl of Norfolk's release in August 1570. Plans for him to marry the captive Queen of Scots were revived (even though she was still married to Bothwell, who was now a prisoner in Denmark), and in early 1571 Mary wrote to the papal agent, Roberto Ridolfi, urging him to solicit help from the Pope, Philip II, and the Duke of Alva. She wanted those potentates to know of "the evil entreatment which I undergo of my person, and other indignities and affronts to which I am subjected, the jeopardy in which I stand of my life, menaced as I am with poisoning and other violent deaths." Mary criticized her cousin for feigning a "willingness to entertain the suggestion of my liberation, to amuse herself at my expense," and accused her of having been "many a time . . . on the point of compassing my death."[47]

Although Mary did not sanction Elizabeth's being deposed or assassinated, this letter was still treason. There was other evidence against her, for she had embroidered a cushion as a gift for the Duke of Norfolk, and the design showed a hand clipping off a barren vine so that the fruitful vine might flourish. The message was clear: She, the fertile Queen of Scots, should supplant her childless cousin.[48] As the plot developed, it was agreed that the Duke of Norfolk would arrest Elizabeth and place Mary on the English throne. When the English queen heard of this, she was said to be "mightily incensed."[49] She had been informed of the plot in late summer 1571 and had lashed out at the French ambassador when he came to pay court to her a few weeks later. Evidently ignorant of the plot himself, he had made the blunder of suggesting that Elizabeth should improve the conditions under which her cousin was kept. "The Queen burst into a most ferocious rage at this and dwelt very strongly upon the evils which she said were being brought upon this country by the queen of Scotland," the ambassador reported to the French king. "She afterward went on to speak of the plots which she and the duke of Norfolk were weaving jointly with your Majesty to turn her off the throne . . . She screamed all this out with so much vehemence that almost everybody in the palace could hear her."[50]

Elizabeth was so infuriated with her cousin that it was widely expected she would put her to death. Norfolk was convicted of treason in January 1572 and executed later that year. Many of her councillors urged her to mete out the same treatment to Mary, notably her spymaster, Sir Francis Walsingham, who declared: "So long as that devil-ish woman lives neither her Majesty must make account to continue in quiet possession of her crown, nor her faithful servants assure them-selves of the safety of their lives."[51] Another courtier remarked: "The Queen of Scotland has been the most dangerous enemy in the world to our Queen."[52] But Elizabeth demurred. No matter how much of a danger her cousin had become, she was still an anointed queen, and Elizabeth flinched from putting her on trial. She therefore delayed enough for matters to settle down once more, and Mary was left to languish in captivity. Elizabeth did, however, insure that she was placed under much stricter supervision. Mary complained bitterly at this, but her cousin scolded her for "filling a long letter with multitude of sharp and injurious words" and added, rather patronizingly, "It is not the manner to obtain good things with evil speeches."[53]

Although Elizabeth refused to punish Mary in the way her council-lors wished, the Ridolfi plot had put paid to any lingering plans she may have had to restore her cousin to the Scottish throne. She there-fore formally recognized James VI as King of Scots. In confirming Eliz-abeth's suspicions that she still hankered after the English crown, Mary had forfeited her liberty for good.

CHAPTER 11

"That She-Wolf"

Little had been heard of Margaret Douglas since 1571, when her husband, the Earl of Lennox, was assassinated in Scotland. She seemed to have settled into a quieter life at her house in Hackney, east London, with regular visits to court. But as ever with this formidable matriarch, a fresh plot to further her dynastic ambitions was never far away. In 1574 she resolved upon one last throw of the dice. Although her grandson, James, was due to inherit the Scottish throne when he came of age, Margaret knew from bitter experience how fast the political tide could turn there. She therefore alighted upon another potential path to the crown of England: her only surviving child, Charles. He had always been rather neglected by his mother, who had favored his elder brother, Henry, and referred to Charles as her "greatest dolour." But now that he had reached the age of eighteen, she appreciated that he could be useful in the marriage market. The countess resolved to find him a bride of noble blood, knowing that this would significantly bolster the chances of her dynasty succeeding Elizabeth, who, aged forty and still unmarried, looked increasingly unlikely to produce any heirs of her own.

On the pretense of wishing to take Charles to visit his infant nephew, Margaret sought the Queen's permission to journey to Scotland. Elizabeth was immediately suspicious, all too well aware that there was usually something sinister behind her cousin's requests for leave of absence. She was convinced that the countess intended to pay a secret visit to her daughter-in-law on her journey northward. Perhaps she wished to give Margaret enough rope to hang herself, for she eventually agreed to her request on the strict condition that she should go nowhere near Chatsworth House, where Mary was being held by the Earl and Countess of Shrewsbury. Margaret assured her that she still hated her daughter-in-law and had no intention of visiting her. She added, for good measure: "for I was made of flesh and blood and could never forget the murder of my child . . . for if I would, I were a devil."[1]

Shortly after embarking, Margaret had an apparently spontaneous change of plan when she received an invitation from the Countess of Shrewsbury to break her journey at Rufford, one of her estates in Derbyshire. Margaret knew Bess well from the days when they had attended court, and she readily accepted the invitation. Bess, who was every bit as ambitious as her former companion, had a daughter, Elizabeth Cavendish, who was just a year older than Charles and still unmarried. From her perspective, the Countess of Lennox's son was an ideal candidate, and she was keen to bring the marriage to pass. The chances are that the two women had concocted the whole scheme long before Lady Margaret left London. Events soon unfolded to their satisfaction. Far from having to be forced to marry, Charles and Elizabeth fell in love at first sight, which made their alliance seem all the less contrived. After only a few days at Rufford, the pair were married with great haste. Neither Margaret nor Bess had seen fit to seek the Queen's permission first, even though the 1536 act made it treason not to.

It was only a matter of time before Elizabeth found out. She reacted with predictable fury, railing against her cousin and Bess for such a blatant act of defiance. She was not fooled by their protestations that the couple had fallen in love of their own accord and that they had not had the heart to prevent the union. Her greatest venom was reserved for Margaret, who had yet again proved disloyal. Convinced that an elaborate plot lay behind the marriage and that Mary, Queen of Scots, was somehow involved, she ordered Margaret and her son to come to London immediately. Upon arrival, they were initially confined to the countess's house at Hackney. Meanwhile, Bess and her daughter were to remain under guard at Rufford until the Earl of Huntingdon, who was leading the inquiry, had completed his investigations.

The Queen's anger at these two troublesome old women increased when she learned that Elizabeth Cavendish was already pregnant. Any child from their union would in theory have a claim to the throne, however distant, thanks to Charles's royal blood. The last thing Elizabeth needed was yet another potential claimant. Determined to exact revenge, she promptly threw Margaret into the Tower.

This was Margaret's third spell in that fortress, and, by now used to the routine, she wasted no time in pleading her case. She insisted that there had been no plot to marry their two children and that by the time she discovered her son's love for Elizabeth Cavendish, "he had entangled himself so that he could have done no other" than marry her.[2] Claiming to be suffering "great unquietness and trouble . . . with the passing these dangerous waters," the countess begged to be set at liberty. Reflecting upon her various incarcerations in the Tower, she added a final lament: "Thrice have I been sent into prison not for matters of treason, but for love matters. First, when Thomas Howard, son of the Duke of Norfolk, was in love with me; then for the love of Henry Darnley, my son, to Queen Mary; lastly for the love of Charles, my younger son, to Elizabeth Cavendish."[3]

Elizabeth punished the other miscreant rather less harshly. She issued a sharp reprimand to Bess and ordered that she and her daughter continue to be held at Rufford until the Earl of Huntingdon had completed his investigations there. This comparative leniency may have been thanks to the intervention of Bess's husband, whom the Queen greatly esteemed. Or perhaps by now Elizabeth and Bess had developed an understanding. Elizabeth knew of Bess's fierce dynastic ambition, but she also knew that she was unlikely to flout the Queen's authority. Indeed, she had proved her loyalty by spying upon Mary, Queen of Scots. She therefore soon forgave her for this latest transgression, calculating that such a show of leniency was more likely to win Bess's loyalty than a strict punishment.

If the Queen had been inclined to show any such clemency toward her much-imprisoned cousin, she soon changed her mind when she learned that the countess had been corresponding regularly with Mary, Queen of Scots, throughout her incarceration. In an astonishing about-face, Margaret was now all kindness and affection toward the woman who just a few years previously she had denounced as a wicked murderess. Sending Mary little tokens of her esteem, she signed her letters: "Your Majesty's most humble and loving mother and aunt." It is likely that Mary's son, James, was responsible for this rapprochement. He represented their joint hopes for the future, the best chance that both women had to see their bloodline succeed to the English throne. Margaret was therefore anxious not to do anything that might distance herself from him. But it did not suit Elizabeth's purpose at all that these two women were back on good terms, and she took a dim view of Margaret's sending the beleaguered Scottish queen letters and gifts from the Tower. She could do little to prevent Mary's subsequently altering her will to include her brother-in-law, Charles, and restoring to Margaret "all the rights she can pretend to the earldom of Angus."[4]

But it seems that the English queen had no taste for detaining her cousin in the Tower this time, and by the end of 1574 Margaret was permitted to rejoin her son and daughter-in-law at Hackney. Her granddaughter, Arbella, was born shortly afterward. Margaret doted on the pretty red-headed child. She was one of the few comforts left to her. Heavily in debt, she was worried about her son, whose health had always been delicate and was now seriously declining. When he died in 1576 at the age of just twenty, his mother fell into a "languishing decay."

Perhaps realizing that her own life was slipping away, Margaret was determined to do everything she could to safeguard the future of Arbella and her "sweet jewel," James VI. She wrote to the latter, assuring him that he was her chief hope for the future. She also worked tirelessly to try to claim the earldom of Lennox for Arbella, but her efforts proved in vain. Pursuing her dynastic ambitions to the last, she died on March 9, 1578, aged sixty-two.

Elizabeth accorded her old adversary a funeral worthy of a countess of royal blood, and she was buried at Westminster Abbey in Henry VII's chapel. Even in death, Margaret had wished her status to be recognized, for in her will she had left instructions for an elaborate tomb to be erected. However, having died in impoverishment, she had not been able to pay for it. The Queen agreed to meet the expense out of her own coffers, but it proved so great that the work was left half finished. Only when James VI ascended the English throne did he order its completion. Margaret would no doubt have been heartened by this act of reverence on the part of her grandson, the man who had, at last, brought her hard-fought dynastic ambitions to fruition.

It is unlikely that Elizabeth felt much regret at the death of the Countess of Lennox. Margaret had never ceased to scheme against the woman whose claim to the throne she always insisted was weaker than her own. All the various plots in which the countess had been involved over the years had given Elizabeth more than enough ammunition to have her executed for treason. And yet she had refrained from doing so; indeed, there is very little evidence to suggest that she had even contemplated it on any of the occasions when Margaret had been her prisoner. Perhaps, after all, she had a grudging respect for this formidable matriarch whose pride and ambition was so all-consuming that even three spells in the Tower could not dampen it.

The following month saw the death of another of the Queen's troublesome cousins. Lady Mary Grey had finally been set "att free leberty" in 1572, following the death of her husband, Thomas Keyes, the previous year. With a modest income of £20 per year from her inheritance, together with a further £80 that the Queen had grudgingly agreed to give her, she had set herself up in a house of her own in Aldersgate, in the heart of the city of London.[5] The Queen's longest-serving attendant, Blanche Parry, had taken pity on the girl, and it was probably thanks to her intervention that Elizabeth had allowed Mary back at court in 1576. She visited there several times over the next two years and was invited to attend the Christmas celebrations at Hampton Court, when she presented her former royal mistress with "four dozen buttons of gold, each containing a seed pearl, and two pairs of sweet gloves"—a generous gift that must have stretched her meagre income. In return, the Queen gave her a silver cup.[6]

But Mary Grey was not destined to enjoy Elizabeth's favor for long. She died on April 20, 1578, aged just thirty-four. She bequeathed various tokens to friends and kinswomen, including a "little gilt bowlle with a cover to it" for Blanche Parry, and "my juell of unicornes horne" to the Duchess of Suffolk.[7] She left nothing to the Queen, and to the end of her days she signed herself "Mary Keyes." It was a small act of defiance in a battle of wills that Elizabeth had easily won, but it showed that Mary had inherited at least some of the characteristic Tudor stubbornness.

The deaths of Mary Grey and Margaret Douglas afforded Elizabeth little respite from the plots and scandals that seemed to be forever erupting at her court. By far the greatest of these had already started to gain ground. It involved the countess's goddaughter and the man whom the Queen regarded as her exclusive property: Robert Dudley, Earl of Leicester.

Lady Douglas Sheffield was the younger sister of Charles Howard, second Baron Howard of Effingham and first Earl of Nottingham. Her unusual first name may have been intended to honor Lady Margaret Douglas, although it is not clear why her last name instead of her first was used. Born around 1542, Douglas had been appointed a maid of honor to Elizabeth by the time of her coronation in January 1559. Her service in this capacity had been fleeting, however, for in 1560 she had left court to marry John Sheffield, second Baron Sheffield, who hailed from a distinguished noble family in Lincolnshire. Lady Douglas was by then sufficiently esteemed by her royal mistress for the latter to present her with a wedding gift. The couple had two surviving children, but their marriage was comparatively brief, for Lord Sheffield died in 1568, aged just thirty.

Douglas returned to court soon afterward and was appointed a gentlewoman extraordinary of the privy chamber. This was one of the less prestigious posts in Elizabeth's household, for it was essentially a stand-in, to be called upon when the regular, salaried members were sick or otherwise absent. However, Douglas was present enough to draw the attention of Robert Dudley. At the time of her return to court, she was about twenty-six years of age and, by all accounts, strikingly attractive—"a lady of great beauties," according to an account written by her seventeenth-century descendant Gervase Holles.[8] Leicester was one of a number of admirers of the young widow, and with her family connections, she could have had the pick of them all. However, her fancy was very much taken by Sir Robert, and she encouraged his flirtation.

Holles alleged that their affair had started as early as 1566, when Dudley accompanied the Queen on progress to the Midlands. According to his account, among the places where they stayed was the Earl of Rutland's Belvoir Castle, where Lord and Lady Sheffield came to pay their respects. He claimed that the "fayre young" Douglas "shone as a star in the Court, both in regard of hir beauty and the richness of her apparell," and Sir Robert, "seeing hir and being much taken with hir. perfections, he made his address of courtship to hir and used all the art that might be (in which he was maister enough) to debauch hir."

Holles went on to claim that Douglas and Leicester fell so passionately in love that they resolved to do away with Lord Sheffield. Before they left Belvoir, Leicester "undertooke the charge of it," and subsequently wrote to his new lover, "wherein after many amorous expressions he tolde hir he had not been unmindfull of removing that obstacle which hindered the full fruition of their contentments." According to the account, Lord Sheffield subsequently discovered the affair when he intercepted one of Leicester's "amorous" letters, and rode to London to confront his rival. When he died there a short while later, Holles was in no doubt that Leicester had poisoned him in an act of despicable "villainy."[9] But there was at this time a concerted campaign to blacken the earl's character in order to prevent his marrying the Queen, and Sheffield's death was one of several that were laid at Leicester's door.[10]

In fact, apart from Holles's account, there is little evidence that the affair between Leicester and Douglas had even begun by the time of Lord Sheffield's death. It is more likely that they only became involved after Douglas had returned to court in 1568, and most accounts date the start of the affair to the early 1570s. In a letter he wrote to his lover, Leicester himself recalled the time "after your widowhood had began, upon the first occasion of my coming to you."[11] By 1573, the affair had become a source of gossip at court. In May, Gilbert Talbot wrote to his father, the Earl of Shrewsbury: "There are towe sisters nowe in ye Courte that are very farr in love with him, as they have bene longe; my Lady Sheffield and Frances Haworthe [Howard]; they (of like stryving who shall love him better) are at great warres together."

Frances was Douglas's younger sister, who would then have been around nineteen years old, whereas Douglas was in her midtwenties. Leicester was a notorious flirt and was no doubt flattered by the attentions of two such beautiful young women, but at this time his attentions were very much focused upon securing the main prize at court: the Queen herself. In the same letter, Talbot reported that the earl "is very muche with her Majestie, and she sheweth the same great good affection to him that she was wonte; of late he hath indevored to please hir more than heretofore." However, when the quarrel between Douglas and her sister grew so fierce that it attracted the Queen's notice, she immediately demanded to know why they were so at odds. When told that it was because they were fighting over her own favorite, Elizabeth was gravely displeased. "The Quene thinketh not well of them, and not the better of him," remarked Talbot, adding that "by this meanes there is spies over him."[12]

The reference to Douglas having "bene longe" in love with Leicester suggests that their affair might have begun some time before her quarrel with her sister brought it to Elizabeth's attention. The evidence suggests that Douglas had already become romantically involved with Leicester by at least 1570. In the early stages of the affair, she seemed content to be Leicester's mistress, but for her, at least, attraction was mixed with love, and she no doubt hoped for more. Her lover felt rather differently. Gratified though he was by having one of the most beautiful ladies at court as his mistress, he was by no means prepared to sacrifice his favor with the Queen for her sake. When Douglas began to put pressure on him to marry, he wrote her a long letter reiterating what he had apparently told her at the start of their affair—that he could never take her as his wife. Although he assured her: "I have, as you well know, long both liked and loved you," he explained that even by having an affair, he was risking "the ruin of my own house" if Elizabeth found out, and that if he took it any further, it would be "mine utter overthrow." He continued: "If I should marry, I am sure never to have favour of them that I had rather yet never have wife than lose them."[13] He therefore offered her two choices: either to continue as his mistress, accepting that she could never be more, or to let him help her find a suitable husband from among her many admirers at court.

It seems that Douglas chose the former option, and they continued to meet "in a friendly sort and you resolved not to press me more with the matter," as Leicester later recalled.[14] However, she was far from satisfied with the arrangement, and her bitterness at Sir Robert's refusal to marry her soon caused another row between them. Although she is often portrayed as a pushover, Douglas was a woman of some intelligence and spirit, prone to stormy outbursts and mood swings. It seems that she and Leicester separated for some months after this spat, Douglas having rejected her lover's attempts to sweet-talk her into obeisance, crying out: "The good will I bare you had been clean changed and withdrawn."[15] They were reconciled some time after Douglas had quarrelled with her sister in May 1573. Perhaps the jealousy at her sister's advances toward Leicester had made her realize that she was still in love with him herself.

It is also possible that Leicester had won Douglas back with a promise of marriage. Now aged forty, he was increasingly conscious of the need to produce an heir, and told his friend Lord North how much he wanted to have children with some "goodly gentlewoman."[16] Whether he had promised it or been persuaded by his lover, by the autumn of that year he had agreed to marry her. According to Douglas's later testimony, the wedding took place some time between November 11 and December 25, 1573, at her family's house at Esher, Surrey. There were at least three witnesses, although when it was later called into question, none of them would admit to having been present. Neither was there any documentary evidence, for Douglas would later claim that this had been stolen by her servants at Leicester's behest.

Although it can never be proven beyond doubt unless further evidence comes to light, the testimony of Lady Sheffield, together with the events that followed, do suggest that the wedding was more than just a figment of her imagination. What is certain is that almost exactly nine months after it was alleged to have taken place, Douglas gave birth to a son, who was christened Robert, after his father. At the time, Leicester was away on progress with the Queen in Bristol, and when a messenger arrived with the news, he immediately wrote to congratulate Douglas on giving him a son who "might be the comfort and staff of their old age." He signed the letter "Your loving husband," which in itself would have been ample proof of their marriage, but this letter—like the other documentary evidence—was subsequently lost. Even though he denied the marriage, Leicester would acknowledge the boy as his, referring to him as "my base son." The fact that he later left the bulk of his estates to him may have been a tacit admittance that he was not illegitimate after all.

Douglas had retired from court before her pregnancy became too obvious. For a time, she lived quietly at Esher and Leicester House, her husband's London home, and was content to keep their marriage a secret. Her servants addressed her as "Countess," although for discretion's sake she refrained from using this title outside the privacy of her own home and continued to be referred to as "the Dowager Lady Sheffield."[17] Although there were rumors, Leicester had apparently succeeded in keeping the whole matter from his royal mistress and enjoyed even greater favor than he had before. Apparently choosing to forget the inconvenient fact that he already had a wife and son, he stepped up his attempts to marry Elizabeth. In 1575 he staged a spectacular series of entertainments for her when she visited his Warwickshire home, Kenilworth Castle, and besieged her with proposals. But while the Queen revelled in his attentions and at times seemed to be giving serious consideration to marrying him, she remained tantalizingly aloof and refused to commit.

In his frustration, Leicester turned his attentions to another lady at court who had caught his eye. Lettice Knollys was the wife of Walter Devereux, first Earl of Essex, and she and Leicester had enjoyed a prolonged flirtation. By now tiring of Douglas, and increasingly besotted with Lettice, Sir Robert decided to rid himself of the former. As Holles (who did not believe that they had married) rather brutally remarked: "According to the nature of all men who think basely of their prostitutes, after he had used hir body sometime and got a base sonne . . . of hir, [he] rejected hir."[18]

The manner of her rejection was no less brutal. In 1578, some five years after their wedding had allegedly taken place, Leicester summoned Douglas to meet him in the gardens of Greenwich Palace. Having seen precious little of him since his affair with Lettice Knollys had begun, it was no doubt with some trepidation that she made her way to meet him. Upon arrival, she found that there were two other gentlemen in attendance, whom Leicester had commissioned to act as witnesses. Without preamble, he told her that their marriage was over and that she was released of all obligations to him. He then offered her £700 if she would deny all knowledge of the marriage and surrender custody of their young son. According to her own account, written a quarter century later, Douglas burst into tears and refused to countenance such a proposal. Leicester then lost his temper and shouted that the marriage had never been lawful anyway, spitefully adding that he "would never come at her again." Devastated by his sudden rejection, and hardly able to come to terms with what he demanded, Douglas asked for more time to think about it. She eventually capitulated, no doubt realizing that if Leicester openly disavowed her, then her reputation would be in tatters and both her own and her son's future ruined. It must have been a painful decision to relinquish custody of the boy, her sole source of comfort now that her husband had rejected her, but she appreciated that he would stand a much greater chance of advancement under

Leicester's patronage. According to her later account, however, she had given in more out of fear for her own safety than concern for her son. She claimed that her hair had recently begun to fall out, and she was therefore terrified that Leicester had plotted her death with "some ill potions."[19] There is little evidence to support her claim, and it is more likely to have been an attempt to damage the reputation of her former husband.

Despite the antipathy that now existed between them, Leicester honored his promise to find Douglas another husband. The year after their separation, she married Edward Stafford, a diplomat, who was shortly afterward appointed Elizabeth's envoy to France. He took his wife with him, which must have gratified Leicester, who seemed to have neatly avoided the Queen's ever finding out about his relationship with Douglas. His optimism would soon prove spectacularly unfounded.

The lady who had so beguiled Leicester that he had cruelly spurned Douglas Sheffield was also a member of Elizabeth's privy chamber. Lettice Knollys, the daughter of Elizabeth's faithful servant and cousin, Katherine Knollys, was appointed a gentlewoman of the privy chamber in 1559, at the age of eighteen or nineteen. Her first period of service coincided almost exactly with that of her rival, Douglas Sheffield, for in December 1560 she married Walter Devereux, later Earl of Essex, and withdrew from court the following year. Over the next few years, she gave birth to five children in close succession. She still occasionally attended court and was certainly there in 1565, when heavily pregnant with her son, Robert, because the Spanish ambassador observed that Sir Robert Dudley was paying court to her.

The ambassador described Lettice as one of the best-looking women of the court, and contemporary portraits of her prove that he was not exaggerating. Like her great-aunt Anne Boleyn, she had seductive dark eyes and a bewitching charm that beguiled many of the male courtiers. Her auburn hair, smooth, pale skin, and pouting lips marked her as one of the most attractive ladies of her age, and like the Queen she had great physical vigor. Although de Silva described her as being a close favorite of Elizabeth, his comment was wide of the mark.

Lettice had neither the personal attributes nor the desire to become an intimate of her royal mistress. Having inherited a good deal of the Boleyn arrogance, she viewed Elizabeth more as a cousin than queen and refused to show her the deference that was required of her ladies. Like Elizabeth, Lettice was vain, demanding, and possessive, and the two women would probably have clashed even if it had not been for the scandal that erupted over Lady Knollys's affair with the Earl of Leicester. For her part, although she had been fond of Lettice on account of her being the daughter of two of her most faithful servants, Elizabeth soon came to resent her. Jealous of her youth and beauty and affronted by her arrogance, she found her an irksome presence at court. Little wonder that when she heard of Lettice's flirtation with Sir Robert Dudley in 1565, she flew into a rage and upbraided her favorite for disloyalty. He duly abandoned the flirtation, but not for long.

As in the case of his relationship with Douglas Sheffield, exactly when Dudley began an affair with Lettice Knollys is not certain. Although both parties would later insist that it was after she was widowed, rumor said otherwise. Her marriage to the Earl of Essex had brought forth many children, but on a personal level the couple had been ill matched, and Lettice was thought to be relieved when her husband went to serve in Ireland in 1573, remaining there for more than two years. It was probably during this time that the long-standing flirtation between Lettice and Leicester deepened into a full-blown affair. Leicester sent her a gift of deer from Kenilworth Castle in 1573, and invited her and her sister Anne to hunt there the following year. By 1575, Lettice had become such a regular fixture at Kenilworth that she was invited to the entertainments held there in the Queen's honor in the summer of 1575. It is possible that while Leicester tried to court Elizabeth, he was also secretly bedding his new mistress.

By the time of Essex's return in November 1575, the affair was "publicly talked of in the streets."[20] The rumors reached such proportions that it was claimed Lettice had already borne her lover two children, including her son, Robert, who would later become the second Earl of Essex and the Queen's great favorite.[21] Before long, Essex himself had heard the whispers, and his antagonism toward Leicester, whom he had never liked, intensified into "great enmity." He chose not to confront the matter, however, and returned to Ireland in high dudgeon the following year. When he died shortly afterward, it was rumored—yet again—that Leicester had poisoned him, although the cause of death was more likely to have been dysentery.

Essex had died embittered against his wife and lamenting "the frailty of women."[22] He ordered that his children be transferred to the guardianship of the Earl of Huntingdon, a relation of his, rather than be corrupted by their mother. He took further revenge upon his wife by leaving her almost destitute, for she was obliged to take over the management of his estates, which were heavily in debt. An attempt to secure some maintenance from her royal mistress failed miserably: Elizabeth was loath to come to the assistance of a woman who had shown her neither loyalty nor respect.

The Earl of Essex's death made little difference to Leicester. Although he was greatly enamored of his beautiful widow, he had no intention of marrying her and seemed content to let the affair take its course. Besides, he was probably already espoused to Douglas Sheffield and found the prospect of bigamy distasteful—even though he put such delicacy aside when seeking the Queen's hand in marriage. But if he thought Lettice would be content to continue as a mere mistress, he was gravely mistaken. As one recent historian has observed, she was "a woman of infinitely stronger character than the lightweight Lady Douglas."[23] Ambitious and still strikingly attractive, Lettice had no intention of wasting her assets by continuing to play the part of a sordid secret, when she could have any man she wanted. Moreover, she needed to find a way out of the financial predicament in which her late husband had left her. Although it would have been an arguably easier—and safer—strategy to fix her attentions on one of her many other admirers at court, she was still strongly attracted to Leicester, and the fact that he was the Queen's favorite no doubt added extra spice to their liaison. She therefore began a concerted campaign to marry him.

Lettice's strategy was altogether more effective than that of her rival, Douglas. Rather than using tearful pleas and tantrums to bring her lover to heel, she resorted to the time-honored trick of falling pregnant. Knowing that Leicester was desperate for a legitimate heir and tired of the woman whom he may or may not have married some five years before, she made sure that she represented an irresistible alternative. Upon learning of her condition, Leicester seems to have capitulated without a struggle, and according to some accounts, he married her in a secret ceremony at Kenilworth some time in the spring of 1578.[24] But Lettice was still not satisfied. As yet, she was in no better a position than Douglas, whom Leicester may also have married in secret. Her new husband could now choose between his two wives or deny them both, as his pleasure dictated. She therefore demanded that he break all ties with Douglas, which he duly did at Greenwich a short while later.

Although she was now Leicester's sole wife—in his eyes, if not in those of the law—Lettice was determined to make sure that she would not be spurned in the same way as Douglas and therefore forced Leicester to agree to a formal wedding ceremony, complete with witnesses. As Holles later observed, Lettice had "served him [Leicester] in his owne kinde every way." Realizing that in his new wife he had met his match, Leicester acceded to her wish and arranged for another ceremony to take place in September at his house in Wanstead, Essex. Both he and Lettice had accompanied the Queen on her progress to the east of England that summer, and Wanstead had been assigned as one of the last stopping-off places before the court returned to London. Three days before Elizabeth and her entourage were due to arrive there, Leicester secured a temporary leave of absence, ostensibly to prepare his home for the royal visit.

Accompanied by his friend, Lord North, he rode ahead to Wanstead on September 20. He immediately sought out his chaplain there, Humphrey Tyndall, and confided "that he had a good seazon forborne marriadge in respect of her Majesties displeasure and that he was then for sondry respectes and especially for the better quieting of his own conscience determined to marry the right honourable Countesse of Essex."[25] The arrangements were duly put in place, and the witnesses assembled. They included the bride's father, Sir Francis Knollys, who, being "acquainted with Leicester's straying loves," was determined to insure that the wedding would be legal and binding.[26] The ceremony took place early the following morning in a small gallery of the house and was attended by a handful of witnesses, including Lettice's brother, who stood by the door to keep lookout. The bride was said to have worn "a loose gown," which was a coded reference to her pregnant state.[27] Just two days later, the Queen arrived and a magnificent banquet was staged in her honor. In the festivities that followed, during which her favorite was as attentive as ever, nobody would have guessed that he had just committed what his royal mistress would view as the ultimate betrayal.

For a time, it seemed that Leicester would be able to hush up this marriage in the same way as he had the one before. Although he had given in to Lettice's demands for a second, formal ceremony, he had done so on condition that it must remain a secret, because "it might not be publiquely knowne without great damage of his estate." His new wife continued to attend the court, and at the Christmas celebrations that year, she presented the Queen with "a greate cheyne of Amber slightly garnishead with golde and small perle."[28] If she had been pregnant at the time of her marriage, then she either miscarried the baby or it was delivered stillborn, for there is no further reference to it.

But it was almost inevitable that the secret would soon come out. The Earl of Sussex had heard of it as early as November 1578 and took great pleasure in informing the French ambassador, eager to discredit his principal rival at court. Remarkably, according to some accounts, it took as long as a year for the Queen herself to find out. The most commonly cited source is her biographer, William Camden, who claimed that it was the Duke of Anjou's agent, Jean de Simier, who revealed the truth to her.[29] The result was explosive. When Elizabeth learned that her despised cousin, Lettice, had stolen her own great favorite, it was too much to bear. Incandescent with rage, she cried that she would send Leicester to "rot in the Tower."[30] When Lettice appeared at court, she lashed out at her, boxing her ears and screaming that "as but one sun lightened the earth, she would have but one Queen in England."[31] She then banished this "flouting wench" from her presence, vowing never to set eyes on her again. At the same time, she resumed her courtship with the Duke of Anjou with far greater enthusiasm than she had before, which may have been Simier's strategy from the beginning.

Any chance that the Queen might soon forgive her favorite and his new wife was destroyed when she discovered that not only had he married Lettice in secret but he might already have been espoused to Douglas Sheffield. Determined to wreak her revenge, she at once ordered an inquiry to determine whether Leicester was a bigamist. The Earl of Sussex was dispatched to interrogate Douglas. Although she had been presented with an ideal opportunity for revenge, Douglas was clearly terrified of the consequences if she admitted how shamefully she had betrayed her former royal mistress. She therefore tearfully insisted that she had "trusted the said Earl too much to have anything to shew to constrain him to marry her."[32] The Queen was irritated by Douglas's refusal to implicate Leicester as a bigamist and made it clear that her former attendant was no longer welcome at court.[33]

Her royal mistress did perform a great service, though, by placing Douglas's son by Leicester under her protection, assuring her "that shee would take great care of him." This may have been more out of spite toward Lettice Knollys than favor toward Douglas, for the boy had been named by Leicester as his heir. The young Robert Dudley would grow up to be a credit to his father. He had inherited Leicester's intelligence and athleticism, and became both an able scholar and a distinguished soldier and explorer. Ironically, had Leicester recognized the validity of his marriage to Douglas, he would have been provided with the legitimate heir he had so longed for.[34]

In the meantime, the woman for whom Leicester had rejected Douglas remained in disgrace with the Queen, who henceforth referred to her as "that she-wolf." She had forgiven Leicester on the understanding that he would pretend that his marriage to Lettice had never happened and could not even bear to hear her name mentioned. However, she could not resist a sideswipe at her rival in a letter she wrote to Leicester a short while later, claiming: "Whosoever professeth to love you best taketh not more comfort of your welldoing or discomfort of your evildoing than ourself."[35]

For her part, Lettice was apparently content to live away from court. In June 1581, she gave birth to a son, Denbigh, much to the delight of her husband. The couple doted on the boy, whom Leicester nicknamed "the noble Imp," and he seemed to complete the domestic harmony they now enjoyed. By 1583, Leicester judged that the Queen had so forgotten the affair that he was now finally able to live with his wife on a permanent basis, rather than contenting himself with occasional visits. It was a serious miscalculation. Shortly after moving into Leicester House, he learned that Elizabeth was furious with him "abowt his maryage, for he opened up the same more playnly then ever before."[36] Soon afterward, he compounded matters by attempting to arrange a marriage between Lettice's daughter, Dorothy Devereux, and James VI of Scotland. Elizabeth railed that she would not have James marrying "the daughter of such a she-wolf," and threatened to destroy Leicester's reputation for good. Undeterred, he and his wife subsequently schemed to marry their son, Denbigh, to Arbella Stuart, herself a potential claimant to the throne.

Arbella's ambitious grandmother, Bess of Hardwick, was instrumental in this. From the moment of Arbella's birth, she had resolved to make an advantageous match for her. In 1577 she had invited the Earl of Leicester to visit her on his way to take the waters at Buxton. The Queen had been fully aware of his plans and condoned them, unaware of Bess's real intentions.[37] Eager to promote the suit of her little granddaughter, Bess had made sure that she presented Arbella to the earl. In praising her virtues, she had planted the idea in his mind that the girl could be betrothed to his own "base son," Robert. Leicester had been quick to appreciate the benefits of such a match. Like the countess, he saw it as a means to secure the throne for his dynasty when he had given up hope of marrying the Queen. By the time he left Chatsworth, a secret agreement had been made to further the match, and Leicester had agreed to persuade his royal mistress that Arbella would thereby have a stronger claim to the throne than Mary, Queen of Scots. Back in London, Elizabeth had no notion that any of this had passed and wrote a gracious letter to Bess thanking her for the hospitality she had shown to her favorite.[38]

Bess and Leicester subsequently abandoned their plans, which was probably due to the earl's marriage to Lettice Knollys the following year. Lettice had not taken kindly to the idea of her new husband's son by Douglas Sheffield being propelled to such greatness by an alliance with a lady of royal blood. She therefore persuaded Leicester to wait until they had sons of their own to offer the countess. When she gave birth to Denbigh in 1581, she and the earl immediately revived the scheme. Bess was only too happy to oblige: Leicester's new son was even better than the first because his legitimacy was beyond question. She therefore began to arrange the finer details, including exchanging portraits of the children, young as they were, in order to put them in mind of such a match. By December 1582, the plan had been discovered. Bernardino de Mendoza informed Philip II: "I understand that Leicester is on the look out to marry his son to a grand-daughter of the countess of Shrewsbury, who is in the same house as the queen of Scots with her grandmother. The most learned lawyers consider that, failing the queen of Scots and her son, this young lady is the nearest heir to the throne."[39] When she heard of this, Elizabeth flew into a rage, provoked beyond measure by the thought that the relentlessly ambitious countess had flouted her professions of loyalty once more by plotting with the Queen's own favorite.

Bess's plans were brought to an end by the untimely death of the "noble Imp" at Wanstead in July 1584. Leicester was at the court in Nonsuch when news reached him that his beloved son had fallen dangerously ill, and he rushed straight to Wanstead, without taking leave of the Queen. When the little boy died shortly afterward, his parents were grief stricken. Denbigh had been Leicester's precious heir, the boy upon whom all his hopes for the future of his dynasty rested. The inscription on his tomb described him as a child "of great parentage but far greater hope and towardness." Given that Lettice was now in her midforties, and Leicester himself was suffering from increasingly poor health, there was little hope of another child.

Leicester stayed at Wanstead for several weeks "to comfort my sorrowfull wyfe." He received letters of condolence from various members of the court, including his adversaries. There is no record of any such comfort from the Queen. Her bitterness against Lettice was still too strong for her to venture such hypocritical sentiments. Even though she had banned her from court, Elizabeth continued to be plagued by her cousin, who had no difficulty in making her presence felt from a distance. The slightest scrap of news about her would be enough to send the Queen into a fury, such as in the summer of 1585, when she heard that Leicester had taken his wife to Kenilworth for a holiday—the very place where he had proposed to his royal mistress a decade before. It would seem that Lettice had tired of living in the shadows, for she began to build up such a magnificent entourage that it rivalled Elizabeth's own. "She now demeaned herself like a princess," an anonymous courtier observed, and "vied in dress with the Queen." The countess continued to parade her magnificence, which incensed her former royal mistress when she heard of it. "Yet still she is as proud as ever, rides through Cheapside drawn by 4 milk-white steeds, with 4 footmen in black velvet jackets, and silver bears on their backs and breasts, 2 knights and 30 gentlemen before her, and coaches of gentlewomen, pages, and servants behind, so that it might be supposed to be the Queen, or some foreign Prince or other," exclaimed the same courtier in astonishment.[40]

Elizabeth's fury upon hearing such reports was as nothing compared to her reaction to another rumor about Lettice that circulated at court the following year. In 1585 Leicester had been sent to the Netherlands to take command of Elizabeth's forces there. The Queen had no doubt been privately satisfied at the thought that it would take him away from her despised rival, but the following February, she learned that Lettice was preparing to join him, "with suche a train of ladies and gentylwomen, and such ryche coches, lytters, and syde-saddles, as hir majestie had none suche, and that ther should be suche a courte of ladies, as shuld farre passe hir majesties court heare." In fact, the rumor proved to be "most falce," but it had done its work. "This informacyon . . . dyd not a lytle sturre hir majestie to extreme collour and dyslike," reported Leicester's agent at court, who described how the

Queen had then declared "with great othes [oaths], that she would have no more courtes under hir obeisance but hir owen, and wold revoke you from thence with all spede." The Earl of Warwick also wrote to warn his brother that "her malice is great and unquenchable," and William Davison added that the rumors "did not a litle encrease the heat of her majesties offence against you." Only after a great effort did Elizabeth's ministers persuade her that the rumors were false, and at length it was reported that they "dyd greatlye pacifie hir."[41] But the "she-wolf" would never be forgiven.

CHAPTER 12

"The Bosom Serpent"

While rumors and plots continued to surround Mary Stuart after the Ridolfi plot in 1572, the years that followed were comparatively quiet, and she seemed to become more resigned to her captivity. Although her movements were now subjected to closer scrutiny than before, she still lived a comfortable life, spending her days reading, embroidering, conversing with her ladies, playing with her numerous pets, and receiving guests. She was also served as a queen within the confines of her various fortresses. Her household included forty-eight servants, and she had privy and presence chambers, a dais, throne, and canopy of estate. She dined in some luxury, regularly being served two courses of sixteen dishes each, thanks to the Queen's generous allowance of £1,000 per year for her food.[1] She also spent extraordinary sums on sumptuous clothes and jewels, eager to make the most of her fading looks. The Earl of Shrewsbury, who had been appointed her guardian, was a kindly man and sympathized with Mary's plight so much that it was rumored he was in love with her.

Although the earl's wife had spied on Mary at first, she gradually began to treat her with more courtesy. Bess was no doubt fascinated by this woman whose legendary beauty and charisma had beguiled so many men. Perhaps she was also unable to resist a little intrigue, knowing how politically important the Scottish queen was. She therefore made an attempt to befriend her. Mary was all too happy to oblige, desperate as she was for company, especially in the form of a woman from whom she could glean information about her cousin Elizabeth. Bess and Mary would spend many hours sitting together at their needlework, gossiping about the English court. Although very different in character, they were united by a fierce dynastic ambition. This was reflected in the embroidery that each woman produced during their conversations, which was laden with symbolism. Mary's emphasized her royal blood and the strength of her claim to the English throne, while Bess identified herself with a number of powerful mythological female figures, such as Penelope and Lucrece.[2]

The captive Scots queen would later claim that Bess grew so affectionate toward her that she declared "that had I been her own Queen she could not have done more for me." She apparently also swore her allegiance to Mary and even promised to help her escape. By contrast, Bess showed no loyalty toward Elizabeth. According to Mary, far from honoring and respecting the English queen, Bess felt only contempt for her. In the now famous "scandal letter" that she wrote to her cousin in 1584, Mary related various stories that Bess had told her about life in Elizabeth's court. The countess had apparently laughed at the notion that Elizabeth was the Virgin Queen, claiming that she was so insatiable that she had seduced a host of men. She had bedded the Earl of Leicester numerous times and had forced Sir Christopher Hatton to have sex with her. She had tried to entice the Duke of Alençon into bed by wearing nothing but a chemise, and had kissed his envoy, Simier, taking "various indecent liberties with him."[3]

Turning to the subject of what it was like to serve this licentious queen, Mary said that Bess had declared: "She would never return to Court to attend you [Elizabeth], for anything in the world, because she was afraid of you when you were in a rage, such as when you broke her cousin Scudamore's finger, pretending to all the court that it was caused by a fallen chandelier." Far from revering their sovereign, most of the ladies in Elizabeth's service played tricks on her and poked fun at her behind her back. According to Mary's account, Bess had been "doubled over with laughter" as she had recounted how her daughter, Mary Talbot, had mocked her royal mistress every time she made a curtsey, "never ceasing to laugh up her sleeve" at her. Meanwhile, Bess and her late friend, Margaret Douglas, Countess of Lennox, had not dared to look at each other when they were in attendance on the Queen "for fear of bursting into gales of laughter."[4]

There is a ring of truth to this latter story, for Elizabeth had let it be known that as her countenance "shone with a blinding light like the sun," none of her ladies should look her in the face for fear of being dazzled by its brilliance. When Bess and her daughter visited court in 1578 and were admitted to the Queen's presence, they had been seized by a silent fit of giggles at the ridiculous notion and had been unable to look at each other in case they laughed out loud.[5]

However, the rest of Mary's account can be given little credence. It is not substantiated by any other source and was probably based upon exaggerated half-truths and hearsay. Moreover, it was written at a time when Mary had fallen out spectacularly with the Countess of Shrewsbury and was therefore trying to do everything she could to discredit her. Fortunately for Bess, Lord Burghley intercepted the letter before the Queen saw it. Whether Elizabeth would have given it any credence if she had read it can never be said for certain, but the allegations it contained were so insulting that even if she had not believed it all, she would probably still have harbored a lingering resentment toward Bess. Meanwhile, when Bess herself heard of it, she fiercely denied everything that Mary had said, and the Privy Council eventually accepted her innocence.

The traumas of her former life in Scotland, together with the strain of her ongoing captivity in England, were beginning to take their toll on Mary's health. She suffered from rheumatism and a chronic pain in her side, her hair turned prematurely grey, and she put on a considerable amount of weight. Despite the efforts she made with her appearance, she bore little resemblance to the beautiful princess who had been the most desirable bride in Europe just a few years before. In November 1582, some fourteen years after her flight to England, she wrote a long and embittered letter to her cousin, listing everything she had suffered at Elizabeth's hands and demanding "satisfaction before I die, so that all differences between us being settled, my disembodied soul may not be compelled to utter its complaints before God." She shared her laments with anyone who would listen, complaining to the Spanish ambassador of the "implacable vengeance with which this Queen was treating her by depriving her of her liberty."[6]

Mary was provoked not just by what she termed "ma longe captivité," but by the disappointment of her hopes that her son, James, would come to her rescue.[7] Her son was a stranger to her, having been raised by others since his earliest infancy and retained in Scotland after her escape to England. No doubt influenced by the regent, Moray, a bitter enemy of Mary, he evidently felt little loyalty toward his mother and was guided more by self-interest. Thus when in 1583 Mary opened negotiations for a joint sovereignty with her son, he sided with Elizabeth and rejected the scheme. The English queen derived some satisfaction from knowing that this prince, of whose birth she had been so jealous, had shown more loyalty to her than to his own mother. "If the half of that good nature had been in his mother that I imagine to be in himself he had not been so soon fatherless," she declared.[8] By July 1586, James had proved this "good nature" by concluding an alliance with the English queen that brought him an annual pension of £4,000 and all but severed ties with his mother for good.

That her son should so readily turn his back on her was a bitter blow for Mary, and she lamented to her cousin: "Alas! Was ever a sight so detestable and impious before God and man, as an only child despoiling his mother and her crown of royal estate?"[9] In her grief and disappointment, she railed against the English queen and her regime. This occasioned some alarm among Elizabeth's courtiers and ministers, who were anxious lest Mary should involve herself in another plot against their royal mistress. In July 1584, George Gilpin, Elizabeth's envoy in the Netherlands, urged that "the Queen of Scots devises be nearly looked to, who will never cease to practise to compass her wicked purpose."[10] By this time, Elizabeth herself was said to be "utterly distrustful" of this "bosom serpent," as Mary became known, and paid little heed to her occasional halfhearted reassurances that she desired nothing but "perfect amity" between them.[11] She was right to be suspicious, for the previous November, her agents had uncovered a conspiracy—known as the Throckmorton plot—to place Mary on the English throne with the aid of an invasion by Spanish and French troops. Although Mary was almost certainly involved in the plot, there was no direct evidence against her, so Elizabeth was unable to act. However, the following year she passed the Act for the Security of the Queen's Royal Person, which bound Englishmen to "prosecute to death" any "pretended successor" who rebelled against her.

In January 1585, Mary was removed from the custody of the Earl of Shrewsbury on the grounds that he had become too sympathetic to his captive. The closeness that existed between them had caused a serious rift between the earl and his wife. Although Mary and Bess had enjoyed a close friendship for a while, this had soon fallen apart when Mary had realized how obstructive the countess could be toward her own dynastic ambitions. The principal cause of discord between them had been Bess's attempts to make an advantageous marriage for Arbella. That the claim of Mary's own son should be so threatened by the countess's schemes filled the Scottish queen with bitter resentment. She wrote a furious letter of complaint to the French ambassador, urging him: "I would wish you to mention privately to the Queen that nothing has alienated the Countess of Shrewsbury from me more but the vain hope which she has conceived of setting the crown of England on the head of her little girl, Arbella."[12]

By the time Mary was removed from the Shrewsburys' custody, she and Bess were barely on speaking terms. Whether she truly believed it or not, the countess was outspoken in her complaints that her husband was sleeping with the Scottish queen and that the latter had borne him at least one child, possibly "several." The Earl of Shrewsbury had certainly developed a soft spot for his beguiling captive, but it is unlikely that he did anything about it. By now well into his sixties, he was beset by ill health and probably incapable of conducting a torrid affair with anyone. But thanks to Bess, rumors of their affair were now rife. Enraged by this slur on her reputation, Mary complained of "the insolence of this vulgar-minded woman" and demanded that the countess revoke her accusations. "She is marvellously grieved with the Countess of Shrewsbury for the foul slanders of late raised upon her by the said countess," reported Henry Sadler to Walsingham, "which, having touched her so near in honour and reputation abroad, she says she can no longer sustain, but trusts that her majesty will suffer her to have justice."[13] In the end, Elizabeth herself was forced to intervene and ordered an investigation into the rumors. A short while later, Mary was able to report: "The Countess of Shrewsbury (I thank God) hath been tried and found to her shame in her attempt against me."[14]

Although the Queen had been obliged to silence Bess on this occasion, she nevertheless sympathized with her predicament. If the earl had not committed adultery with Mary, Queen of Scots, then he had been overly friendly toward her at a time when his attitude toward his wife was one of increasing disdain. Whereas before he had been attracted by Bess's strength of character and ambition, now he found her domineering manner irritating and complained to a friend: "It were no reason that my wife and her servants should rule me and make me the wife and her the husband."[15] He also resented the amount of time Bess was spending on creating a magnificent home at Chatsworth, and argued with her over the income from their estates. Meanwhile, his own behavior was becoming ever more erratic, leading some observers to doubt his sanity.

An example of this came in July 1584, when Shrewsbury suddenly threw his wife out of her beloved Chatsworth, claiming that he could no longer bear the sight of her. The countess could scarcely believe what had happened. Although their relationship had begun to unravel, she had not anticipated such a drastic reaction. Shocked and distraught, she resolved to leave Derbyshire with all haste and throw herself on the mercy of the Queen. It must have been with great relief that, upon her arrival at court that summer, she found Elizabeth inclined to sympathize with her cause. Her royal mistress had never approved of marital disagreements, believing that having entered into that sacred union, couples ought to do their best to live in harmony. Bess was equally steadfast on the subject of marriage. True, she had had three husbands before Shrewsbury, but she had served each faithfully and had only been separated from them by death. It was inconceivable that she might seek a way out of her predicament by divorcing the earl.

The Queen now found herself in the unenviable position of referee between the warring couple. Shrewsbury insisted that she should punish his wife for her slanders and banish her from court. Elizabeth was still very fond of the earl, who, unlike his wife, had always proved the most loyal of servants. On the other hand, she disapproved of his treatment of Bess, with whom she felt a measure of female solidarity. Loath though she was to interfere in the dispute, she eventually succumbed to Bess's "piteous and lamentable" pleas to "take the order of this cause into her own gracious hands" and launched a commission of inquiry to examine the claims of both sides, making it clear that the purpose of this was to preserve their marriage, not end it.[16]

It may have been thanks to Bess's persuasions that the Queen decided to remove Mary Stuart from Shrewsbury's custody. Bess had told her royal mistress "that, so long as the Queen of Scots was in the hands of the earl of Shrewsbury, she would never be secure, as he was in love with her."[17] When he heard of this, the earl pretended indifference and, having recently come to court, kissed the Queen's hand "for having, as he said, freed him from two devils, namely the queen of Scotland and his wife."[18]

While the earl continued to rant and rave against his wife, Bess maintained a dignified composure, responding to his increasingly outlandish claims with reason and patience. This strategy served her well, for in the end it was she who triumphed. In the summer of 1586, the Queen summoned both parties to her presence and announced the verdict of her commission. She declared that she could "not suffer in our realme two personnes of your degree and qualitie to live in such a kinde of divided sort," and therefore ordered that they be reunited. The Countess of Shrewsbury was completely vindicated, and the earl was instructed to take her back into his house. Shrewsbury pretended compliance, and it was reported that both he and Bess "shewed themselves very well content with her Majesty's speeches, and in good sort departed together, very comfortable to the satisfaction of all their friends."[19] Once back at their estates, however, they lived virtually separate lives, and the earl declared that he would "neither bed with her nor board with her."[20]

If Mary had derived any satisfaction from witnessing the very public spat between her former custodians, she did not have long to enjoy it. Now under the guardianship of Sir Amyas Paulet, an austere man with a good deal less sympathy for her than Shrewsbury had demonstrated, her imprisonment became even more onerous. This fresh misery prompted her to undertake a course of action so dangerous that if it failed she would lose everything—her life included.

In the summer of 1586, Anthony Babington, a Catholic gentleman who had become acquainted with Mary through his service to the Earl of Shrewsbury, masterminded a plot to assassinate Elizabeth and place Mary on the throne, assisted by Spanish forces. Walsingham soon heard of it through his spies and determined to lay a trap for the Scottish queen. A channel of communication was established for Mary, whereby she would send coded letters hidden in beer barrels to the conspirators. Little did she know that all of these were intercepted by Walsingham, who was waiting patiently until he had enough evidence to condemn her. All she had to do was mention Elizabeth's death, and it would be treason. He did not have to wait long. On July 17, Mary wrote to Babington endorsing his suggestion that the English queen be "despatched" by a group of noblemen. "Sett the six gentlemen to work," she urged. She had as good as signed her own death warrant.

The following month, Babington and his fellow conspirators were arrested and interrogated. Under torture, they confessed to the whole plot—and, crucially, Mary's complicity in it. Surely now Elizabeth would have no choice but to finally put her cousin to death. In the immediate aftermath, it certainly looked as if she would do so. She ordered Amyas Paulet to tighten the security around his captive and deprive her of her former luxuries. Paulet was only too happy to oblige, and Elizabeth praised him for his efforts in keeping "so dangerous and crafty a charge." She also sent him a letter full of fury at her traitorous cousin, demanding that he "Lett your wicked murtheress know how with hearty sorrow her vile deserts compelleth these orders, and bid her from me ask God forgiveness for her treacherous dealings towards her saver of her life many a year, to ye intolerable peril of her owne." Referring to all the times Mary had tried to persuade her to show some female solidarity, Elizabeth scoffed that her cousin's actions were "far passing a woman's thought, much less a princess."[21]

Elizabeth also wrote to Mary herself, lambasting her for showing such base ingratitude for all the help she had offered her over the years. "You have in various ways and manners attempted to take my life and bring my kingdom to destruction by bloodshed," she began. "I have never proceeded so harshly against you, but have, on the contrary, protected and maintained you like myself."[22] As well as committing her fury to paper in her letter to Paulet, she also railed against Mary's treachery to her courtiers and councillors in London and took particular exception to anyone whom she perceived as having favored her cousin in the past. "Well, what do you think of your Queen of Scotland?" she demanded of one hapless ambassador. "With black ingratitude and treachery she tries to kill me who so often saved her life. Now I am certain of her evil intent."[23]

Yet however great Elizabeth's anger, and however much she might rant to her courtiers and ministers about her cousin's treachery, she still shrank from deciding her fate. Only after intense pressure from Burghley and Walsingham did she agree, with considerable reluctance, to appoint a commission for Mary's trial. This took place at Fotheringay Castle, Northamptonshire, in October 1586. Although Mary defended herself with skill and dignity, the verdict was never in question. Toward the end of the month, the trial was transferred to the Star Chamber in Westminster, and on October 25 the verdict was pronounced. Mary, Queen of Scots, was found guilty of conspiring toward "the hurt, death and destruction of the royal person of our sovereign lady the Queen."

Parliament was convened to decide upon the sentence, and on December 4 it was proclaimed that Mary should be put to death.

Yet still it required Elizabeth's sanction. No execution could go ahead unless she signed the warrant herself. And still she wavered. This was partly because of a justifiable concern that putting Mary to death would threaten England's international security, for France or Spain might well launch a revenge attack. Mary herself had written to Philip II upon hearing of the commission's verdict, urging him to avenge her death by invading England and taking the crown for himself. There was also Scotland to consider. Mary's son had already let it be known that he would "no ways keep friendship if his mother's life be touched."[24]

But aside from these diplomatic matters, there was another, more pressing concern—one that above all else made Elizabeth flinch from signing the warrant. Mary was no straightforward traitor: she was an anointed queen. She was also of Tudor blood. To put such a woman to death would set a shocking and dangerous precedent, not to mention plaguing Elizabeth's conscience more than anything else she had done. "What will they now say that for the safety of her life a maiden Queen could be content to spill the blood even of her own kinswoman?" she lamented.[25] Katherine and Mary Grey had been punished for their treachery, but they had at least absolved Elizabeth of guilt by dying of natural causes. The same had been true of Margaret Douglas. Mary, Queen of Scots, on the other hand, had been a constant thorn in Elizabeth's side for almost thirty years, and the threat she posed to her security had increased tenfold after her flight to England. It would be too much to hope that if Elizabeth deferred her execution in favor of continued imprisonment, Mary would die at a convenient moment before causing any further trouble. She had already proved far too great a danger for that.

There was another reason why Elizabeth found the idea of putting Mary to death so abhorrent. She had apparently finally grasped the point that her cousin had made so often about female solidarity. Although she had scoffed at this in her letter to Paulet, it apparently weighed increasingly upon her mind, especially after Mary referred to it in a letter she wrote in December, stressing "our sex in common."[26] Her councillors were evidently aware of this, for it was one of the main issues they tackled in a paper that set forth the necessity of having Mary executed. "What compassion is to be had of her who has transgressed the bounds of that modesty and meekness that her sex and quality prescribes . . . shall any difficulty be alleged in executing her that so heinously has gone about to procure the murdering of the Lord's anointed, a lady, a Queen, a virgin?"[27]

This tactic of praising Elizabeth's virginity in contrast to Mary's licentiousness showed just how far the two women's roles had been reversed. Not so long before, Mary had been hailed as a shining example of womanhood; Elizabeth as an abomination. Yet, ultimately, while Mary had confirmed all the stereotypes about women's incapacity to rule, Elizabeth had confounded them. Mary had proved to be a woman first, a queen second; Elizabeth had proved the opposite and had triumphed because of it. In the end, her virgin state had been widely accepted not as an anomaly but as a virtue, a symbol of the discipline and self-control that had so obviously been lacking in Mary's character.

No matter how persuasive her ministers' arguments, they did little to ease Elizabeth's dilemma. Perhaps the agony of her decision was intensified still further by the horror she still felt over her own mother's execution. That she should condemn a kinswoman to such a death must have filled her with dread. This was undoubtedly the greatest crisis of her reign so far, and she almost broke down under the stress. There was no reprieve from her councillors, who plagued her daily—hourly, even—with demands for her cousin's death. In November, while Elizabeth was seeking refuge at her favorite palace of Richmond, a parliamentary delegation arrived to put yet more pressure on her to conform to their wishes. Her response made it clear how tormented she was by the dilemma she faced. "Though my life hath been dangerously shot at," she told them, "yet I protest there is nothing hath more grieved me, than that one not differing from me in sex, of like rank and degree, of the same stock, and most nearly allied to me in blood, hath fallen into so great a crime." Referring to their calls for her to sanction Mary's execution, she claimed: "I am so far from it that for mine own life I would not touch her," and went on to utter what have become some of the most famous words in her oft-quoted speeches. "I assure you," she told the astonished delegation, "if the case stood between her and myself only, if it had pleased God to have made us both milkmaids with pails on our arms, so that the matter should have rested between us two; and that I knew she did and would seek my destruction still, yet could I not consent to her death." Dismissing her statesmen, she told them: "I must desire you to hold yourselves satisfied with this answer answerless."[28]

Elizabeth had made it clear that if Mary were to repent, then she would spare her life—something she claimed to wish "with all my heart."[29] Her cousin stubbornly refused to oblige and continued to insist upon her innocence. This effectively forced Elizabeth's hand, for she knew that it was now a choice between Mary's life and her own. "I am not so void of judgement as not to see mine own peril; nor yet so ignorant as not to know it were in nature a foolish course to cherish a sword to cut mine own throat." Yet still she desperately sought a way to save both their lives, and lamented: "I am right sorry [this] is made so hard, yes, so impossible."[30]

Elizabeth's decision was made yet more "impossible" by a long and impassioned letter that Mary wrote to her in December. Clearly trying to shock her cousin into sparing her life, Mary brought her face to face with the reality of ordering her execution. "When my enemies have slaked their black thirst for my innocent blood, you will permit my poor desolated servants altogether to carry away my corpse, to bury it in holy ground, with the other Queens of France, my predecessors, especially near the late queen, my mother." She also made reference to Henry VII—"your grandfather and mine"—and begged that she might be permitted to send "a jewel and a last adieu to my son."[31]

The Earl of Leicester noted that the letter "wrought tears" from his royal mistress, who pleaded that the "timerousnes of her [own] sex and nature" made the dilemma she faced all the more agonizing. For all this show of womanly weakness, however, Elizabeth's favorite rightly predicted that Mary's letter "shall do no further [damage] herein."[32] Perhaps, after all, the Queen wanted to prove them wrong about women's inability to rule and make it clear that she really did have the "heart and stomach of a king." Although Elizabeth kept her councillors—and her cousin—waiting for a further month after receiving this letter, by the end of January she had apparently decided that death was the only option for Mary. Still shrinking from sanctioning her execution, she urged Paulet to "ease her of this burden" by secretly putting Mary to death with poison or some other means. This would absolve Elizabeth of responsibility—and also clear her of blame in the eyes of the world. Paulet was horrified at the suggestion and utterly refused to carry it out.

With apparently no other option left to her, on February 1 Elizabeth finally signed her cousin's death warrant. The controversy surrounding this action would tear her council apart and has been the subject of intense debate ever since. The central question is whether Elizabeth, as she later claimed, signed the warrant but ordered her secretary William Davison not to issue it until she gave the order. Melville certainly believed that this was the case and noted in his memoirs that she had given Davison "express command" not to deliver it.[33]

However, Davison's own account, made during his incarceration in the Tower, casts doubt upon this theory. He described the occasion in detail, saying that he had brought the warrant to Elizabeth along with several other papers, and that she had signed it, apparently without a care. She had known full well what it was, however, for she had asked her secretary if he was sorry for it. He had replied that he was, but that it was necessary: "Which answere her highnes approving with a smiling countenance, passed from the matter, to aske me what ells I had to signe and theruppon offering unto her some other warrants & instructions touching her service, yt pleased her with the best disposition and willingnes that might bee to dispatch them all." According to Davison, the Queen had then specifically ordered him to have the warrant sealed before giving it to the lord chancellor with a special order to "use it as secretly as might bee." She had also told him to inform Walsingham that she had at last given in to pressure by him and others and ordered Mary's execution, joking that "the greeife therof would goe neare (as she merrilie sayd) to kill him."[34]

The apparently blasé way in which the Queen had signed the warrant made Davison confident that she was not about to change her mind. Nevertheless, he waited two days before presenting it to her councillors. In the meantime, he received further proof of Elizabeth's resolve, for the day after signing the warrant, she told him that she had been troubled with a dream that the "Scottish Queene" had been executed. She had apparently "bin soe greatly moved with the newes" that she had railed against Davison in her dream with such "passion shee could have done I wott [know] not what." However, she told all this to her secretary "in a pleasant and smyling manner," and when he asked her what it meant and "whither having proceeded thus farre, shee had not a full and resolute meaning to goe through with the sayd execution according to her warrant," she confirmed, "with a solemne oath in some vehemencye" that she had no regrets.[35]

The following day, Davison presented the warrant to a group of councillors. They agreed that it should be implemented straightaway, and it was duly dispatched to Fotheringay. Upon receiving it, Paulet, no doubt relieved at having received no further orders to have Mary secretly assassinated, immediately set about making preparations for the execution. Mary herself took the news of her fate calmly, with "a stable and stedfast countenance," determined to set herself up as a Catholic martyr by declaring that she was being put to death for her faith, rather than for any treachery. She spent the night before her execution praying devoutly, a crucifix in her hand, and consoled her weeping ladies by telling them "how signal a mercy God was showing her in rescuing her from the power of so bad a woman as the queen of England."[36]

On the morning of February 8, 1587, Mary, Queen of Scots, mounted the scaffold in the great hall of Fotheringay Castle. She was barely recognizable from the beautiful woman who had so beguiled the world. An eyewitness described her as: "round shoulder'd, of face fat and broad, double chinned and hazel eyed; borrowed hair."[37] But still she had that presence and enigma that drew all eyes to her. She was ever one for theatrical gestures: when her ladies took off her outer gown, it revealed an underdress of scarlet, the color of martyrs. She then proclaimed her status as an anointed queen and, one last time, stressed the responsibilities that she shared with her cousin as a fellow sovereign, woman, and "sister." Turning to her executioner, she pardoned him and told him she was "glad that the end of all her sorrowes was so neare." She also told her women to cease their "whininge and weepinge," urging them instead "to thanke God for resolutenes."

When Mary lowered her head onto the block and gave the signal that she was ready for death, the executioner "struck at her neck" with his axe, but missed and instead sliced into the side of her face. "Lord Jesus, receive my soul!" Mary exclaimed, at which the executioner again hacked at her neck but still did not sever it. It was only with the third blow that Mary's head finally fell upon the scaffold. When the executioner stooped to pick it up, it came away in his hands, and he was left holding only her wig. In the increasingly macabre farce, Mary's little dog then scurried from where he had been hiding under her dress and "laid itself down betwixt her head and body, and being besmeared with her blood, was caused to be washed, as were other things whereon any blood was."[38]

Mary's desire to set herself up as a martyr had succeeded. As news of her execution spread across Europe, she was portrayed as a brave Catholic princess who had died at the hands of that wicked heretic the Queen of England. One account referred to her as "that sweet saint and martyr," while another observed: "She finished her happy and blessed martyrdom, to the comfort of all true Catholics, and to the shame and confusion of all heretics." Meanwhile, a Spanish envoy reported: "Our Lord will have taken [Mary] into heaven, seeing that she died a martyr."[39]

When the news reached the court in London, Lord Burghley ordered that it should be concealed from the Queen until the evening. Accompanied by Sir Christopher Hatton, vice chamberlain of the council, he then went to seek out his royal mistress, no doubt apprehensive about what her reaction would be. Upon being told that her cousin had been executed, Elizabeth was apparently at first not able to comprehend it. "Her words failed her," claimed William Camden. "She was in a manner astonished."[40] She remained that way for the rest of the evening, giving no indication of the storm that was about to come. The next morning, she flew into a rage so fierce that her councillors had never seen the like. One observer tremblingly noted that she was in such "heate and passion" that she screamed out against the execution "as a thing she never commanded or intended." She then set about "casting the burthen generallye uppon them all," but, as Davison lamented, "chiefly upon my shoulders."[41]

Elizabeth's storms usually passed as swiftly as they had arrived. This one did not. As the days passed, it seemed only to grow in intensity, and in her near-hysterical fury, she threatened to throw all her councillors in the Tower for such blatant defiance of her orders. In the meantime, she "commanded them out of her sight."[42] Even her closest adviser, Lord Burghley, was said to be in "deep disgrace" for many weeks after the event. Ten of the offending councillors were ordered to appear before the lord chancellor, the Lord Chief Justice, and the Archbishop of Canterbury in order to justify their actions. It was Davison who bore the brunt of the Queen's rage. Determined to create a scapegoat for the whole affair, Elizabeth stripped him of his office, sent him to the Tower, and subjected him to a full interrogation by the Star Chamber. He was subsequently fined £10,000, a sum far beyond his means, and ordered to stay in prison for as long as Her Majesty pleased. While few people believed that he was truly to blame, a rash of propaganda was put forward casting him as the villain.[43]

As well as lambasting Davison and her other councillors, Elizabeth also effected a show of extreme sorrow at the death of her cousin. Her first biographer noted: "She gave her selfe over to griefe, putting her selfe into mourning weedes, and shedding abundance of teares." An ambassador at court concurred that she had "taken to her bed owing to the great grief she suffered through this untoward event."[44] Elizabeth's "great grief" was viewed with some skepticism by her enemies abroad. Philip II declared: "It is very fine for the Queen of England now to give out that it was done without her wish, the contrary being so clearly the case."[45] Indeed, her show of anguish may have been largely for the benefit of such potentates: One of the reasons why she had held back from ordering the execution for so long was because she

feared an international backlash. In the immediate aftermath of
Mary's death, it seemed that her fears were justified. The English ambassador in Paris noted with some alarm: "Truly I find all men here in
a fury, and all that love not her Majesty in a great hope to build some
great harm to her upon it." He added that the French king "took it
very evil" when he heard the news.[46]

Worse was expected from James VI, for quite apart from the political implications, it was his mother who had been executed. It was reported that he took her death "very heavily."[47] Elizabeth wrote an
impassioned letter to him, full of grief at Mary's execution and putting
the blame squarely upon others. "I would you knew, though felt not,
the extreme dolour that overwhelms my mind for that miserable accident which, far contrary to my meaning, has befallen," she wrote. "I
beseech you that as God and many more know how innocent I am in
this case."[48] James was all too easily appeased. Shortly after receiving
this letter, it was reported that if the Queen persisted in declaring her
innocence, "the King shall love her and honour her before all other
princes." He went on to publicly excuse her of all blame in "the late execution of his mother, and layeth the same upon her Council." The
swiftness with which James forgave Elizabeth sparked a great deal of
criticism. Melville scornfully remarked that "the blood was already
fallen from his Majesty's heart." By contrast, this news of his pardon
"did wonderfully content her Majesty, who desireth nothing more
than to have it generally conceived that she had least part in the action."[49]

It had certainly been politically expedient for Elizabeth to deny any
involvement in Mary's death, but the near hysteria of her reaction suggests that she was more than just a consummate actress. At least some
of the fury she directed against her hapless councillors may have been
inspired by resentment at the fact that, for the first time in her reign,
they had forced her into making a decision. Until then, she had always
succeeded in giving them "answers answerless" and procrastinating
just long enough for the issue at hand to be resolved or superseded by
another, more pressing, concern. But with Mary, she had been placed
under relentless pressure, not just in the weeks following her trial and

sentence, but throughout the nineteen years of her imprisonment. As Elizabeth herself told her councillors accusingly at the time: "You have laid a hard hand on me."[50] If she did not really believe that they had issued the warrant unlawfully, then she did blame them—indirectly—for Mary's death. Had she been left to her own devices, she would undoubtedly have settled upon more of a compromise, dangerous though that might have been to her own security. As it was, she had been forced into taking a course of action that would plague her for the rest of her life. It was the first occasion upon which she—a female sovereign—had allowed herself to be bullied by her male councillors. She resolved that it would never happen again.

It is also possible that Elizabeth's show of remorse at her cousin's execution was at least partly genuine. When she heard of the manner of Mary's death, she was horrified at the lack of dignity that had been accorded to her. Mary had been deprived of her servants as she made her way to the scaffold, and she had been "led like a lamb to the butchery." The many accounts of the blundering executioner would also have appalled Elizabeth, who was already sensitive enough to female sovereigns being put to death, thanks to the experience of her mother and stepmother, both of whom, like Mary, had represented carnal femininity. Little wonder that she claimed to have nightmares about it for many years to come. What appalled Elizabeth most, however, was that she had put a fellow sovereign to death. She had an almost superstitious revulsion about doing so, convinced that vengeance would be wreaked upon her by either an earthly or heavenly force. Fear for her own future now mingled with remorse.

For all her professed anguish at the death of Mary, Queen of Scots, Elizabeth failed in the most basic duty toward her cousin. By July 1587, some five months after the execution, Mary's corpse still lay rotting in Fotheringay Castle. The stench, which was already "noisome," had become intolerable in the summer heat, so that none wished to enter the room where it was kept. At the end of that month, Elizabeth finally gave orders for her cousin's body to be moved to Peterborough Cathedral for burial. Although she made a show of ordering a lavish funeral, with full royal honors and great pomp, it was too little, too late. Furthermore, while it was not customary for royalty to attend funerals, the woman whom the English queen chose to represent her as chief mourner was not even the highest-ranking member of her entourage. This fact, together with the apparent neglect over ordering the funeral in the first place, seemed like a deliberate insult to the late queen. Only after Elizabeth's death were Mary's remains reinterred at Westminster Abbey by her son in a rather belated show of filial loyalty. She was buried in the chapel directly facing that of her English cousin—in death, as in life, on opposing sides.

CHAPTER 13

Gloriana

If Elizabeth had hoped that by putting her cousin Mary to death she would be extinguishing her most dangerous rival for good, then she was mistaken. In a sense, the woman whom the Duke of Norfolk had referred to as the English queen's "competitress"[1] was an even more threatening adversary now than she had been in life. In executing her cousin, Elizabeth had given her an almost mythical status. Mary had made sure that in the eyes of the world she died a martyr, not a traitor. She had called upon the Catholic powers to avenge her death, and the very next year, the greatest of them took her at her word.

In May 1588, Philip II launched his armada against England, ostensibly in Mary's name, even if he had other motives. This was the greatest threat that Elizabeth had faced in her reign, and that England had faced since the Norman invasion more than five hundred years before. Of course, it was not entirely due to Mary, but her death had given Philip the excuse he needed finally to destroy this troublesome heretical queen who had so long defied him. Just as she had when agonizing over Mary's fate, so Elizabeth was again forced to set aside her customary prevarication and lead her country decisively into battle.

When the English queen emerged victorious, it transformed her image into one of invincible majesty that became ever more idealized as her reign progressed. Like Mary, she had attained mythical status— but in her own lifetime. In so doing, she had been compelled to abandon the hesitation and doubt born of youth and insecurity, and instead embrace a more confident, resolute vision of herself. It was in finally resolving the complex question of what to do with Mary, Queen of Scots, and in dealing with the dramatic repercussions that followed, that Elizabeth was able to mature into the Gloriana of legend.

The defeat of the Spanish Armada won Elizabeth greater popularity than ever before. Ever one to capitalize upon public opinion, she made sure that her subjects were in little doubt that the victory was thanks to God's favor toward her, and ordered commemorative medals to be struck that carried the words: "God blew with his wind and they were scattered." Even her most implacable adversaries were forced to acknowledge the greatness she had attained. Shortly after the victory, Pope Sixtus V exclaimed: "She is only a woman, only the mistress of half an island, and yet she makes herself feared by Spain, by France, by Empire, by all!"[2] Elizabeth had apparently conquered not just Philip II's armada but also the widely held prejudices against her ability to rule as a "mere woman." In a marked contrast to his earlier remark, Lord Burghley declared: "There was never so wise woman born as Queen Elizabeth."[3] She had come to reign as absolutely as her father, dominating her ministers and brooking no opposition to her will. Toward the end of her reign, one contemporary marvelled that it was "a rare thing that a woman sitting in council amongst the gravest and best experienced men of her time should be able to examine and individually to control their consultations."[4]

Emphasizing her divine status had enabled Elizabeth to transcend what she perceived to be the weakness of her sex. It had also won her widespread acclaim and loyalty among her subjects. She claimed that it was because she had been chosen by God that she, a "mere woman," was able to rise above her natural weaknesses and rule over a kingdom. The divine nature of her status was stressed time and again in her speeches. When pressured by Parliament to take a husband, she told them that even though "the weight and greatness of this matter" might be considered too much for her feminine understanding, "yet the princely seat and kingly throne wherein God (though unworthy) hath constituted me . . . boldeneth me to say somewhat in this matter."[5]

By presenting herself as God's chosen queen, Elizabeth ensured that no subject—whether male or female—could oppose her will. She told another of her parliaments: "I am your anointed queen, I will never be by violence constrained to do anything." She later wrote a prayer of thanks to God "for making me (though a weak woman) yet thy instrument to set forth the glorious gospel of thy dear son Christ Jesus."[6] She claimed that it was thanks to God's protection that she had survived the perils of her half sister's reign to become queen herself. "It is thou who hast raised me and exalted me through thy providence to the throne . . . pulling me from the prison to the palace, and placing me a sovereign princess over the people of England." Later, in her famous "Golden Speech" to Parliament in 1601, she again drew attention to the contrast between her frailty as a woman and her power as a divinely appointed sovereign, declaring: "Shall I asribe anie thing to myselfe and my sexelie weaknes I were not worthie to live then and of all most vnworthie of the mercies I have had from god."[7]

As her reign progressed, Elizabeth would invoke her divine status to devastating effect. She became not just God's representative on earth but also a divinity herself: the peerless Virgin Queen whose powers set her apart from ordinary mortals. Presenting herself in this way was a stroke of genius, for it rendered her maidenly state something to be celebrated and admired. She was the Virgin Mary here on earth, appointed by God to bring her people to glory. Even in her own lifetime, myths and legends began to surround her, thanks in no small part to the adulatory poems and prose that were written in her honor. She was "Gloriana," "a most royal queen and empress" to Edmund Spenser, as well as "Diana," "Cynthia," "Astraea," and "Belphoebe," "a most virtuous and beautiful lady." The portraits of her became ever more fantastical, emphasizing her ethereal nature as an eternally youthful goddess ruling over her adoring subjects.

But this was not the only strategy the Queen employed to secure her people's loyalty. In rallying her troops at Tilbury when the onset of the armada was imminent, she had given one of the most famous speeches of her reign, telling the assembled masses: "I know I have the body of a weak and feeble woman, but I have the heart and stomach of a King, and a King of England too."[8] On many other occasions, she would promote herself as a king or a prince in the speeches she gave to her subjects. "I have the heart of a man, not of a woman, and I am not afraid of anything," she once declared.[9] Whenever she needed to stamp her authority upon overbearing councillors or ambassadors, she would often invoke a more specific male image by comparing herself to her father, Henry VIII. In one of her speeches to Parliament, she told them: "And though I be a woman, yet I have as good a courage, answerable to my place, as ever my father had."[10]

At the end of Elizabeth's reign, Sir Robert Cecil reflected that she had been "more than a man, and, in truth, sometimes less than a woman."[11] The Queen did indeed have a number of characteristics that sixteenth-century society understood to be distinctly male. For a start, she had an exceptionally sharp mind and was the intellectual superior of most of her male contemporaries. She was also remarkably brave in the face of adversity. "She had more valour in her than was fit for a woman," remarked one of her contemporaries with a mixture of admiration and disapproval.[12] Elizabeth's remarkable energy and vigor were also viewed as masculine traits. She was a tireless horsewoman and would hunt and hawk with as much enthusiasm as her male courtiers. Unlike many women who followed the chase, she shot as well as rode, and did not flinch from cutting the throat of a deer when it had been cornered. Even in old age, she would still enjoy brisk early morning walks in the gardens of her palaces, tiring out many of her young female attendants in the process. It was said that she preferred the male steps of the galliard, a popular dance of the time, because these involved athletic leaps. She was often boisterous. When amused, she would laugh as uproariously as a man, and at other times would storm through her apartments, slapping and beating her ladies.[13]

Nevertheless, the fact that Elizabeth was a highly sensual woman who revelled in her male courtiers' attentions suggests that she was far from being the one-dimensional "honorary male" that some commentators have claimed. Although she often bemoaned the weakness of her sex, this was more in terms of the limitations it threatened to place on her power. In other ways, she celebrated her femininity and used it to devastating effect in bringing her courtiers to heel.

One of her most feminine traits, at least in the eyes of her male courtiers, was her tendency to delay and procrastinate when faced with pressing issues. The question of her marriage was the most notorious example of Elizabeth's delaying tactics, but there were many others. Whenever she was under pressure to make a decision, she would excuse herself on the basis of her "feminine weakness," and would drive her councillors to distraction as she vacillated between one option and another. Indecision was seen as a typical female failing, so this confirmed their fears about having a queen to rule over them. Few of them realized that Elizabeth was merely playacting. The experience of her childhood and the example of Mary's reign had taught her the danger of inflexibility. Catherine of Aragon had stubbornly refused to submit to the king's will, and had died miserable and impoverished as a result. Mary Tudor had doggedly pursued her vision of restoring Roman Catholicism to England without realizing that in so doing she was losing the goodwill of her people. Elizabeth had learned that single-minded principles brought little but grief and suffering in the end; it was far better to be pragmatic and alter one's policy according to the ever-shifting tides of politics.

What her ministers took to be indecision or procrastination was actually political shrewdness and a determination to wait and see which way the tide would turn. The Queen was also deliberately unpredictable and made it impossible for her councillors to judge what she was really thinking. "Hir wisest men and beste counsellors were oft sore troublede to knowe hir wyll in matters of state: so covertly did she pass hir judgemente," recalled Sir John Harington.[14] The wisdom of this policy was proved time and again. By keeping her various suitors in play with "answers answerless," she navigated England through the turbulent waters of international politics. And by refusing to act precipitately against Mary, Queen of Scots, she was rewarded when that same woman was delivered into her clutches after fleeing from Scotland. Few, if any, of her male contemporaries seemed to realize that this was a deliberate ploy. "It is very troublesome to negotiate with this woman, as she is naturally changeable," complained the Count de Feria.[15]

More positively, Elizabeth was capable of showing a very tender kindness and sympathy for her subjects. She wrote a heartfelt letter of condolence to her friend Lady Margery Norris, whom she affectionately called "My Old Crow," upon the death of the latter's two sons in the Irish wars. "Harm not yourself for bootless help, but shew a good example to comfort your dolorous yoke-fellow [husband]," she began. "Nature can have stirred no more dolorous affection in you as a Mother for a dear Son, than a gratefulness and memory of his service past, hath wrought in Us his Sovereign."[16] She showed a similar sensitivity when the Countess of Huntingdon's husband died, and also wrote a kind letter to Elizabeth Drury upon the death of hers, assuring her: "a queen for her love, who leaves not now to protect you when your case requires care."[17]

During the early part of her reign, Elizabeth had drawn more attention to her masculine than her feminine attributes. As it progressed and she grew more confident as a queen regnant, she allowed both sides of her character to shine through. She also interchanged them as the situation demanded. She was the *belle dame sans merci* to her courtiers, an infuriatingly indecisive and parsimonious woman to her councillors, a mother to her people, and the Virgin Queen to all. But when it suited her, she would suddenly invoke her "kingly" majesty in order to show courage in the face of battle, assert her intellectual and political acumen, and lambast her recalcitrant councillors if they refused to do her will.

In her Golden Speech of 1601, Elizabeth claimed "the glorious name of a King" but added that she had the "royall authoritie of a queene."[18] This interchanging of imagery was what made her such a brilliant monarch. It set her apart from all her predecessors—kings and queens alike—and created a unique identity that would assume iconic status both during her own lifetime and over the centuries that followed.

The armada was a pivotal point in this development of the so-called Cult of Elizabeth, giving her the confidence to experiment with and promote her image. Her confidence was matched by that of her kingdom: England had faced down the might of Spain and begun to establish itself as a world power. But the year 1588 was also one of immense sorrow for the Queen. Just a few short weeks after her victory over Philip II's fleet, her closest male favorite, Robert Dudley, Earl of Leicester, died. Elizabeth was grief stricken at the loss of the only man whom she had truly loved. For the rest of her life, she kept in a locked casket by her bed a note that he had written her shortly before his death. The Queen inscribed it "His last letter."[19] In the days immediately after his death, she kept to her rooms, unable to face her court or council, and she would often fetch a tear at the mention of his name.

By contrast, Leicester's widow was preoccupied by more material concerns. Although he had in theory left Lettice well provided for with an annual income of £3,000 per year, together with £6,000 worth of plate and furniture, he had bequeathed her the unenviable task of acting as executor to his estates. "The question is whether she would be wise in the circumstances to accept the office," observed one of her associates.[20] She would have done well not to, for in the ensuing years, she was beset by a series of costly legal battles over her late husband's estates. The most serious of these involved Robert Dudley, his son by Lady Douglas Sheffield, whose inheritance of Kenilworth caused Lettice considerable difficulties, for the adjoining manors had been bequeathed to her.

The greatest problem Lettice faced, though, was how to settle Leicester's debts. These were substantial, totaling some £50,000, of which £25,000 was owed to the Crown. It was in vain that she pleaded with the Queen to release her from these debts. This was Elizabeth's chance for revenge, and she grabbed it with both hands, ordering a minute examination of Leicester's estate so that she could recover every last penny that was owed to her. She pursued her old rival relentlessly over the ensuing months until the latter was forced to sacrifice a large part of her jointure, including Leicester House, as well as a number of her jewels.[21]

But the countess "was of a light, easy, healable nature" and refused to be bowed by either Elizabeth's persecution or her own increasing penury.[22] This fact was proved when, in July 1589, she married again, barely ten months after Leicester's death. Her new husband, Christopher Blount, was a friend of her son, Robert, and was some seventeen years her junior. The marriage shocked society, and even Lettice's adoring son lamented her "unhappy choyce." She defended her action by claiming that she needed the support of a husband in her troubles, but few people believed her.

For the next few years, Lettice retreated into the background—much to Elizabeth's relief. The execution of Mary, Queen of Scots, the previous year had removed another rival. But Elizabeth had learned that there would always be others to take their place. In late summer 1588, a young woman arrived at court who would plague the English queen for the rest of her days.

From the moment of her birth, Arbella Stuart was destined to become a rival to Elizabeth. Although James VI of Scotland was technically the first in line to the throne, many believed that Arbella's claim was stronger because she had been born on English soil, whereas James was an alien and as such was disqualified from inheriting the crown or possessing land in England. Furthermore, Arbella was favored by Catholic nobles with Spanish sympathies, because James had recently forged an alliance with France. She therefore constituted a threat not just to Elizabeth but also to the ambitions of James and his mother, Mary, Queen of Scots.

Arbella's mother, Elizabeth Cavendish, had died in 1582, which had caused her grandfather, the Earl of Shrewsbury, to "humblie and lowlie beseech her majestie, to have pyttie uppon her poore Orphanntt

Arbella Stewarde."[23] The young girl had been transferred to the care of her maternal grandmother, Bess of Hardwick, and had been brought up in the latter's Derbyshire home, Chatsworth. Bess had given her an education befitting a royal princess, and Arbella had shown early promise, quickly learning a wide range of languages, including Latin, Greek, and Hebrew. Bess noted proudly that she was "very apte to learne, and able to conceive what shall be taught her."[24] In 1583 at the age of seven, Arbella had written a letter to the chancellor, Sir Walter Mildmay, in order to show off her "learning." Mildmay had been so impressed that he had told Sir Francis Walsingham that she was "a very proper child, and to my thinking will be like her grandmother, my old Lady Lennox." He added that "the little lady" had included a request in her letter that Mildmay should present "her humble duty to her majesty with her daily prayer for her majesty."[25]

This latter request had no doubt been prompted by Arbella's ambitious grandmother, Bess, who was grooming the young girl as a successor to the English throne and wished Elizabeth to acknowledge her as such. But the Queen had proved extremely reluctant to name an heir in the past, and she refused formally to recognize the girl's claim, just as she had that of James VI. Instead she sought to limit Arbella's influence from the beginning. When the girl was two years old, her English lands and revenues were seized by the Crown. James VI had used a similar ploy by refusing to grant her the earldom of Lennox with its accompanying lands. Together, these two facts meant that the young girl was dependent upon the Queen's generosity.

However, Arbella had a formidable advocate in the person of her grandmother and guardian, Bess of Hardwick. Determined to improve the girl's fortunes, Bess pestered the Queen to grant her granddaughter an allowance befitting her status. In May 1582, only a few months after taking over her care, she wrote to Lord Burghley on behalf of "my dearest jewel, Arbella" and asked him to persuade his royal mistress to contribute £600 per year toward the girl's upbringing, £400 of which was the allowance that Arbella's mother, Elizabeth, had been given during her lifetime. "So trust I you will consider the poor infant's case, who, under her majesty, is to appeal only to your lordship for succour in all her distresses." In order to persuade him to help, Bess emphasized her granddaughter's credentials as an heir to the throne. "I have specyall care not only such as a naturall mother has of her best beloved chyld, but much more greatter a respect how she ys in bloude to her majesty, albeyt one of the poureste as depending wholly on her majestys gracyous bountye and goodnes."[26]

Eventually, Elizabeth gave in to Bess's insistent requests and agreed to grant Arbella an annual income, although at £200 this was rather less than her grandmother had hoped for. The Countess of Shrewsbury also fought to restore Arbella's Lennox inheritance. Even though she failed to secure the earldom for her granddaughter, she referred to her as "Countess" and instructed all her servants to do the same. She also ordered her other grandchildren to curtsey whenever they met Arbella, in order to emphasize the superiority of her status. When she introduced the girl to Mary, Queen of Scots, she is alleged to have taunted the latter by saying that her claim to the throne was inferior to Arbella's. Little wonder that Arbella grew up with a firm conviction that she was destined to be queen. As the Venetian ambassador, Scaramelli, later recalled: "She has very exalted ideas, having been brought up in the firm belief that she would succeed to the Crown."[27]

Although Bess had doted upon her precocious young granddaughter, and the latter had no doubt revelled in her attentions, as Arbella grew up she felt increasingly suffocated by her grandmother's domineering nature and the fact that she insisted upon directing every element of her life. When she began to show signs of independence, Bess was determined to quell them, and reprimanded her "in despiteful and disgraceful words . . . which she could not endure."[28] Being raised as a pampered but restricted princess had not had a beneficial effect upon Arbella's character. She began to display a dangerous mixture of arrogance and intemperance, prone to romantic fancies and paranoia. In later years, she would become so unstable that some historians have suggested she may have suffered from mental illness.[29]

But the Countess of Shrewsbury was blind to Arbella's faults—or perhaps thought that she could correct them in due course. She therefore continued to groom her as an heir to the throne and was delighted when Arbella received an invitation to court in 1587. Convinced that Elizabeth intended to use the occasion to formally name the young girl as her heir, Bess made excited preparations for her departure and no doubt coached her granddaughter in exactly how to behave when she was presented to the Queen. Arbella duly made her way to Theobalds, Lord Burghley's Hertfordshire home, where the court was on progress. She was then twelve years old, and, by all accounts, very pretty. The Venetian ambassador described her as being "of great beauty, and remarkable qualities, being gifted with many accomplishments."[30]

Bess's preparations seemed to have paid off, for the visit was a tremendous success, although the initial signs had not been promising. Bess's son Charles, who was also present, reported: "Her Majesty spoke to her [Arbella] but not long."[31] However, Elizabeth soon changed tack and began to show Arbella great favor by inviting her to dine in the presence chamber and seating her next to herself. This was a considerable honor and one that many courtiers had hankered after for years. Meanwhile, the Queen's host and principal minister, Lord Burghley, "made exceeding much of her," and dined with her and her uncle, Charles, on which occasion he was heard to make a favorable remark about the girl to Sir Walter Ralegh.

For all these pleasantries, Arbella's visit was loaded with significance. Just a few months before, Mary, Queen of Scots, had been executed. As well as removing the strongest rival claimant to the throne, this dramatic event had also cast doubt upon the position of Mary's son, James, as a likely successor. By showing his chief rival, Arbella, such favor, Elizabeth therefore seemed to be indicating that she intended to name her as heir. This was apparently confirmed by a remark the Queen made to the French ambassador about Arbella shortly after the visit: "Look to her well: she will one day be even as I am."[32] However, as was her custom, Elizabeth stopped short of formally acknowledging her claim.

Arbella had been intoxicated by her brief glimpse of the glittering world of the court, and it was with considerable reluctance that she returned to the stultifying atmosphere of Chatsworth. However, her grandmother was well pleased with her and immediately began pestering for another invitation to court, eager to push home their advantage. It came the following year, and Arbella was delighted to learn that this time she was to be presented at the court in London.

She arrived in late summer 1588. The court was even more spectacular than usual because there was a host of celebrations to mark the defeat of the Spanish Armada. Arbella's visit began at Greenwich and continued at Whitehall when the court moved there toward the end of the summer. The Queen invited her to attend as a member of the privy chamber, and she was much feted by other members of the court. However, whereas last time Arbella had been humbled and delighted by Elizabeth's favor, this time she seemed to expect it as her right. The Venetian ambassador recalled that she "displayed such haughtiness that she soon began to claim the first place; and one day on going into chapel she herself took precedence of all the Princesses who were in her Majesty's suite; nor would she retire, though repeatedly told to do so by the Master of Ceremonies, for she said that by God's will that was the very lowest place that could possibly be given her." Elizabeth was furious when she heard of this, and "in indignation, ordered her back to her private existence without so much as seeing her before she took her leave, or indeed ever afterwards."[33]

Arbella's own account of the visit, written some years later, suggests that there may have been another reason for her expulsion from court. She was thirteen years old by the time of this second visit, and was blossoming into an attractive young woman. A portrait painted at around this time shows her to have had reddish fair hair, a heart-shaped face, large dark blue eyes, and a small pouting mouth. Her royal blood made her even more appealing to the ambitious young men of the court. Chief among them was the Earl of Essex, son of Lettice Knollys, who was rapidly rising to prominence. When Elizabeth caught her talking to the earl "in a friendly fashion" in the privy chamber, she lashed out in a jealous rage. "How dare others visit me in distress when the Earl of Essex, then in highest favour, durst scarcely steal a salutation in the privy chamber," Arbella complained petulantly, "where, howsoever it pleased her Majesty I should be disgraced in the presence at Greenwich and discouraged in the lobby at Whitehall."[34]

Little wonder that the Queen, who was now in her midfifties and unable to use her looks to retain Essex's interest, should be so provoked by this arrogant and impertinent girl, whose pretty auburn coloring perhaps reminded her of her younger self.

But it is also possible that Elizabeth was acting out of more than mere jealousy when she dismissed Arbella from court. This wily queen was ever cautious not to commit to one claimant or another, preferring to play them off against one another. Conscious that she had boosted Arbella's status by making much of her last time, she no doubt resolved to do the opposite now in order to keep everyone guessing. The ploy worked: many of Arbella's former supporters dismissed her as an irrelevance in the race for the English throne. One foreign observer noted that "small account" was made of her in Scotland and Spain "by reason she was not Catholic," while others thought she was the weaker claimant because of her sex. One contemporary protested: "A woman ought not to be preferred, before so many men."[35]

This did not stop Elizabeth from using Arbella as a pawn in her political power games. As early as 1585, she had considered marrying her to James of Scotland, who had expressed "an affectionate favour and good will" toward his young cousin. But she had swiftly abandoned this plan because she realized that it would make the succession too certain. Far better to keep the two main claimants on opposing sides. In 1587, the year of Arbella's first visit to court, Elizabeth was contemplating offering her as a bride to Rainutio Farnese, son of the Duke of Parma, in order to neutralize the threat then posed by Spain. When the political tide turned again the following year, it was rumored that Arbella would marry the king of Denmark's son instead and that the couple would inherit the throne as man and wife upon Elizabeth's death.[36]

In 1587 James VI had entered the fray by demanding "that the Lady Arabella be not given in marriage without the King's special advice and consent." His preferred candidate was Lord Esmé Stuart, to whom he had given the Lennox title and whom he looked set to name as his heir. He made it clear that even if this match did not come about, he wanted "to have the bestowinge of hir." Elizabeth rebuffed his demands. "Hir majeste wold know how she sholde be bestowed," insisted her envoy.[37] Realizing that he would get nowhere with the Queen, James changed tack and wrote to Arbella directly, emphasizing their "naturall bonde of bloode" and praising her virtues. "I cannot forbeare to signifye to you hereby what contentment I have receaved hearing of your so vertiouse behaviour . . . it pleaseth [me] most to sie soe vertiouse and honourable syouns arise of that race whereof we have both our discent."[38]

As James stepped up his efforts to control Arbella, Elizabeth responded in like manner. Over the next fifteen years, the girl's name would be linked with practically every eligible suitor in Europe. As far as the English queen was concerned, this arrogant young woman was a commodity to be dispensed with at her pleasure.

CHAPTER 14

"Witches"

Sir Walter Ralegh, one of the Queen's greatest favorites during the 1580s, denounced her ladies as being "like witches, capable of doing great harm, but no good."[1] Having fallen into disgrace over an affair with one of the Queen's ladies, he had attempted to use them as intermediaries with his royal mistress, but this had failed to regain Elizabeth's good opinion. The latter may have made a show of banning her ladies from meddling in political affairs at the beginning of her reign, but it had soon become clear that this was little more than a front. Just as her attendants were keen to exploit the unrivalled access they had to the Queen, so was she quick to appreciate the advantages of controlling her court through their networks. Before long, Elizabeth's women were perceived to have so much influence that many of the great men at court sought their intervention. As one courtier wryly observed: "We worshipped no saints, but we prayed to ladies in the Queen's time."[2]

With a wealth of evidence to prove the influence wielded by the women of Elizabeth's household, it is perhaps surprising that their role has been so overlooked.[3] The assumption has often been that because they inhabited the private world of the Queen, they were kept quite separate from the political. But this in turn underestimates the political importance of personal relationships in her court. It was a tightly knit world in which almost all the occupants—female as well as male—were related by ties of blood, marriage, or friendship. Women may have been barred from holding political office, but as the stories of some of the female protagonists at court will show, they played an integral part in both the political and personal lives of their sovereign. They were the only members of court who were guaranteed close access to the Queen throughout the reign. While her male courtiers and councillors hung about in the public rooms beyond, her ladies would spend hours alone with her, exchanging gossip, sharing her innermost thoughts, and attending to her person.

As her reign progressed, Elizabeth increasingly used her ladies as a source of information about affairs at court. They were much better placed to pick up gossip from the attendants of male courtiers, councillors, and ambassadors, and as their presence was a good deal less obtrusive than the Queen's, they could listen in on conversations and report the contents back to their mistress. They also had extensive connections through husbands, fathers, children, stepchildren, and godchildren, as well as servants, retainers, and other associates. Far from confining their activities to domestic pursuits, these women took an active interest in state affairs and made every effort to find out the latest developments. So effective were their networks that even the most important officials, such as Lord Burghley or his son Robert Cecil, were often extremely disconcerted to find that the Queen was already well acquainted with matters of which they had hoped to keep her in ignorance.

Being in constant attendance upon their royal mistress also gave the ladies of her household an unparalleled ability to assess her mood and judge the best time to present petitions. Sir John Harington recalled an occasion when one of Elizabeth's ladies had "come out of her presence with an ill countenance, and pulled me aside by the girdle, and said in a secret way, 'If you have any suit today, I pray you put it aside; the sun doth not shine!'"[4]

Men soon came to realize that sending messages to the Queen via her ladies was a surer guarantee of success than addressing her directly. Lord Burghley had resorted to this means after being banished from court for his involvement in the execution of Mary, Queen of Scots. His friend Lady Cobham assured him: "If you will write I will deliver it. I do desire to be commanded by you."[5] She proceeded not just to speak favorably of him to the Queen but also to keep him informed of everything that passed at court during his enforced absence.

By the late 1580s, Blanche Parry, the Queen's longest-serving attendant, had become an unofficial personal secretary to her mistress, drafting letters for her and checking the contents of others before they were sent out. This was an expected part of her role as chief gentlewoman of the privy chamber, for as such she was in charge of the Queen's personal papers, while the Privy Councillors and other officials were responsible for her majesty's state documents. But Blanche was more than just a secretary. It was apparently well known at court that sending official material to the Queen via her chief gentlewoman was an effective way of ensuring a positive result, because Blanche would often append her own comments in favor of the sender.[6]

Mistress Parry soon gained a reputation as one of the most influential women at court. Her cousin Rowland Vaughan claimed that she was one of a "trinity of Ladies able to work Miracles," and that in "little Lay-matters," she would "steal opportunity to serve some friends' turns."[7] These "friends" were numerous. Blanche was constantly besieged by requests for assistance during her long service in the Queen's household. She could not possibly accede to them all, and the evidence suggests that she prioritized those from among her own relations.

Despite knowing the influence she could wield, Blanche always made it clear that Elizabeth was the most important person in her life. In her will, she referred to her as "my dear Sovereign lady and mistress," and this reflected the awe and reverence with which she always treated her. Such constancy and devotion formed a marked contrast to

the fickleness and backbiting of court life. As such, it was a vital stabi-
lizing influence for Elizabeth, who came to rely on it more and more
throughout her long reign as her other favorites fell away through
death or disgrace. Blanche was also a precious link with the past. By
the 1580s, she was the only person in Elizabeth's life who could remember her mother, Anne Boleyn, and who shared her private remi-
niscences of her childhood. Blanche had been with Elizabeth for as
long as she could remember—her constant companion and loyal confidante. Little wonder, then, that when, on February 12, 1590, Mistress
Parry died after fifty-seven years of diligent service, the Queen was
bereft.

Blanche had remained in good health for almost all of her eighty-
two years, and it was only toward the end that she had been troubled
by illness, receiving ever more frequent visits from the Queen's
apothecary. Even though her eyesight grew so poor that she was practically blind, she had remained in active service. Her death came as a
shock to the Queen and the court as a whole. Tradition has it that Elizabeth was with Blanche in her final hours. Within days, news of her
death had spread across the kingdom. On February 17, George Talbot,
Earl of Shrewsbury, received a letter informing him: "On Thursdaye
last Mrs Blanshe a Parrye departed; blynd she was here on earth, but I
hope the joyes in heven she shall se."[8] Another letter of the time noted
the "great sorrow" that was shown by the Queen and her ladies.

Elizabeth was determined to honor her late servant's memory, and
she ordered a funeral befitting a baroness. Blanche was interred in St.
Margaret's Chapel, Westminster. A marble effigy was later erected, on
which an inscription recorded that she had been "Beneficial to her
kinsfolke and countrymen." A marble figure of Blanche formed part
of this elaborate monument. It is the most detailed likeness of her that
exists today. Her features are striking, with high cheekbones, a small,
pursed mouth, and piercing, intelligent eyes. Her face is framed by
neatly curled hair, which is kept in place by a French hood. She is
dressed in the finest clothes of black satin, with a high ruff around her
neck, and the Queen's livery is prominently displayed on her dress. At
her throat is a bejewelled necklace, which, along with the books she is

holding, serves as a reminder of her former positions in the royal household.

Blanche's will indicates that she was a wealthy woman by the time she died. She was able to make numerous bequests to friends and relatives, including jewels, money, and land amounting to several thousand pounds.[9] She also left provision for a magnificent tomb to be erected at her home church of Bacton in Herefordshire. The epitaph attests to the personal sacrifice that her service to Elizabeth had entailed:

So that my tyme I thus dyd passe awaye
A maed in Courte and never no man's wyffe
Sworne of the quene Ellsbeth's bedd chamber allwaye
Wythe maeden quene a maede dyd ende my lyffe.

The tomb was also clearly intended as a final testament to her love for the Queen. It depicts Elizabeth, resplendent in an elaborate gown and jewels. Kneeling beside her is her old servant, Blanche. Her expression is one of adoration, as if worshipping the Virgin—only in this case, it is the Virgin Queen.[10] The monument may be the earliest example of Elizabeth being represented as an icon. If so, it paved the way for hundreds of allegorical paintings and images and helped define Elizabeth as Gloriana, the unassailable, godlike queen whose name would reverberate down the centuries. In death, as in life, Blanche Parry had served her mistress well.

A year earlier, Elizabeth had lost another of her long-standing attendants. Like Blanche, Lady Fiennes de Clinton had been at the heart of court affairs for more than a half century. Although in theory her role was restricted to service in the privy chamber, in reality her influence spread a good deal further. The intrigue and backbiting of the Elizabethan court honed her political skills, and there are several references to her shrewdness in the "business matters" in which the Queen had forbidden her ladies to meddle.

A contemporary observed that the Queen confided in Lady Clinton because she was a lady "in whom she trusted more than all others."[11] The intimacy between the two women had strengthened as the reign progressed. The Queen seemed to take genuine pleasure in Lady Clinton's company, and they shared both humor and intellect. As well as spending many hours with her in the privy chamber, the Queen would also attend private supper parties with her away from court. One of these, in May 1587, was hosted by Lady Clinton at her London residence. The Queen had decided on a whim that they should go there, so her friend was obliged to rustle up a suitably lavish supper at short notice.[12]

That Elizabeth should involve Lady Clinton in her most private affairs is a testament to the degree of trust that existed between them. The closeness of their relationship was remarked by many at court, and it gave Lady Clinton a great deal of influence. Before long, she was besieged by petitions from ambitious place-seekers or those who had fallen from the royal favor. There are various recorded instances of her furthering these requests with the Queen,[13] but for the most part, she only did so when it suited her own interests or those of her family.

When Lady Clinton died in March 1589, her royal mistress was heartbroken. The "Fair Geraldine" had served her faithfully for more than thirty years and had been her childhood companion before that. Her intellect and shrewdness had won her the respect and admiration of the Queen, who missed her wise counsel and affection. Elizabeth ordered a magnificent funeral to be conducted at Windsor, where her cousin's body was interred in the royal chapel, next to that of her second husband.

Other ladies were quick to step into the void created by the deaths of Lady Clinton and Blanche Parry. Principal among them was Anne Dudley, Countess of Warwick. Anne had been one of Elizabeth's favorite maids of honor before her marriage to Robert Dudley's brother, Ambrose, Earl of Warwick, in 1565. She had then been promoted to gentlewoman of the privy chamber. The Countess of Warwick proved extraordinarily diligent in her duties. Her husband once told Sir Francis Walsingham that she had "spentt the cheffe partt of her yeares both painfully, faythfully, and servycably, yea after soche sortt as without any dishonour to her maiestie any kinde off wage nor ytt any belmyshe to her powre sellff."[14] Elizabeth appreciated her efforts and found her service indispensable. She placed an ever-greater reliance upon Lady Anne's steadfast loyalty and support, and as the years wore on, she treated her more as a friend and confidante than as a servant. The countess's niece would later claim that her aunt was "more beloved and in greater favour with the Queen than any other woman in the kingdom, and no less in the whole Court and the Queen's dominions, which she deserved. She was a great friend to virtue."[15]

As a result of the high esteem in which she was held by Elizabeth and Leicester, the countess received more requests for favor than any other lady of the privy chamber. It was said that she was "a helper to many petitioners and others in distress," and the records certainly bear this out.[16] Lady Anne's networks were extensive, both in England and abroad, and she made the most of the aristocratic connections her marriage gave her. She was kept abreast of international affairs through her contacts with English ambassadors and envoys. Lord Hunsdon, for example, sent her secret reports from Hesse, which was then a province of Prussia, after taking up the ambassadorship there. Foreign ambassadors were also well aware of her influence, to the extent that when she fell ill in the late 1590s, it was reported as far afield as Venice, Italy.[17]

Anne would mostly use her position at court to further the suits of her friends and family—particularly her nephew by marriage, Sir Robert Sidney. The Queen had appointed him governor of Flushing, a key strategic post in the Netherlands. Although this was a great honor, it was an unwelcome one for Sir Robert, who did not relish the prospect of spending long periods of time away from his wife and family. He also knew that Elizabeth's troops in the Netherlands were poorly provided for, and that many were running up huge debts in order to furnish themselves and their men with the necessary supplies and munitions. He therefore sent frequent requests for leave, which the Queen, realizing his worth, was loath to grant.

In October 1595, after ten years' service, Sir Robert's wife pleaded with Lady Warwick to intervene on their behalf with the Queen. The countess advised that "as yett it is no tyme to move yt," for her royal mistress was deeply concerned with the ever-growing Spanish presence in the Netherlands and feared that Philip II intended to launch another armada from there. However, toward the end of that year, a tragic event in Anne's family gave her the perfect opportunity to press her nephew's suit once more.

Lord Huntingdon, the husband of Sir Robert's other aunt, Katherine Hastings, had fallen gravely ill. The couple were held in some esteem by Elizabeth. Katherine was another member of the Dudley family, being the youngest sister of the Queen's late favorite. Lord Huntingdon, meanwhile, was descended on his mother's side from Margaret Pole, Countess of Salisbury, daughter and sole heiress of George, Duke of Clarence, brother of King Edward IV. In the early days of their relationship, when Elizabeth was still establishing her regime, his royal blood had caused "some jealous conceit" of Huntingdon and his wife, but she had soon come to appreciate their unswerving loyalty. In 1572 she had appointed him the earl president of the Council of the North, one of the most prestigious posts outside of the court because the council was designed to keep the volatile northern counties under control.

Although Katherine spent long periods of time at their estate in Ashby de la Zouche, Leicestershire, she also made regular visits to the court, and by the early 1590s she had become a more or less permanent feature there. It is not clear whether the Queen gave her an official role in her household, but she was certainly in attendance most of the time. By 1595, she was sufficiently in favor for the Earl of Huntingdon to thank the Queen for being so gracious "to my poor wife, which I can no ways in any sort do anything to deserve."[18] Accordingly, her friends and relatives began seeking preferment through her intervention with the Queen. In September that year, Rowland Whyte, Sidney's agent at court, urged his master: "I pray you wryt to my Lady Huntingdon by every passage, for 'tis looked for, and desire her favour to obtaine your leave to return to see her, which will much advance yt; for the Queen is willing to give her any contentment that may comfort her."[19]

On December 16, 1595, news reached court that the earl was "left sick in York." Ever the pragmatist, Elizabeth dispatched the Earl of

Essex there, "to see things ordered in those parts till a trusty President can be found, if God should call him away." She also insisted that the news be kept from the Countess of Huntingdon, anxious not to worry her unnecessarily. Three days later, a messenger arrived with the sorrowful tidings that the earl had died. Still Elizabeth was careful to conceal the truth from his widow, knowing the great love that she had borne the earl. The court was then away from London, but in her anxiety to prevent anyone but herself from breaking the news to Katherine, she set off at once for the capital. "The Queen is come to Whitehall on such a sudden that it makes the world wonder," observed Rowland Whyte, "when it is but to break it unto her herself." He noted that she had taken great care to prepare the ground with the countess.[20]

Later that day, Elizabeth journeyed to Katherine's London home, where she at last relayed the devastating news that the earl was dead. "I am not able to deliver unto you the passions she fell into and which yet she continues in," reported Whyte to his master. Near hysterical in her grief, Katherine had wept and wailed long into the night. Elizabeth was so concerned about her friend that she resolved to visit her again the following day. "The Queen was with my Lady Huntingdon very privat upon Saturday," observed Whyte, "which much comforted her."[21]

With no children of her own to support her, the Countess of Huntingdon fervently wished to see her nephew, Sir Robert Sidney, who was still serving in the Netherlands. As she fretted over his absence, she fell seriously ill, and those closest to her despaired of her life. "It is not possible Lady Huntingdon should continue long, so weak she is," observed one. Upon hearing of this, the Queen at once granted Sidney permission to leave his post when pressed to do so by the countess's sister-in-law, Lady Warwick. "Lady Warwick used Lord Huntingdon's death as an excuse to persuade the Queen to let Robert Sidney come home on leave to comfort his aunt," reported Whyte.[22]

Such opportunism did not always bring results. In January 1597, Sidney distinguished himself at the Battle of Turnhout, in which the combined Anglo-Dutch forces triumphed over the might of Spain.

Seeing this as an ideal opportunity to win him some more leave, the Countess of Warwick wrote to assure him: "I doubt not butt that it will be easelie satisfied . . . wherof I would be as gladd, as I will be of the effecting of anything you shall desire."[23] However, while the Queen expressed her pleasure at Sidney's valiant service, she would not be persuaded to accede to her attendant's request.

Undeterred, Lady Warwick petitioned her royal mistress on many other occasions during the years that followed. In March 1597, Rowland Whyte reported to his master that she had assured him: "If I had any occasion to use her to further any of yours I should come boldly unto her."[24] Perceiving the Earl of Essex's rise to prominence at court, Lady Warwick increasingly used his influence to try to sway the Queen. Together they were a force to be reckoned with: Essex was Elizabeth's favorite male courtier, and Lady Warwick her closest female friend and confidante. While Essex helped to further the countess's suits and recommended her to others seeking Elizabeth's favor, so she intervened on his behalf when he was out of favor with his royal mistress. This happened in 1599, when he had been banished from the court for flouting the Queen's orders in Ireland. Lady Warwick sent him a message assuring him that if he came to Greenwich, where the court was then in residence, she would contrive an opportunity to let him into the palace gardens when the Queen was in a good mood, so that he could plead her forgiveness in person.

As a close associate of the Earl of Essex, Lady Warwick alienated Sir Robert Cecil, whom the Queen had appointed as her principal secretary in 1596. The countess's niece, Lady Anne Clifford, recalled: "Sir Robert Cecil and the House of the Howards . . . did not much love my Aunt Warwick."[25] Becoming ever more embroiled in the faction-ridden politics of the 1590s, Lady Warwick was forced to assume greater secrecy in her dealings on behalf of Sir Robert Sidney, knowing that Cecil would counter any attempt to have him recalled. In his correspondence with her and his master, Rowland Whyte took to using code in order to disguise both Lady Warwick's name and those of the principal players at court. Thus, in January 1598, he informed Sidney that his aunt had agreed to deliver his request for leave to the Queen, but only after Cecil—"200"—had left court, because she knew he would interfere.[26]

The increasingly tenuous nature of Lady Warwick's influence with the Queen was demonstrated the following year, when the latter took offense at Sir Robert Sidney's tarrying too long in England after finally being granted permission to visit his family. In August his steward wrote to warn him that Elizabeth was "much displeased" that he was still on English soil, despite reports that he had set off for the Netherlands some weeks before. The Countess of Warwick had informed Whyte of this and warned him that her royal mistress suspected Sir Robert of staying in the hope of seeing his ally, the Earl of Essex. Both she and Whyte strenuously denied this, but Elizabeth would not be placated, and a week later the countess reported: "the Queen is still full of the delays you made, and wishes you to write at the first opportunity," adding that she herself would deliver his letter. Sir Robert duly wrote to his sovereign, but as well as begging her forgiveness for delaying his return, he also petitioned her for the post of Lord President of Wales, which would enable him to return to his beloved estate of Pembroke. His aunt delivered the letter to Elizabeth and helped her to read it. While it worked the desired effect of winning the Queen's pardon, she stopped short of granting him the much-coveted post.[27]

The Queen's "good opinion" of Sidney, expressed in this letter, did not last for long. By the end of the year, Whyte was writing to warn his master: "cc [Lady Warwick] finds 1500 [the Queen] more cold towards you than was wont, and could not tell what to make of it."[28] When in October 1600 Lady Warwick presented yet another request for Sidney's leave of absence, the Queen gave one of her famous "answers answerless," telling her: "Well, well, he shall come over, but I will see further yet." This so incensed the countess, who was tired of being rebuffed by her royal mistress, that she was said to be "passionately troubled."[29]

This episode proved that it was Elizabeth herself who held the reins of power. She would not be duped by her ladies any more than by her ministers. They might wield influence, but only with her sanction. If any of them got above their station, they would be slapped down at once. By 1599, it seemed that Lady Warwick's power was on the wane.

In supporting the Earl of Essex, she had backed the wrong horse. He fell spectacularly from favor that year and would resort to ever more extreme—and ultimately fatal—measures to try to regain his sovereign's good graces. The countess appealed in desperation to her old adversary, Sir Robert Cecil. "Your help is sought for and found," she wrote, "now let it be obtained for one that hath lived long in Court with desert sufficient, being coupled with others." In a rather unconvincing show of self-deprecation, the countess insisted that she did not by nature have "much of the fox's craft or subtlety and as little of the lion's help; having lost friends almost all, no face to crave, no desire to feign."[30]

Perhaps Cecil worked some good on behalf of his former rival, or perhaps Elizabeth simply missed her old friend, for Lady Warwick was soon restored to her favor. When the countess fell very ill in the summer of 1599, it was noted that her royal mistress was "more than usually subject to fretting and melancholy."[31] The following year, Elizabeth showed great honor to Lady Warwick's family by attending the wedding of her niece, Lady Anne Russell, to Lord Herbert, at Blackfriars.

Meanwhile, the Countess of Warwick's sister-in-law, the widowed Katherine Hastings, had risen to prominence at court. It was observed that the Queen would keep her by her side for many hours of the day. This prompted Katherine's nephew to seek her intervention in his perpetual quest for leave from the Netherlands. At first she was reticent to use her favor with Elizabeth in this way. She was also mindful of Robert Cecil's growing power and, being more cautious than her sister-in-law, did not wish to alienate him. "I found a fear in her to speak for you," reported Whyte. "All see 200 [Cecil]'s power and fear to displease."[32] He was more hopeful a few months later, when he wrote to Sidney: "her access is good, and she very gracious with her Majesty." By February 1598, it was noted that "Lady Huntingdon is at Court and with her Majesty very privat twice a day." Even so, she still gave little indication that she was willing to intervene on her nephew's behalf. "I see no fruit come to you by it," complained Whyte, "though none so fit as herself to do it . . . I cannot see what good she doth for her frends."[33]

Although Whyte concluded that the Countess of Huntingdon's influence must be very limited, it is at least equally possible that she was reticent to further her nephew's requests because she knew how much Elizabeth needed to keep him in his vitally important post. The countess had also seen at firsthand how little her sister-in-law, Lady Warwick, had benefited from persisting in furthering his suits: indeed, it had diminished her standing at court. Therefore, only when she judged both the case and the timing to be right would she agree to intervene with her royal mistress on his behalf. In the meantime, she grew so impatient with Rowland Whyte's frequent pleas for assistance that she refused to grant him an audience. "Lady Huntingdon has been at Court these 7 days," he whined to his master in May 1600. "I am made a stranger unto her." Two months later, he was still complaining that "she admittes me not to her presence."[34]

Perhaps it was Katherine's refusal to exploit her position that endeared her to Elizabeth even more. Being constantly besieged by petitions from her ladies, it must have made a refreshing change that Lady Huntingdon was apparently content to spend time with the Queen for the pleasures this brought her rather than for any more material benefits. By summer 1600, she had supplanted even Lady Warwick in Elizabeth's affections, and it was noted that "She governes the Queen, many howres together very private."[35] The use of the word *governes* is interesting. It is tempting to take it literally and conclude that Lady Huntingdon had somehow succeeded where no woman—or man— had before by making Elizabeth subservient to her will. More likely, however, is that it was the Queen's time, not her person, that Katherine dominated. She would not have been able to do so unless it was Elizabeth's professed desire.

The idea that Elizabeth came to prefer the company of those ladies who did not try to use their positions for their own gain is borne out by looking at the remaining members of her close entourage during these later years of her reign. They included Katherine Howard, Countess of Nottingham. Although Katherine had had five children during the course of her long marriage, she had always returned to her post as soon as possible after each birth and carried out her duties with extraordinary diligence. Elizabeth rewarded her faithful service by giving her ever-greater responsibilities. She was among a select group of ladies who were entrusted with the care of the royal jewels, including those given as New Year's gifts to the Queen. These were as extensive as they were valuable. On one New Year's Day alone, Katherine took charge of three heavily ornate necklaces, four precious stones, two bracelets, twenty pairs of aglets, two jewel-encrusted bodkins, a bezoar stone (a stone from the liver of animals, thought to be a cure for various diseases), a bejewelled gown, and an orange taffeta waistcoat covered in gold lace.[36] She herself was not stinting in the gifts she presented to her royal mistress. Knowing Elizabeth's fondness for animals, she once gave her "a jewell of golde being a catt and myce playing with her garnished with small diyamonds and perles," and on another occasion a gold greyhound with a diamond-studded collar, and a gold and ruby dolphin.[37]

Around 1572, Katherine had been promoted to first lady of the bedchamber, one of the most exalted offices in Elizabeth's household. As well as being in constant personal attendance upon the Queen, she supervised her royal mistress's extensive wardrobe, and the Spanish ambassador noted that she also presided over the table of the ladies of the privy chamber. Her prestigious position, combined with Elizabeth's obvious affection for her and their ties of kinship, placed Katherine in an extremely influential position. Yet it is testament to her loyalty to the Queen that, for the most part, she chose not to exploit this for her own gain. In stark contrast to the numerous petitions received by the likes of Anne and Katherine Dudley, the contemporary sources contain only a handful of references to Lady Howard's involvement in patronage.

Katherine was one of the most faithful and longest serving of all Elizabeth's attendants. She was also one of the few women left at court who had known Elizabeth before she became queen, and she represented the traditional values of loyalty and selfless devotion that her royal mistress had come to prize so highly. The same was true of Mary Radcliffe, who had served the Queen for more than thirty years and had never lost her good opinion. As a result, she enjoyed some prestige, but it is to her credit that she too chose never to exploit it.

By contrast, Helena Snakenborg had suffered a temporary loss of favor in 1576 when she had married Thomas Gorges, a groom of the privy chamber, even though the Queen had refused her permission. Helena had been distraught when her royal mistress had dismissed her from court, and she had written to beg forgiveness for "a poure, desolat, and banished creture."[38] Elizabeth soon forgave her, and even when she was obliged to leave court again two years later in order to give birth to her first child, her royal mistress was so far from resenting it that she agreed to act as godmother, and presented Helena with a beautiful silver-gilt bowl at the christening. Helena repaid the Queen's trust by refusing to exploit her position to further the causes of friends or associates, and she came to enjoy a great deal of influence because of it.

Elizabeth would have need of companions such as Helena Gorges, Katherine Howard, and Katherine Dudley during the later years of her reign. They represented a precious link with the past in a court that was increasingly looking to the time when this aging queen would no longer be at its apex.

CHAPTER 15

"Flouting Wenches"

Sir Francis Knollys, the aged vice chamberlain of the Queen's household, found his duties at court ever more tiresome in the later years of the reign thanks to the boisterous antics of the ladies and maids of honor whose chamber adjoined his own. An old man in need of his sleep, he complained that they would "frisk and hey about in the next room, to his extreme disquiete at nights, though he had often warned them of it."[1] Elizabeth herself was growing increasingly impatient with the young girls who served her, preferring the company of her faithful old friends and servants. As she grew into old age, she became ever more intolerant of their audacious antics and often "swore out [against] such ungracious, flouting wenches," making them "cry and bewail in piteous sort."[2] Her anger was perhaps born of frustration at losing her grip on the formerly strict moral standards at her court, and of bitterness that she herself was no longer young and desirable, no matter how much her fickle courtiers might flatter her so.

More often than not, Elizabeth's outbursts at the "flouting wenches" of her household were prompted by her discovery that they had married or had affairs in secret. This had always been a source of friction between Elizabeth and her women, but as her reign progressed, it became an increasingly frequent occurrence. It seemed that almost every year, there were scandals involving clandestine seduction, unwanted pregnancies, or elopements. By the late 1590s, this had become so prevalent that one courtier disapprovingly noted: "Maides of the court goe scarce xx [20] wekis with child after they are maryed, wherein man hath lybertie of conscience to play the knave."[3]

One of the most notorious scandals of the 1590s involved Elizabeth (Bess) Throckmorton. Bess had entered the Queen's service as a gentlewoman of the privy chamber in 1584, at the age of nineteen. The Throckmorton family had long been connected to the court, although not always to their advantage. Bess's mother, Anne, had been imprisoned with Lady Jane Grey and had suffered mental torture from witnessing her fate. Her father, Sir Nicholas Throckmorton, had had a checkered relationship with the Tudors. Under Henry VIII he had been appointed to the household of his cousin, Katherine Parr, where he had become acquainted with the young Princess Elizabeth. However, during Mary I's reign, he had been implicated in the Wyatt rebellion, only narrowly escaping with his life and fleeing to exile on the Continent. He returned upon Elizabeth's accession, and she demonstrated her regard by appointing him to several important posts, including that of ambassador to France. It was there that he became acquainted with Mary, Queen of Scots, and he was so beguiled by her that he would often take her side against his royal mistress, which earned him a sharp reprimand. In 1569 he was suspected of involvement in the Duke of Norfolk's plot to marry the Scottish queen and was thrown into prison. Although he was subsequently released, he never regained Elizabeth's trust.

Sir Nicholas died in 1571, when Bess was just six years old. If she understood little of his fall from grace at that time, she would have learned his fate as she grew into adulthood. Later events would prove that it had not imbued her with the same sense of caution that her royal mistress had learned from her own childhood experiences.

Bess's appointment to the privy chamber was a prestigious one for a girl of her age. She was one of the few salaried members of Elizabeth's household and was also granted bouge of court (food from the royal kitchens), as well as three of her own servants. Bess soon proved an asset to the Queen. She was already well used to the etiquette of the court, having been introduced there at the age of seven. As a result, she needed little training for her role in the privy chamber. Elizabeth appreciated the speed at which Bess learned her duties, and she soon became one of her most trusted ladies.

Intelligent, witty, passionate, and forthright, Bess Throckmorton was by all accounts also something of a beauty. A description of her written some years later praised her as "a fair handsome woman," whose "charms" beguiled many men at court.[4] She had an exquisite sense of style and embellished the black-and-white gowns that she and her companions were obliged to wear with as many jewels as she could get away with under the Queen's jealous scrutiny.

Among Bess's many admirers was Sir Walter Ralegh, the man who had succeeded Leicester as the Queen's great favorite. Born in 1554, Ralegh was some twenty-one years Elizabeth's junior, but he paid court to her like a lover, showering her with romantic poems and letters, all praising her beauty and allure. Ever the gallant, he treated his royal mistress like a precious jewel and went out of his way to fulfil her every desire. According to popular legend, on one occasion he threw down his cloak over a "plashy place," so that the Queen might walk over it without getting her feet wet.[5] Ralegh's looks were those of the eponymous romantic hero. Tall, dark, and handsome, he was also athletic and brave, having embarked upon various daring voyages to far-flung corners of the globe. Elizabeth was delighted by his attentions, and before long their relationship was a source of much gossip at court. In December 1584, a foreign visitor was astonished to see the obvious intimacy that existed between them, and described how the Queen had pointed "with her finger at his [Ralegh's] face, that there was a smut on it, and was going to wipe it off with her handkerchief; but before she could he wiped it off himself."[6]

Elizabeth was not the only woman at court to be beguiled by this handsome adventurer. Many of her ladies fancied themselves in love with him. A natural flirt, Ralegh encouraged their attentions with tales of daring escapades in faraway lands. Like Leicester, he was discreet in his flirtations and no doubt bedded many more women at court than he admitted to. He certainly had the opportunity. As captain of the gentleman pensioners, he was sworn to protect the Queen's ladies and had a key to their chambers. He was always careful to keep any liaisons from the Queen in order to maintain the pretense that she was the sole object of his devotion. But when the seductive young Bess Throckmorton caught his eye, he was so desperate to have her that he abandoned his accustomed discretion—with catastrophic results.

The antiquarian John Aubrey provides a salacious account of Ralegh's seduction of a "maid" at court, who was almost certainly Bess. He wrote that Sir Walter "loved a wench well: and one time getting up one of the maids of honour aginst a tree in a wood . . . who seemed at first boarding to be something fearful of her honour, and modest, she cried, 'Sweet Sir Walter, what you me ask? Will you undo me? Nay sweet Sir Walter! Sweet Sir Walter!' At last, as the danger and pleasure at the same time grew higher, she cried in the ecstasy 'Swisser Swatter! Swisser Swatter!' "[7]

It is not clear exactly when the affair began, but it was likely to have been in 1590 or 1591, by which time Bess had fallen in with the Earl of Essex's circle of wild friends, which included his sister, Penelope, and Sir Walter Ralegh. They would hold wild parties at Essex's house, long after the Queen had retired, donning raunchy clothes and eating suggestively shaped food "to stir up Venus." Their debauched gatherings formed a stark contrast to the strictly controlled etiquette of the court, and Bess revelled in them. No matter how late she stayed out, she would always be back in her chambers by six o'clock the following morning, ready to serve her mistress.

This carefree existence came to an abrupt end in July 1591, when Bess discovered herself to be with child. In panic, she fled to her lover and begged him to marry her. No doubt aghast at the prospect of their secret being discovered, Sir Walter nevertheless did the honorable thing and made Bess his wife. Theirs had been more than just a thoughtless affair, for they seemed to share a genuine love for each other. Even so, Ralegh was taking a considerable risk in marrying the girl, knowing how his royal mistress would react if she found out. Meanwhile, Bess continued with her duties in the privy chamber, hiding her swelling stomach as best she could. She finally secured a leave of absence from court at the end of February 1592, when she was more than eight months pregnant, and went to the house of her brother Arthur, who had enlisted the services of a midwife.

Bess may have succeeded in concealing her pregnancy from the Queen, but rumors had begun to circulate about her relationship with Ralegh, and her sudden departure from court set tongues wagging even more. Sir Walter had also absented himself and was at Chatham docks supervising preparations for an expedition to Panama. When he heard that his secret was being whispered throughout the court, he wrote with some alarm to Robert Cecil, who had been no great friend to him in the past, and denied any involvement with Bess. "I mean not to come away, as they say I will, for fear of a marriage and I know not what," he assured him. "And therefore I pray believe it not, and I beseech you to suppress what you can any such malicious report. For I protest before God, there is none on the face of the earth that I would be fastened unto."[8]

Thus betrayed by her husband, Bess was forced to endure the ordeal of childbirth alone. On March 29, she gave birth to a son. A messenger was dispatched straightaway to Chatham, and Ralegh responded by sending £50 to his wife. However, he continued to prepare for his voyage, and two days later he travelled to Portland in Dorset to gather men and munitions. In his absence, Essex attended the secret baptism that Bess had arranged for her son, who was christened with the curious name of Damerei, after one of Ralegh's Plantagenet forebears, the royal dynasty that preceded the Tudors. She was back at court just four weeks later and, with some audacity, resumed her duties to the Queen as if nothing had happened.

But as had so often been proved in the past, no secret could remain hidden for long at court. In late May the scandal broke, and the Queen discovered her gentlewoman's betrayal. The scale of Bess's treachery was staggering. Not only had she married Elizabeth's closest favorite, but she had also broken the vows of loyalty that she had sworn upon entering the privy chamber. Her illusions about Ralegh's ardent affection shattered, the Queen lashed out at the perpetrators, her fury stoked by humiliation at being so deceived. Bess's first reaction was one of defiance. She collected her baby son from his wet nurse and went to her husband's London home, Durham House. He joined her there shortly afterward, and they enjoyed a brief day together as a family before Elizabeth's men came to haul Bess away to the custody of Sir Thomas Heneage, one of the Queen's most trusted spymasters. Both she and Ralegh, who remained under arrest at Durham House, were subjected to fierce interrogation, along with Bess's brother and his wife.

Ralegh desperately tried to clamor his way back to favor by sending urgent messages to the Queen—his "nymph" and "goddess"—lamenting his misery at being deprived of her presence and assuring her of his undying love. Upon hearing that Elizabeth was leaving for Nonsuch Palace, he wrote to Cecil: "My Heart was never broken till this Day, that I hear the Queen goes away so farr of, whom I have followed so many Years, with so great Love and Desire, in so many Journeys, and am now left behind her in a dark Prison all alone . . . I, that was wont to behold her riding like Alexander, hunting like Diana, walking like Venus, the gentle Wind blowing her fair Hair about her pure Cheeks, like a Nymph, sometime siting in the Shade like a Goddess, sometimes singing like an Angell, sometimes playing like Orpheus: behond the Sorrow of this World!" Apparently despairing of life now that he was deprived of his sovereign's presence, he begged Cecil to "Do with mee now therfore what yow list," for he was "wery of life."[9]

Such melodrama did Ralegh few favors. Having learned the extent of his betrayal, which had involved not just a secret courtship but also a marriage and birth, the Queen was in no mood to be seduced by his overblown romantic sentiments. As she tried to make up her mind what to do with him, rumors began to circulate that he and his new wife would be thrown in the Tower as common traitors. On July 30, 1592, Sir Edward Stafford, husband of Lady Douglas Sheffield, wrote to Anthony Bacon about "the discovery of Sir Walter's having debauched that lady," and told him: "If you have . . . anything to do with Sir Walter Ralegh, or any love to make to Mrs Throckmorton, at the Tower tomorrow you may speak with them."[10]

The rumors proved to be true. On August 7, 1592, Elizabeth committed both Sir Walter and Bess to the Tower. The scandal was the talk of the court. The poet Edmund Spenser wrote of it in *The Faerie Queene,* casting Bess as "Amoret" and Sir Walter as "Timias," who suffered at the "wrathful hand" of "Belphoebe." Ralegh himself resorted to poetry in an attempt to win back the Queen's favor, but to no avail. He and Bess continued to languish in that fortress with no prospect of release.

But while Sir Walter lamented his wretchedness, Bess remained unrepentant. She never showed any remorse for her actions and indeed seemed to revel in the fact that she had married one of the most desirable men of the court. Rather than seeking the Queen's pardon, she tried to use her network of contacts to secure her release. The fact that she signed all her letters to them "Elizabeth Ralegh" is an indication of her defiance. When Elizabeth heard of this, she was highly affronted and resolved to keep her wayward gentlewoman in the Tower for perpetuity. By contrast, she soon forgave Ralegh. On September 15, he was released from the Tower and permitted to go to Dartmouth, where one of his fleets had recently returned bearing riches from the Azores. Although technically still a prisoner, Ralegh was permitted to share in the spoils. As ever, the Queen had proved much more inclined to forgive the man than the woman.

Meanwhile, Bess remained a prisoner in the Tower, abandoned by her friends and unacknowledged by her husband, who was more concerned with clawing his way back to favor. In October 1592, London was hit by an outbreak of the plague, and it may have been this that killed her young son, Damerei, who subsequently disappears from the records. By now, Elizabeth's heart was so hardened toward Bess that she showed no sympathy at this tragic event. It was not until two months later, on December 22, that she finally granted her release.

Bess expected to be invited back to court, but her confidence was borne more of arrogance than insight, for the Queen made it clear that she would never again be admitted to her presence. Lady Ralegh was therefore obliged to join her husband at his Dorset estate of Sherborne, where they enjoyed a life of peaceful domestic harmony for the next few years and had another son, Walter, in 1593. Four years later, Sir Walter was finally allowed back to court, but his wife was bitterly disappointed to find that her banishment still stood. She continued to besiege her friends and contacts there with requests for them to speak to the Queen on her behalf. She soon realized, though, that for every person who did so, there would be at least another who stoked Elizabeth's antipathy against her. In 1602 she complained bitterly to Robert Cecil: "I understand it is thought by my Lady Kelldare that you should do me the favour to let me know how unfavourably she hath dealt with me to the Queen . . . I wish she would be as ambitious to do good as she is apt to the contrary."[11]

At the end of Elizabeth's reign, Lord Henry Howard, an enemy of the Raleghs, gloatingly remarked that although "much hath been offered on all sides to bring her into the Privy Chamber of her old place," the Queen was still determined to exclude her from it. "His [Ralegh's] wife, as furious as Prosperpina,[12] with failing of that restitution in court which flattery had led her to expect, bends her whole wits and industry to the disturbance of all motions."[13] According to Howard, in her fury Bess had sought revenge by conspiring with the Queen's enemies and holding secret meetings with the associates of James VI in an attempt to ingratiate herself with the future king of England.[14]

Bess Throckmorton's betrayal was one of many to occur in the Queen's household during the later years of her reign. Anne Vavasour, a gentlewoman of the bedchamber, fell pregnant by the Earl of Oxford and actually gave birth in the maidens' chamber. Her baby's cries gave the game away, and when news of it reached the Queen, she ordered that Mistress Vavasour be conveyed at once to the Tower, even though she was still recovering from the birth. For some considerable time afterward, it was reported that "Her Majesty is greatly grieved with the accident."[15] A later controversy involved Mary Fitton, a maid of honor whom some believe was the "dark lady" of Shakespeare's sonnets.[16]

She would steal out of her apartments at night disguised as a man in order to meet her lover, William Herbert, eldest son of the Earl of Pembroke. Her misdemeanor was discovered when she was "proved with chyld," and she was banished from court.[17]

Having endured so many scandals, Elizabeth became deeply embittered against any of her ladies who dared to have affairs or marry in secret. Her punishments, even for minor misdemeanors, became ever more severe as she desperately tried to regain control. Her godson, Sir John Harington, noticed that she "doth not now bear with such composed spirit as she was wont; but . . . seemeth more forward than commonly she used to bear herself towards her women; nor doth she hold them in discourse with such familiar matter, but often chides for small neglects."[18] Leicester's "base son," Robert Dudley, was exiled from court in 1591 for merely kissing Mistress Cavendish, a lady of the household. Meanwhile, Elizabeth Bridges and Elizabeth Russell received a sharp reprimand for stealing a glimpse of the Earl of Essex as he played at sports. "The Queen hath of late used the faire Mrs Bridges with words and blowes of anger, and she with Mrs Russel were put out of the coffer chamber," reported Rowland Whyte in April 1597.[19] They were permitted to return three days later, but had to endure their royal mistress's sour looks and sharp reproofs for many weeks afterward. A short while later, when the Queen suspected Lady Mary Howard of having an affair with Essex, she lashed out in fury at her maids, reducing them to tears. "She frowns on all the ladies," remarked Harington in October 1601.[20]

The fact that Essex was already married had done nothing to restrain his licentiousness with the Queen's ladies. This marriage itself had been the source of some scandal. The earl had secretly married Frances, widow of Sir Philip Sidney, much to the Queen's anger. Although she forgave him sooner than she did his new wife, Mistresses Bridges and Russell had found to their cost that she was still highly sensitive to any transgressions involving her favorite. As well as still smarting over his secret marriage, the Queen was painfully aware that age had ravished her own looks, and her insecurity led to a jealous possessiveness. Perhaps she also knew that for all his flattery, Essex secretly mocked her as "an old woman . . . no lesse crooked in minde than in body," much to the amusement of the "Ladies of the Court, whom he had deluded in love matters."[21]

Essex's remark was due to more than mere mockery. He had once burst into the Queen's bedchamber unannounced and was aghast to find her stripped of her courtly finery, her grey hair and deeply wrinkled face rendering her virtually unrecognizable from the carefully constructed image she presented to the world. Elizabeth was no less horrified, and it was said that the episode played as great a part in the earl's downfall as his failed insurrection some time later.

It is little wonder that the Queen took such care to shield her true appearance from all but her closest ladies. The ritual of dressing her had become increasingly elaborate as age began to overtake her. She had originally worn wigs that matched her own coloring, but as she grew older, these were used to conceal her greying hair.[22] At the same time, ever more layers of makeup were applied to complete the so-called "mask of youth." Her face, neck, and hands were painted with ceruse (a mixture of white lead and vinegar), her lips were colored with a red paste made from beeswax and plant dye, and her eyes were lined with kohl. Ironically, most of these cosmetics did more damage to the skin than aging ever could. Ceruse was particularly corrosive, and one contemporary observed with some distaste: "Those women who use it about their faces, do quickly become withered and grey headed, because this doth so mightily drie up the naturall moisture of their flesh."[23] But Elizabeth insisted that she continue to be adorned with this and other dangerous cosmetics, and only ever let her closest ladies see what lay beneath.[24]

This was more than mere vanity on the Queen's part; it was essential that a sovereign be presented in as magnificent a style as possible in order to emphasize his or her divinely appointed status and authority. One contemporary described Elizabeth as "most royally furnished, both for her persone and for her trayne, knowing right well that in pompous ceremonies a secret of government doth much consist, for that the people are naturally both taken and held with exteriour shewes."[25] For Elizabeth, it was also imperative that she appeared as youthful and attractive as possible—on the surface at least—in order both to uphold her marriage prospects and to show no sign of bodily weakness that her enemies could seize upon.

At the end of each day, in the privacy of her bedchamber, Elizabeth's ladies would remove her dark red wig, jewels, and other accessories, and gently wipe off the thick makeup that covered her face, bosom, and hands. Thus divested of her queenly adornments, Elizabeth would become the private woman once more. As the gulf between her public persona and what lay beneath grew ever greater, so did the fierceness with which she guarded these secluded hours with her ladies.

Despite all her efforts, the onset of old age was becoming increasingly obvious to everyone at court. Sir John Harington noticed that Elizabeth had started to let herself go, and described her as being "much disfavourd, and unattird." The Venetian envoy, Scaramelli, agreed that she who had once been such a leader of fashion was now sadly out of touch: "Her skirts were much fuller and began lower down than is the fashion in France," he reported, adding: "her hair is of a light colour never made by nature."[26] As she tried desperately to stop the "mask of youth" from slipping, Elizabeth appeared as a grotesque parody of her former self. Reporting on his visit to court in 1597, Monsieur de Maisse sniggered that she was "strangely attired" in an elaborately decorated dress that was so low cut that "one could see the whole of her bosom," which he added was "somewhat wrinkled."

On another occasion, de Maisse reported with a mixture of amusement and disgust that "she often opened this dress and one could see all her belly, and even to her navel." Her hair, which had long since turned grey, was covered by a "great reddish-coloured wig, with a great number of spangles of gold and silver, and hanging down over her forehead some pearls, but of no great worth." Meanwhile, her face "appears very aged . . . and her teeth are very yellow and unequal, compared with what they were formally, and on the left side less than on the right. Many of them are missing so that one cannot understand her easily when she speaks quickly." In desperation, Elizabeth tried to maintain the pretense, fooling herself and others that she was still the most desirable woman in Europe. "When anyone speaks of her beauty she says she was never beautiful, although she had that reputation thirty years ago. Nevertheless, she speaks of her beauty as often as she can." John Chapman, who had served in Lord Burghley's household, also saw through the Queen's attempt to "dazzle" her subjects by her ever more outrageously ostentatious clothes, so that by "those accidental ornaments [they] would not so easily discern the marks of age and decay of natural beauty." Lorenzo Priuli, the Venetian ambassador in France, was more brutal, describing Elizabeth as being of an "advanced age and repulsive physical nature."[27]

Elizabeth's fading looks were mirrored in the loosening of her grip on the affairs of the court, where the formerly strict moral standards now began to decay rapidly. As new generations of young ladies joined her household, they were frustrated by what they perceived to be the Queen's old-fashioned attitudes, and they were unwilling to make the sacrifices of their predecessors in order to serve her faithfully. "Now there was much talk of a Mask which the Queen had at Winchester, & how all the Ladies about the Court had gotten such ill names that it was grown a scandalous place, & the Queen herself was much fallen from her former greatness and reputation she had in the world," reported the Countess of Warwick's niece.[28] Meanwhile, William Fowke observed: "The talk in London is all of the Queen's maids that were," and related how the real cause of Mrs. Southwell's absence from court was found to be not a "lameness in her leg," as she claimed, but her having fallen pregnant by a "Mr. Vavisor."[29] A contemporary verse poked fun at the widespread licentiousness that now existed within Elizabeth's household:

> Here lyeth enterred under this Mound,
> A Female of Sixteen yeares old.
> More men than yeares have been upon her
> And yet she died a Maid of Honor.[30]

Another of the Queen's ladies who put sexual gratification ahead of service was Elizabeth Vernon, a cousin of the Earl of Essex. She succeeded in attracting one of the most notorious rakes at court, Henry Wriothesley, third Earl of Southampton. Strikingly attractive, with bright blue eyes, long auburn hair, and a lithe figure, the earl had many admirers at court, both male and female. In a veiled reference to his sexual ambiguity, Thomas Nashe paid him the following dubious compliment in 1594: "A dere lover and cherisher you are, as well of the lovers of Poets, as of Poets themselves." The earl was a great patron of poets and playwrights (including William Shakespeare), and many of his closest companions were male. Nevertheless, he also displayed a fondness for ladies at court, and in the late 1590s he began an affair with the alluring Elizabeth Vernon, a maid of honor who was then, like him, in her midtwenties.

From the beginning, it was clear that the passion was more on her side than his, and the earl soon tired of her cloying affection. In January 1598, Rowland Whyte reported that there had been "some unkindness" between them.[31] Seeking a way out of the relationship, the earl declared his intention to travel abroad, at which "his faire mistress doth wash her fairest face with to many tears." He secured Robert Cecil's agreement that he should accompany him on a mission to France, and when he left a few weeks later, Mistress Vernon was described by Rowland Whyte as "a very desolate gentlewoman that hath almost wept out her fairest eyes."[32] With shrewdness born of many years at court, Whyte had already guessed that her grief was due to more than just pining for her absent love. As he rightly predicted, the fair Elizabeth had fallen pregnant.

Mistress Vernon attempted to conceal her "grave condition" for as long as possible. Eventually, though, her increasing girth caused tongues to wag. Ever one for gossip, Sir John Chamberlain gleefully observed: "Some say that she hath taken a venue [a thrust, in fencing terms] under the girdle, and swells upon it," adding: "yet she complains not of foul play but says the Earl of Southampton will justify it."[33] When news of his mistress's pregnancy reached the earl, he was predictably reluctant to return home and do his duty. Elizabeth Vernon had anticipated this, however, and persuaded the Earl of Essex to intercede with him on her behalf. Essex proved as good as his word, and arranged for Southampton to be conveyed back to England in the strictest secrecy. He then provided his own house in London as a venue for the wedding. That Elizabeth Vernon was desperate to marry, despite the many examples of other ladies at court falling foul of the Queen in this way, suggests that she—like they—wished her child to be legitimate.

Soon after the ceremony, which took place in August, the Earl of Southampton returned to France, while his new wife stayed on for a while at Essex House. She had found a trusty protector in its owner, for when the time came for her to be delivered, he sent her to stay with his sister, Lady Rich, who was well used to hushing up scandals, having had various extramarital affairs of her own. On November 8, 1598, Elizabeth was delivered of a baby girl, whom she christened Penelope after her hostess.

By now, the affair was one of the worst-kept secrets at court, and it soon reached the Queen's ears. When she was told of the "Lady of Southampton and her adventures," she was furious at being deceived yet again. "Her patience was soe muche moved that she came not to ye Chapple," reported one of Essex's servants. It was bad enough that another of her ladies had been embroiled in such a scandal, but what made it worse was the involvement of the Queen's chief favorite, whose enemies were quick to inform her that he had organized the wedding. "She threatenethe them all to the tower, not only the parties but all that are partakers of the practize," reported the same servant, adding with some irony: "I now understand that the Queen hathe Commanded that there shalbe provided for the nouille [new] Countesse the sweetest and best appointed lodgings in the fleet [prison]."[34]

Meanwhile, Sir Robert Cecil had written to Southampton on the Queen's behalf, ordering him back to England "with all speed."[35] "I must now put this gall in my ink, that she knows that you came over very lately, and returning again very contemptuously; that you have also married one of her maids of honour without her privity, for which with the circumstances informed against you, I find her grievously offended; and she commands me to charge you expressly (all excuses set apart) to repair hither to London, and adventure your arrival without coming to Court, until her pleasure be known."[36] Southampton wrote at once to his friend the Earl of Essex, lamenting "her Majesty's heavy displeasure conceived against me." He added, rather optimistically: "My hope is that time (the nature of my offence being rightly considered) will restore me to her wonted good opinion."[37]

His confidence was misplaced. Upon returning to London, Southampton was consigned immediately to the Fleet prison, where his new wife was languishing in one of the most unpleasant lodgings. They remained there until Elizabeth considered that they had learned their lesson. Still hopeful of being restored to the Queen's household, after their release the new Countess of Southampton went to court with her husband to request an audience. They joined the throngs of people waiting to petition the Queen as she made her way to chapel, but Elizabeth walked straight past them as if they were invisible. After waiting for a further two hours, the Countess sent a message to her former royal mistress via Lady Scudamore "that she desired her Majesty's resolution." Elizabeth angrily retorted "that she was sufficiently resolved but that day she would have a talk with her [the Countess's] father."[38] She duly spoke to Sir John Vernon and instructed him to take his daughter home in disgrace.

The Queen refused to allow her former maid of honor ever to return to her service, and the latter therefore spent the rest of the reign flitting between her father's and husband's estates, and the London home of their ally, the Earl of Essex. Unusually, neither did her husband ever recover the Queen's favor. In June 1599, his friends at court informed him that she was still "possessed with a very hard conceit" against him. When Essex found him employment in Ireland as general of the horse, Elizabeth "sharply chided" her favorite and told him to remove the earl forthwith, for she had "taken displeasure against Southampton, because he had without acquainting her, contrary to that which noblemen were wont to doe, secretly married Elizabeth Vernon."[39]

Essex had also made an unsuccessful attempt to rehabilitate his mother, Lettice, the Queen's old adversary. Even though Lettice was still in theory a member of Elizabeth's household, having never been formally dismissed, the Queen made it plain that, highly though she esteemed Essex, his mother would never be forgiven. Lettice, who had retired to her house at Drayton Bassett in Staffordshire, declared herself to be on strike until such time as her royal mistress might relent. However, in private she complained to her son that this exile only added "greater disgrace" to her situation.[40]

After languishing there for two years, the countess was heartened by news from her friends at court that "her majesty is very well prepared to hearken to terms of pacification," and assured her son that she would journey there with all haste, even in the depths of winter.[41] She duly made her way to Essex House in January 1598 and awaited a summons. Meanwhile, the court was buzzing with anticipation at the thought of witnessing the reunion of these two old adversaries. "The greatest newes here at Court is an expectation that my Lady Lester shall come to kisse the Queen's hands," reported Rowland Whyte, adding that "yt is greatly labored in, and was thought shuld have bene yesterday, but this day a hope is yt wilbe." Four days later, he wrote again to say that "her Majesty will not yet admyt my lady his [Essex's] mother to come to her presence, having once given some hope of yt."[42]

The Queen was in no hurry to summon back to court the woman whom she still referred to as the "she-wolf." In fact, she seemed to enjoy keeping her on tenterhooks. On March 1, it was arranged that Elizabeth would meet Lettice at the home of Lady Chandos. A "great dinner" was duly prepared, and the countess hastened to the house with "a faire jewell of £300" to present to her royal mistress. However, "upon a soddain she [Elizabeth] resolved not to goe, and soe sent word." Essex was furious when he heard of this slight to his mother and went at once to the Queen, not stopping to change out of his "night gown." For once, he found his way barred, and even when he tried to access her apartments by "the privy way," he failed to gain admittance and was forced to go back to his bedchamber.[43]

Just as Lettice had given up hope and was preparing to leave for her country estates, she finally received the long-awaited summons. The meeting between these two great rivals was courteous but brief. "My

Lady Lester was at Court, kissed the Queen's hands and her brest, and did embrace her," reported Whyte, noting that Elizabeth had kissed Lettice in return before drawing the audience to a close. The countess was far from forgiven, however, and the Queen had no intention of repeating the encounter. When Lettice, who was "very contented" by the event, tried to push home her advantage by expressing a wish "to kiss the Queen's hands" again, her request was summarily dismissed.[44] Within days, Elizabeth was referring to the countess with "some wonted unkind words," and made it clear that she would never again grant her an audience. When Essex pleaded with her to relent, she snapped that she had no wish "to be importuned in these unpleasing matters." Thenceforth, if ever he sulked or proved disobedient (which was often), she would say that he had inherited his difficult nature from his mother.[45]

The countess either did not realize that her royal mistress had reverted to her former antipathy or set little store by it, because two years later she was again petitioning to meet her. This time, it was on her son's behalf, for he had fallen from favor as a result of having flouted the Queen's orders while serving in Ireland. Lettice tried everything to try to persuade Elizabeth to receive her and even sent a "most curious fine gown" worth £100. Elizabeth admitted that she "liked it well, but did not accept nor refuse, only answered, that things standing as they did it was not fit for her to desire what she did."[46] The countess did not give up. She came to stay with her daughter, Lady Penelope Rich, in order to be closer to the court and her son, who was by now under house arrest. In February 1600, the two women moved to a house that overlooked Essex's prison, but when Elizabeth heard of this, she was incensed. "The Earl of Essex is little spoken of at Court," reported Whyte. "Mislike is taken that his mother and friends have been in a house that looks into York garden where he uses to walk, and have saluted each other out of a window."[47] Undeterred, Lettice subsequently moved to her son's own house in the hope of seeing him when he was released, but this only made Elizabeth more determined not to grant him his liberty.

Although Essex was eventually released and restored to some measure of favor with the Queen, he was highly aggrieved by his diminished influence at court, which his enemies (in particular, Sir Robert Cecil) had orchestrated in his absence. His nature was marked by a dangerous combination of arrogance and intemperance, and, in 1601, blind to the consequences, he launched a rebellion to oust his enemies from the council. His friend and stepfather, Christopher Blount, was among the rebel force, along with several other high-ranking noblemen. But it failed to gain more widespread support and was easily quashed by the royal forces. Although Essex claimed that his revolt had not been against the Queen but her ministers, he was convicted of high treason and sentenced to death. While he languished in the Tower, his mother was at her house in Drayton Basset, frantic with worry. She adored her "Sweet Robin" with a clinging, possessive love that bordered on the incestuous.[48]

Lettice left no trace of her feelings upon hearing of her son's rebellion and conviction. There are no touching last letters to her son, or to her husband, who had also been condemned to death. Neither did she go to plead with the Queen for his life. She had apparently finally realized just how much Elizabeth despised her and had no wish to make matters worse. When the axe severed the heads of her son and husband on February 25, 1601, she might have reflected, in the torment of her grief, that her old rival had finally wreaked her revenge. Yet it was a revenge that destroyed Elizabeth just as much as her despised adversary, for she had loved Essex with a fierce possessiveness that rivalled his mother's, and the agony of having to order his death was said to have hastened her own. In the long-running battle between the two women, it was not clear who had won.

The Queen and Lettice would never meet again. After her son's death, the countess had no wish to return to court, even if she had been welcome. Robert Cecil later recalled that she had been "long disgraced with the Queene" and their rift would never be healed. When Lettice died in 1634 at the remarkable age of ninety-three, at her request she was buried at Warwick "by my deere lord and husband the Earle of Leicester."[49] In death, as in life, she was determined to gain the upper hand over her rival.

It looked certain that the Queen would take the same revenge upon Elizabeth Vernon as she had upon Lettice Knollys. The Earl of Southampton had been arrested for his involvement in Essex's rebellion and sent to the Tower to await his fate. Meanwhile, his wife wrote frantic letters to Sir Robert Cecil, begging him to secure a pardon from the Queen, who now held her husband's life in her hands. Lamenting the "miserable distress of my unfortunate husband," and declaring her "infinite and faithful love unto him," she ventured "humble petitions to His holy anointed [the Queen], prostrate at her feet if it might be, to beg some favour." Bitterly regretting having lost her royal mistress's favor when it might have saved her husband, she described herself as "the most miserable woman of the world . . . And in that through the heavy disfavour of her sacred Majesty unto myself, I am utterly barred from all means to perform those duties and good to him I ought to do."[50]

It seemed that her pleas had fallen on deaf ears, for on February 17, the Earl of Southampton was found guilty of high treason and sentenced to death. Beside herself with panic and grief, the countess again wrote to Cecil. "The woeful news to me of my Lord's condemnation passed this day makes me in this my most amazed distress, address myself to you. I do beseech you to conjure you by whatsoever is dearest unto you that you will vouchsafe so much commiseration unto a most afflicted woman, as to be my means unto her sacred Majesty that I may by her divine self be permitted to come to prostrate myself at her feet, to beg for mercy for my Lord."[51] Still Elizabeth refused to grant her former maid an audience. However, her heartfelt plea may have had some effect, for the Queen eventually commuted Southampton's sentence to life imprisonment.

Back at court, Elizabeth's ladies were growing increasingly impatient with their mistress's authority and showed little respect toward her. Lady Mary Howard was typical of this new breed of courtier. Tired of the strictures that Elizabeth had imposed with regard to her ladies' dress, Lady Mary one day appeared at court in an ostentatious gown made from a rich velvet and "powdered with golde and pearle." Harington recalled that this had "moved manie to envye; nor did it please the Queene, who thoughte it exceeded her owne." Elizabeth was so jealous that a few days later she ordered a servant to steal the gown from Lady Mary's chamber and bring it to her. She duly put it on herself and paraded it in front of her ladies, demanding to know "How they likede her new-fancied suit?" None of them dared to admit that it was "far too shorte for her Majesties heigth." At length the Queen addressed Mary Howard herself, who resentfully snapped that it was "too short and ill becoming." "Why then," Elizabeth purred, "if it become not me, as being too shorte, I am minded it shall never become thee, as being too fine; so it fitteth neither well." According to Sir John Harington: "This sharp rebuke abashed the ladie, and she never adorned her herewith any more." The dress was carefully packed away, never to be seen again while Elizabeth was on the throne.[52]

Far from being "abashed," Lady Mary deeply resented this humiliating reprimand, and thenceforth she refused to carry the Queen's cloak or serve her drinks at mealtimes, as her duties required. When Elizabeth upbraided her for insolence, she "did vent such unseemly answer as did breed much choler in her mistress."[53] It is a sign of how much things had changed that a lady should dare to show such disrespect.

Mary Howard was by no means the only one of Elizabeth's ladies to flout her authority. It was said that they often laughed at her behind her back for "trying to play the part of a woman still young."[54] The Earl of Essex's rebellion in 1601 had seriously destabilized the Queen, making her ever more fearful and paranoid. "These troubles waste her muche," reported Harington. "Every new message from the city doth disturb her . . . the many evil plots and designs have overcome all her Highness' sweet temper. She walks much in her privy chamber, and stamps with her feet at ill news, and thrusts her rusty sword at times into the arras in great rage . . . the dangers are over, and yet she always keeps a sword by her table."[55]

According to another observer: "the court was very much neglected, and in effect the people were generally weary of an old

woman's government."[56] Increasingly, they looked north of the border
to James VI, anxious to ingratiate themselves with the Queen's likely
successor. As Camden noted: "They adored him as the Sunne rising,
and neglected her as now ready to set."[57] Elizabeth was well aware of
this and was tormented that "the question of the succession every day
rudely sounded in their ears."[58]

Although it seemed ever more likely that James would succeed, another potential claimant to the throne was causing the English queen a
great deal of unease. Thanks to her royal blood, Arbella Stuart's name
had been linked to a whole host of suitors during the 1590s and early
1600s. It was even rumored "that Sir Robert Cecil intends to be king,
by marrying Arabella, and now lacks only the name."[59] Arbella had
paid a third visit to court in 1592, accompanied by her grandmother.
Bess of Hardwick had been determined to secure Arbella's place in the
succession, and despite all the previous setbacks, she had confidently
anticipated that Elizabeth would use the occasion to name Arbella her
heir. She had therefore planned their journey south as a triumphal procession, and they had arrived at court in magnificent state that June.
Arbella had again attracted attention at court, but Elizabeth had
stopped short of acknowledging her as the next Queen of England.
Bitterly disappointed, the girl later reflected: "What fair words I have
had of courtiers and councillors, and so they are vanished into
smoke."[60]

Even though her visit to court had not been a success, Arbella was
far from glad to leave. She was becoming increasingly frustrated with
her grandmother's oppressive regime, and the latter no doubt added to
her misery by chastising her for her conduct at court, which seemed to
have well-nigh ruined her hopes for the succession. Once back at
Chatsworth, things went from bad to worse. Now in her late teens, Arbella became increasingly rebellious, and one of Bess's stewards reported that she was neglecting her studies and refusing to go to bed at
the accustomed hour, preferring to stay up late making "merry."[61]

Her grandmother reacted by imposing even more limitations upon
her freedom. She ordered that Arbella was to be kept under strict surveillance at all times, and more often than not, she herself was the cus-

todian. "She was under very strict custody of her grandmother, Lady Shrewsbury," reported Marin Cavalli, the Venetian ambassador in France, "and was never allowed to be alone or in any way mistress of her actions."[62] Being forced to spend so much time with her aging, overbearing grandmother, and having no company of her own age, Arbella grew ever more resentful. James VI would later recall "that unpleasant life which she hath led in the house of her grandmother with whose severity and age she, being a young lady, could hardly agree."[63]

In 1597 Bess took her granddaughter to live at her magnificent new home, Hardwick Hall. Although a change of scene might have provided a temporary relief from her oppression, Arbella was soon just as miserable as she had been at Chatsworth. Blinded by her own ambition, Bess failed to see the effect that her domineering influence was having on the girl. Arbella's behavior became increasingly erratic, but the more she lashed out against her grandmother's strictures, the more severe these became. Even though she was now well into her twenties, Arbella was still treated like a child, and complained of "being bobbed and her nose played withal" if she disobeyed her grandmother. John Starkey, the chaplain at Hardwick, noted with some sympathy that the girl's misery "seemed not feigned, for oftentimes, being at her books, she would break forth into tears."[64]

Eventually it became too much to bear, and in late 1602 Arbella hatched a plan to escape. Frustrated by the many negotiations for her marriage, all of which had come to nothing, she resolved to find a husband for herself. The man that her hopes alighted upon could hardly have been a less appropriate choice. Edward Seymour was the grandson and namesake of the first Earl of Hertford, who had caused such a scandal all those years ago by marrying Lady Katherine Grey. Allying herself to a family that had long been tainted by treachery was a disastrous move. Moreover, Edward Seymour was himself of royal blood, so the Queen was sure to suspect a conspiracy.

It was no easy matter for Arbella to make contact with her intended husband, for her grandmother continued to watch her like a hawk. But a number of Bess's servants sympathized with her plight and hated to see how harshly she was treated. One of them, John Dodderidge, risked his position by agreeing to convey a message from Arbella to the Earl of Hertford, who was then living at his house in London, which proposed that she marry his sixteen-year-old grandson.

When Dodderidge reached the earl's house and handed over his young mistress's urgent message, it did not elicit the response that she had expected. The earl had only recently been in trouble with the Queen for trying to prove the validity of his marriage to Lady Katherine Grey, so he was horrified when he learned of Arbella's scheme, knowing that even to have received her message was enough to implicate him. With "many bitter reprehensions," he chastised Dodderidge for thus coming to him and immediately informed the council. The hapless messenger was duly placed under armed guard and questioned before being sent to the court for further interrogation by Sir Robert Cecil. In the meantime, he managed to send a secret message to Arbella, warning her: "My reception here is contrary to all expectation."[65] She never received it.

When Elizabeth learned of Arbella's plan, she was outraged. This haughty young woman, whom she had never liked, had confirmed all her prejudices by trying to marry a scion of the most traitorous family in England. That her choice of husband was himself of royal blood made it certain, in the Queen's mind, that Arbella had been plotting to seize the throne. In fact, this had probably been far from the foolish young woman's mind: her primary motivation had been to escape her miserable life at Hardwick, and she had probably not thought beyond that. But Elizabeth was convinced that it was part of a greater conspiracy and was determined to punish the perpetrators.

News of the controversy spread like wildfire, and before long it had reached the courts of Europe. In France, Cavalli reported "the uproar, which has happened in England recently, about Arbella." When the perpetrator herself heard that she had been discovered, she "went down on her knees and implored pardon; declaring that she had taken this step to induce the Queen to change her prison, for she knew that any other must be much milder than the one she was in."[66] But Elizabeth's suspicions would not be dispelled so easily, and she dispatched one of her trusted officials, Sir Henry Brouncker, to Hardwick in order to interrogate Arbella. She instructed him to talk to the young woman alone, without the overbearing presence of her grandmother, who would no doubt attempt to answer the questions herself.

In fact, Bess had known nothing of her granddaughter's transgressions. True, she had hoped that the girl might one day inherit the throne, but this was to be achieved by persuading Elizabeth to name her as heir, not by conspiring behind the Queen's back. When Brouncker arrived at Hatfield and told Bess that he was commanded by the Queen to question Arbella, the countess was taken aback, having no idea what on earth her granddaughter could have done. The fact that Elizabeth had ordered her to be kept in ignorance made her panic that she herself was under suspicion, but Brouncker assured her of the Queen's goodwill. "The old lady took such comfort at this message as I could hardly keep her from kneeling," he reported. Bess herself wrote to thank the Queen shortly afterward: "When I considered your Majesty's great wisdom in it, I did in my heart most humbly thank your Majesty for commanding that course to be taken," she wrote. "These matters were unexpected of me, being altogether ignorant of her vain doings, as on my salvation and allegiance to your Majesty I protest."[67]

For once, Bess's loyalty to the Queen won out over her dynastic ambition. She was so "wonderfully afflicted" when she heard of Arbella's plot and "took it so ill" that it was all Brouncker could do to stop her beating the girl. Meanwhile, he informed Arbella that she was under suspicion of treason, and observed: "it seemed by the coming and going of her colour that she was somewhat troubled, yet (after a little pause) she said that the matter was very strange to her; she was much grieved that your Highness should conceive an ill opinion of her." At first Arbella denied having written to the Earl of Hertford and was very "obstinate and wilful" in all of her answers. Eventually, though, she realized that it was futile to maintain the pretense and promised to tell Brouncker everything on condition that he would "promise to conceal it from her grandmother." However, as she began to relate her version of events, her interrogator found that it was "done so confusedly with words so far from the purpose as I knew not what to make of it."

He therefore told Arbella to prepare a written confession instead. This was hardly more coherent. "When I read it, I perceived it to be confused obscure and in truth ridiculous. I told her it was not a letter fit for me to carry, nor for your Highness to read . . . She wrote again and little better than before, which made me believe that her wits were somewhat distracted either with fear of her grandmother or conceit of her own folly."[68]

Brouncker eventually concluded that there was no conspiracy and that the whole sorry episode had been sparked by the wild imaginings of a troubled mind. Bess's reaction made it clear that she'd had neither knowledge nor involvement in it. Desperate to prove her loyalty, upon his departure she "fastened a purse full of gold on me in honour of your Majesty," and it was only with great difficulty that he was able to persuade her to take it back. Nevertheless, the countess was still determined to make Elizabeth realize that she had played no part in the reckless scheme and therefore wrote to her again, declaring her intention to disown "this unadvised young woman." "I am desirous and most humbly beseech your Majesty that she may be placed elsewhere, to learn to be more considerate," she pleaded, "and after that it may please your Majesty either to accept of her service about your royal person or to bestow her in marriage, which in all humility and duty I do crave of your Majesty for I cannot now assure myself of her as I have done." In another letter, she declared: "For my own part, I should have little care how meanly soever she were bestowed so as it were not offensive to your Highness."[69]

The fact that Bess was thus prepared to relinquish all her long-cherished ambitions for Arbella and hand her over to the Queen to do with her what she wished proves the strength of her loyalty. In case her royal mistress remained in any doubt, she added a heartfelt expression of her devotion: "I will not respect my trouble or charge to do your Majesty any service that shall lie in me during life."[70] Her words seemed to work the desired effect, for Elizabeth sent word that she "remains satisfied with your proceedings, nothing appearing in them but fullness of care to prevent inconveniences and desire to accomplish in all things her Majesty's pleasure." She later wrote to Bess in person: "I assure you, there is no Lady in this land that I better love and like."[71] So great was the Queen's trust in her old servant that she gave her responsibility to insure that Arbella did not step out of line again. At her instruction, Bess was effectively required to spy on her own granddaughter and report back anything untoward.[72]

For Bess, this was a mixed blessing. On the one hand, she was overjoyed that Elizabeth had shown such trust in her, but on the other she found the prospect of continuing to live with "this inconsiderate young woman" utterly distasteful. Nevertheless she declared: "What it shall please her Majesty to command me, to the uttermost of my power I will do my best service, though it be to the shortening of my days . . . Even to the last hour of my life I shall think myself happy to do any acceptable service to her Majesty."[73] She would be as good as her word. If the Queen had wished to test her loyalty, she soon had ample proof of it. Within weeks of accepting her commission, Bess was reporting her suspicions that Arbella "had some other like matter in hand." In her paranoid state, the countess seized upon the slightest look or word from her granddaughter as a sign that she was plotting again. Before long, the strain was starting to take its toll. "The old lady groweth exceedingly weary of her charge, beginneth to be weak and sickly by breaking her sleep, and cannot long continue this vexation," reported Brouncker.[74]

Elizabeth continued to suspect that there had been more people involved, but her attitude to Arbella was surprisingly sympathetic. "In the observation of the root from whence this motive sprung in the lady [Arbella], she [the Queen] doth perceive that some base companions, thinking it pleasing to her youth and sex to be sought in marriage, were content to abuse her with a device that the Earl of Hertford had a purpose to match his grandchild with her."[75] Such leniency could hardly have been expected, given the severity of the Queen's treatment of the other rivals to the throne, notably the Grey sisters, when their secret plots were uncovered. Perhaps it was thanks to Bess's influence with Elizabeth that she chose to forgive her granddaughter so easily.

In January 1603, when she heard that the Queen had decided to pardon her misdemeanors, Arbella wrote a letter of heartfelt thanks from

"Your Majesties most humble and dutifull handmaid," urging: "I humbly prostrate my selfe at your Majesties feete craving pardon for what is passed and out of your Princely clemency to signify your Majesties most gratious remission to me by your Highnesse letter to my Lady my Grandmother whose discomfort I shall be till then."[76] This was probably dictated by Bess, who was far more conscious than her wayward granddaughter just how narrowly she had escaped punishment.

Arbella, meanwhile, seemed to be slipping ever further into insanity. Just a few days after writing her letter of humble submission to the Queen, she wrote another declaring that she was betrothed in secret to "some one near and in favour with Her Majesty," but refused to reveal his identity.[77] Elizabeth was taking no chances and sent Brouncker back to Hardwick to interrogate the young woman. Arbella eventually confessed that her secret lover was the already married James VI, whom she had never met. She then wrote a petulant letter to the court, declaring: "It was convenient her Majesty should see and believe what busy bodies, untrue rumours, unjust practices, colourable and cunning devices are in remote parts among those whom the world understands to be exiled from her Majesty's presence, undeservedly."[78] In the meantime, she went on a virtual hunger strike in order to draw further attention to her claims. If she thereby hoped to persuade Elizabeth to invite her back to court, she was sadly mistaken. The Queen merely ordered that the security at Hardwick be increased in order to prevent any further attempts at escape.

Tormented by her confinement and goaded by her grandmother's constant reproofs, Arbella's mental state was hardly strong enough to withstand the shocking event that took place in February, when her old chaplain and tutor, John Starkey, was discovered with his throat cut. It was rumored that he had killed himself because he was plagued with guilt about the part he had played in his protégée's intrigues. Driven mad with fear and grief, Arbella wrote at once to her interrogator, Sir Henry Brouncker. "If you thinck to make me weary of my life and so conclude it according to Mr Starkey's tragicall example, you are deceived," she railed. "I hope it is not hir Majesties meaning nor your delusive dealing, and sure I am it is neither for hir Majestie's honour nor your creditt, I shall be thus dealt withal . . . I recommend my innocent cause and wrongfully wronged and wronging frende to your consideration and God's holy protection . . . For all men are liers."[79]

The more Arbella ranted, the less of a threat she appeared to the court back in London. The chances of this reckless, half-mad woman ever succeeding to the throne seemed more remote than ever. "We are very sorry to find by the strange style of the Lady Arbella's letters that she hath her thoughts no better quieted," the lords of the council wrote to Brouncker on March 14, "especially considering her Majesty's own ready inclination . . . to have taken no other course with her than was expressed by our first joint letter." A few days later, their agent confirmed that Arbella "hath neither altered her speech nor behaviour. She is certain in nothing but in her uncertainty." It was clear that he was growing increasingly exasperated with his charge, for he confided to Cecil that her "wilfulness . . . is much greater and more peremptory than before . . . I find her so vain and idle as I seldom trouble her . . . all her words and actions are so contrary to reason as no man can divine aright of her."[80]

Deranged though she now was, Arbella was lucid in one thought at least, and that was her urgent desire to escape from Hardwick. Brouncker noted that she "desireth liberty," and told his masters: "I persuade her to patience and conformity, but nothing will satisfy her but her remove from her grandmother, so settled is her mislike of the old lady." Shortly after Brouncker wrote this, Arbella made a desperate attempt to flee from her grandmother's clutches. This caused some alarm at the court in London, where it was feared that a conspiracy was afoot. But there is little evidence to support this, and the ill-planned nature of Arbella's attempt suggests that she was the only one involved. Although it was reported that Elizabeth intended to have Arbella imprisoned at Woodstock Palace, just as she herself had been during her half sister's reign, the young woman remained under her grandmother's care at Hardwick.

Arbella was the last of the female claimants to Elizabeth's throne, and like all her predecessors, she ultimately found her royal blood a curse. Not only did she fail to learn from their example, but she disregarded her own previous experience and entered time and again into such reckless courses that her downfall was more due to self-destruction than to any external forces. The irony is that if Arbella had not been so foolhardy as to conspire to marry Seymour and escape from Hardwick during Elizabeth's last months, but had instead bided her time, she would probably have stood a much greater chance of advancement in the next reign. As it was, her highly strung nature combined fatally with her romantic delusions to ensure that she, too, would join the ranks of Elizabeth's tragic rivals.

CHAPTER 16

"The Sun Now Ready to Set"

Although in the later years of her reign Elizabeth had started to lose her grip on the affairs of the court, she enjoyed good health almost until the end. She still rode in the hunt, enjoyed her accustomed long walks "as if she had been only eighteen years old," and danced energetically until well into her sixties. In January 1599, a Spanish visitor to the court observed with some astonishment that after a feast held there one evening, the Queen "was to be seen in her old age dancing three or four gaillards." The following year, she insisted upon going on progress, ignoring the entreaties of her councillors. "Her Majesty byds the old stay behynd, and the young and able to goe along with her," reported Rowland Whyte. Even as late as February 1603, Scaramelli observed with some astonishment that the Queen "is in excellent health, as I hear on all sides, and in perfect possession of all her senses."[1]

But even this formidable monarch could not defy time forever. Her courtiers continued to flatter her, but she no longer paid any heed to their compliments, realizing at last—as they had done some years before—that she was a "crooked old woman." It was perhaps this that hastened the Queen's decline. According to one account, she called for a looking glass for the first time in twenty years, and upon seeing her face "lean and full of wrinkles," she "fell presently into exclayming against those which had so much commended her, and took it so offensively, that some which before had flattered her, dourst not come into her sight." Thereafter, she was "extreame oppressed" with a deep melancholy.[2]

In her grief, Elizabeth sought peace away from society, closeting herself in her privy chamber with just a few favored ladies. "She has suddenly withdrawn into herself, she who was wont to live so gaily, especially in these last years of her life," observed Giovanni Scaramelli, the Venetian envoy to England. In late 1602, Sir John Harington paid a visit to court and remarked sadly that his godmother was "a lady shut up in a chamber from her subjects and most of her servants, and seldom seen but on holy days." He realized that he was one of the few who felt any sorrow at the steady decline of "this state's natural mother," for he added: "I finde some lesse mindfulle of whate they are soone to lose, than of what they may perchance hereafter get."[3]

Elizabeth seemed to have lost all her former lust for life, and it was as if she had decided that it was time to give it up altogether. She would not be cheered by any of her courtiers. When Harington read her some verses that he had written in her honor, she told him: "When thou doste feele creepinge tyme at thye gate, these fooleries will please thee lesse; I am paste my relishe for such matters." Even her favorite cousin, Sir Robert Carey, could do no better. As he later recounted, when he assured her that she looked to be in the best of health, she replied: "'No, Robin, I am not well' . . . and in her discourse she fetched not so few as forty or fifty great sighes . . . I used the best words I could to persuade her from this melancholy humour; but I found by her it was too deep rooted in her heart, and hardly to be removed."[4]

In January 1603, the Queen left the court in Whitehall and moved to Richmond Palace, her "warm box," to which she could "best trust her sickly old age."[5] She was accompanied by a small entourage of her ladies. The gulf between Elizabeth and the wayward young maids of honor who entered her service had grown ever greater, as they fixed their sights upon the next regime. Women such as the Countess of Warwick, on the other hand, belonged to a generation that was quickly passing away. The countess had served Elizabeth for more than forty years and represented a treasured link with the ideals and traditions of Gloriana's heyday. Perhaps it was for this reason that the Queen sought her company above all others during her final illness. The countess's niece, Lady Anne Clifford, recalled that her aunt visited Elizabeth regularly when the latter took to her bed in March 1603. "My Lady used to goe often thither and caried me with hir in the coach," she wrote. The young girl would wait in the outer chambers while Lady Warwick attended her royal mistress, often staying with her until "verie late."[6] Another of the Queen's faithful servants was with her: Helena Gorges (née Snakenborg), whom Elizabeth had taken such a shine to when she had arrived from Sweden almost forty years before. Having been surrounded by her mother's relatives throughout her reign, the Queen also made sure that one of them would be in attendance upon her now. The chosen lady was Philadelphia Scrope (née Carey), sister of Elizabeth's former favorite, the Countess of Nottingham.

The women who accompanied Elizabeth to Richmond hoped that the comforts of that palace would ease their mistress's troubled state, but she continued to slip into a steady decline, and it soon became clear that it would be her final illness. Elizabeth, too, realized this and had apparently resolved to hasten her own end. "Shee refused to eate anie thing, to receive any phisicke, or admit any rest in bedd," reported one of those present. Her attendants began to despair as day after day she turned food and drink away, "holding her finger almost continually in her mouth, with her eyes open and fixed upon the ground, where she sat on cushions without rising or resting herself, and was greatly emaciated by her long watching and fasting." Ever mistress of her own fate, she railed against her physicians when they tried to press their medicines upon her, and those around her began to suspect that she had simply decided to die. "The Queene grew worse, because she would be so, none about her being able to perswade her to go to bed," recalled an exasperated Carey. John Manningham, another visitor to the palace, observed: "It seemes she might have lived yf she would have used meanes; but shee would not be persuaded, and princes must not be forced."[7]

In her increasing delirium, Elizabeth was haunted by strange apparitions. "She told a lady, one of the nearest about her person, that she had seen a bright flame about her, and asked her if she had not seen visions in the night," reported the Countess of Warwick's niece. Unnerved by this, and worn down by the strain of trying to care for their mistress, her ladies also began to imagine things of a supernatural nature. One of them was said to have left the Queen sleeping for a few moments in order to get some fresh air and had been surprised to encounter her mistress a few moments later. Fearing that she was about to receive a severe reprimand for leaving her alone, she hurried forward to present her excuses, and was aghast when the apparition "vanished away." Upon returning to the royal bedchamber, she found the Queen still sleeping. At around the same time, two of her ladies discovered a queen of hearts playing card on the underside of a chair with a nail driven through the forehead, "which the Ladyes durst not then pull out, remembring that the like thing was reported to be used to others for witchcraft."[8]

Sir Robert Carey recalled in his memoirs that as the Queen lapsed ever further into decline, she was tormented not just by ghostly visions but also by memories of real people whom she had known—above all, her women, both past and present.[9] Principal among them during these early days of her illness was Arbella Stuart. In early March, Elizabeth had been greatly troubled by news of Arbella's attempted escape from Hardwick. "The Queen has received information that some dangerous practices have been intended for the violent removing of the Lady Arabella out of the charge of her grandmother," reported the lords of the Council. Another observer noted that "the rumours of Arabella much afflict the Queen," and Scaramelli claimed: "It is well known that this unexpected event has greatly disturbed the Queen . . . as far as health was concerned, her days seemed numerous indeed but not now she allows grief to overcome her strength."[10]

When it became clear that Elizabeth was dying, the rumors about this unlikely successor grew ever wilder. It was said that Arbella had not just escaped but had married one of her many suitors and thereby intended to seize the crown. Such rumors greatly tormented the Queen. "She raves of . . . Arabella, and is infinitely discontented; it is feared she will not long continue," reported one anonymous observer. The Venetian envoy likewise claimed: "Her Majesty's mind was overwhelmed by a grief greater than she could bear. It reached such a pitch that she passed three days and three nights without sleep and with scarcely any food. Her attention was fixed . . . on the affairs of lady Arabella, who now is, or feigns herself to be, half mad."[11] Brouncker confirmed that Arbella was more "wilful" than ever, and that this "ariseth out of a hope of the Queen's death."[12] Scaramelli later reported: "Her conduct is thought to have killed the Queen," and referred to Arbella in a dispatch to his masters as "Omicida della Regina."[13]

But it was the dead, as well as the living, that plagued Elizabeth during her final days, as her mind wandered ever further back over her life. The old feelings of guilt about Mary, Queen of Scots, rose to the fore once again, and the Queen was said to have wailed out in torment at the part she had played in her execution. Sir Robert Carey described how she "shedd many teares and sighes, manifesting her innocence that she never gave consent to the death of that Queene."[14] She was also reminded of another former rival, Lady Katherine Grey, when one of those present at Richmond suggested that she might name Katherine's eldest son as her heir. "I will have no rascal's son in my seat but one worthy to be a king," exclaimed Elizabeth with one last flash of her accustomed fury.[15]

Thoughts of her women inspired more tender feelings in the dying Queen. She was deeply grieved by the death of her old servant Katherine Howard, Countess of Nottingham. She had received news of this in late February, and it was widely believed that this painful reminder of her mortality hastened her own death. "The Queen loved the Countess well, and hath much lamented her death, remaining ever since in deep melancholy that she must die herself, and complaineth much of many infirmities wherewith she seemeth suddenly to be overtaken," observed Anthony Rivers, a contemporary at court. In another letter, he claimed that Elizabeth "rests ill at nights, forbears to use the air in the day, and abstains more than usual from her meat, resisting physic, and is suspicious of some about her as ill-affected." Scaramelli agreed: "The Queen for many days has not left her chamber . . . they say that the reason for this is her sorrow for the death of the Countess."[16]

It was rumored that her sorrow was also due to a deathbed confession that the countess had made to her former royal mistress. The story goes that when the Earl of Essex had been languishing in the Tower after his abortive rebellion, he had resolved to send the Queen a ring that she had once given him and that was secretly understood between them to be a token of their mutual devotion. Looking out of his window, he had spied a passing boy and threw the ring down to him, urging him to convey it with all speed to Lady Philadelphia Scrope, his chief ally in the Queen's privy chamber. In a fatal case of mistaken identity, the boy had instead presented the ring to Lady Scrope's sister, Katherine Howard, whom she closely resembled. Knowing what it signified, Katherine had at once sought out her husband and asked him what she should do. He had instructed her to hide the ring away and say nothing of it to anyone. Thus receiving no sign of Essex's continuing devotion, the Queen duly ordered his execution. The matter had preyed on Katherine's conscience, and it was said that on her deathbed she had confessed everything to Elizabeth, who cried: "God may forgive you, but I never can!" The anguish her servant's revelation had caused was said to have hastened her own death shortly afterward.[17]

At length the Queen was persuaded by the Lord Admiral, widower of her loyal servant, to retire to her bed. Thenceforth her life slipped rapidly away. Sir Francis Vere, commander of the Queen's troops in the Netherlands, heard from his agent at Richmond that "between the coffer chamber and her bed chamber he saw great weeping and lamentation among the lords and ladies, as they passed to and fro, and perceived there was no hope that Her Majesty should escape." Vere lamented: "I never thought to live to see so dismal a day."[18] Although Elizabeth rallied briefly after taking to her bed, she was then seized by a "defluxion in the throat," which left her unable to speak and "like a dead person." Four days later, Scaramelli reported: "Her Majesty's life is absolutely despaired of, even if she be not already dead."[19] On March 23, as she lay senseless, her ladies and councillors gathered around her bed anxiously watching for signs of life, Elizabeth suddenly rallied. With tears streaming down her cheeks, she exhorted her ministers to care for the peace of the realm, and when the Lord Admiral asked her if James VI should be her heir, she lifted her hand up to her head and slowly drew a circle around it to indicate a crown.[20]

That evening, Elizabeth's councillors, Archbishop Whitgift, and everyone who had thronged into the royal bedchamber departed. Only her ladies remained. They watched over her as she drifted between waking and sleeping. Between two and three o'clock the following morning, their royal mistress breathed her last. "Her Majesty departed this life, mildly, like a lamb, easily like a ripe apple from the tree," reported her favorite chaplain, Dr. Parry.[21] Lady Scrope opened the window of the bedchamber and dropped a sapphire ring to her brother, Sir Robert Carey, who was waiting below. The ring had been given to her by James VI, who had instructed her to send it to him as a sign that the Queen was dead. It was to this new king of England that Sir Robert now rode with all haste.

The Countess of Warwick had feared "some Commotions" upon the Queen's death being announced and her successor being proclaimed king, but in the event, this happened as quietly as her mistress's passing. "This peacable coming in of the King was unexpected of all parts of the people," reported the countess's niece, Anne Clifford.[22] It seemed that in their haste to acknowledge King James, the late queen's beloved subjects had forgotten all about her. For days after her death, her corpse lay at Richmond, wrapped in a cerecloth in a "very ill" fashion. It was said that it was left entirely alone and that "mean persons had access to it."[23] In truth, it had been guarded throughout by Elizabeth's former ladies. Scaramelli reported: "The body of the late Queen, by her own orders, has neither been opened, nor indeed seen by any living soul, save by three of her ladies."[24]

These ladies, who included the Countess of Warwick and Helena Snakenborg, watched over the Queen's corpse as it was carefully placed in a lead coffin, and accompanied it as it was taken at night from Richmond to Whitehall Palace by barge in a sombre torchlit procession. Upon arrival, it was carried into a withdrawing chamber and placed upon a bed of state, "certain ladies continually attending it."[25] They proceeded to watch over it day and night until the funeral three weeks later.

The late Countess of Nottingham's granddaughter Elizabeth Southwell, who had been serving as a maid at court during Elizabeth's final days, later claimed that her former royal mistress had had one last surprise in store for her women. The story goes that Mistress Southwell had been among the ladies watching over the coffin at Whitehall, "and being all in our places about the corpse, which was fast nailed up in a board coffin, with leaves of lead covered with velvet, her body burst with such a crack that it splitted the wood, lead and cere-cloth, whereupon, the next day, she was fain to be new trimmed up."[26] Often though this tale has been repeated, it is almost certainly untrue—just one of many scandalous and grotesque reports put about to discredit the late queen.

After lying in state at Whitehall, Elizabeth's body was taken to Westminster Hall in preparation for her funeral. This took place at Westminster Abbey on April 28, 1603, and it was a fitting testament to one of the greatest queens England would ever know. Crowds thronged the streets, and as Elizabeth's coffin passed by, "divers of the beholders fell a weeping, especially women."[27] The magnificent funeral procession comprised more than 1,000 nobles, bishops, ministers, and courtiers, and 260 "poor women . . . apparelled in black, with linen kerchiefs over their heads." Following directly behind the coffin were the Queen's ladies. Chief among them was Helena Gorges, who is depicted in a contemporary illustration of the funeral wearing a black hood and cloak and clutching a handkerchief to stay her tears. She and the other members of the late queen's household who followed her coffin to its final resting place were as assiduous in serving their royal mistress in death as they had been in life.

Acknowledgments

I have been extremely lucky to have the support of many people, both professionally and personally, during the research and writing of this book. In particular, I would like to thank the inspirational Alison Weir for providing the original idea and for masterminding our series of joint events on Anne Boleyn and Elizabeth I, "The Whore and the Virgin." My thanks also go to my editor, Ellah Allfrey, for suggesting a new approach to the book and for her enthusiasm and guidance throughout. My agent, Julian Alexander, was instrumental in developing the original idea, and his continuing advice and support have been, as ever, invaluable. I was delighted to have the support of Hannah Ross, senior publicist at Jonathan Cape, whose creativity and enthusiasm have insured a fantastic series of events and publicity. I am also grateful to Katherine Murphy for her assistance with editing and picture research and to Geraldine Bear, who did such a good job on the index.

I owe a debt of gratitude to my colleagues at the Heritage Education Trust for their support and interest in the book, notably Gareth Fitzpatrick, John Hamer, and Jean MacIntyre. I have also been very fortunate in having the encouragement of my colleagues at Historic Royal Palaces, in particular John Barnes, Michael Day, David Souden, Sam Brown, and the Interpretation Team. I would like to thank Dr. Edward Impey for generously sharing his research on Amy Robsart and Cumnor Place. Likewise, Charles Lister of Boughton House for all his help with the picture research. It is thanks both to Charles and to Gareth Fitzpatrick that a little-known portrait of Elizabeth as a princess has come to national prominence.

My friends and family have been unfailingly encouraging throughout. I would particularly like to thank my parents, both for their encouragement with this book and for inspiring me with a love of history as a child. My sister, Jayne, and her family have also been very supportive, and I would like to give special thanks to my nieces, Olivia and Neve Ellis.

From the beginning, my friend Honor Gay has shown a limitless enthusiasm for the book and has been an invaluable source of ideas and inspiration during our long country walks. Likewise, Maura and Howard Davies have taken a keen interest in my research and writing, and have always proved willing to offer advice and support on everything from subject matter to subtitles. I owe a huge debt of gratitude to my friend Doreen Cullen for her wisdom and guidance, as well as to my fellow authors Ruth Richardson, Sarah Gristwood, Siobhan Clarke, Christopher Warwick, and Julian Humphrys for offering advice and encouragement. It is thanks to Jean and Gladney Wadsworth that I was able to undertake some of my research by the beautiful Dorset coastline. Finally, my former headmaster, Len Clark, has shown an enduring interest in my writing career, for which I am most grateful.

Notes

Abbreviations

BM MS British Museum Manuscript, British Library
CSPD   Calendar of State Papers, Domestic Series
CSPF   Calendar of State Papers, Foreign Series
CSPS   Calendar of State Papers, Spanish
CSPV   Calendar of State Papers, Venetian
HMC   Historical Manuscripts Commission
L&P   Letters and Papers of Henry VIII
SP   State Papers
TNA   The National Archives

INTRODUCTION

1. Harrison and Jones, *Andre Hurault de Maisse,* 109.

CHAPTER 1: Mother

1. CSPV, VI, ii, 1059.
2. Ibid., VI, ii, 1105.
3. Elizabeth's seventeenth-century biographer, William Camden, claimed that Anne was born in 1507, as did other sources of that time. But this would have made her no more than six years old when she entered Margaret of Austria's service in 1513, an impossibly young age.
4. Ashdown, *Ladies-in-Waiting*, 23–24.
5. Strickland, *Lives of the Queens of England*, II, 572.
6. CSPV, IV, 365.
7. Wyatt, "Extracts from the Life of the Virtuous, Christian and Renowned Queen Anne Boleyn," 423–24.
8. In reality, this was little more than a second nail growing on the side of one of her fingers. Anne was so self-conscious about it that she took to wearing long-hanging oversleeves, which instantly became fashionable among court ladies.
9. Wyatt, "Extracts," 424.
10. Ibid., 441.
11. L&P, IV, II, i, 1467.
12. Ibid., IV, II, i, 1468.
13. CSPV, IV, 287.
14. Bodleian Library MS Don. C42, ff.21–33.
15. L&P, VII, ii, 251.
16. CSPV, IV, 288.
17. Ibid., IV, 57, 288.
18. L&P, IX, 288.
19. Ibid., V, i, 11.
20. Ibid., VI, i, 150.
21. Ibid., VI, i, 356.
22. Ibid., VI, i, 179.
23. Ibid., VI, i, 295.
24. Ibid., VI, i, 300.
25. Ibid., VI, ii, 436, 446.
26. Ibid.
27. CSPS, Mary I 1554–58, XIII, 166.
28. Weir, *Henry VIII*, 137.
29. L&P, VI, ii, 459.
30. The first Queen of England—albeit an uncrowned one—was Matilda, the daughter and sole surviving heir of Henry I. She came to the throne in April 1141, but her claim was disputed by her cousin, Stephen. After a

bitter civil war, Matilda was forced to relinquish the throne in November 1141, having reigned for just seven months.

31. Strickland, *Lives of the Queens of England,* II, 651.

32. BM Harleian ms 543, f.128; and 283, f.75.

33. L&P, VII, i, 465.

34. Wyatt, "Extracts," 441.

35. L&P, VI, ii, 469.

36. Ibid., VII, i, 360.

CHAPTER 2: "The Little Whore"

1. L&P, VII, i, 36.

2. Heath, "An Account of Materials Furnished for the Use of Queen Anne Boleyn."

3. Strype, *Ecclesiastical Memorials, Relating Chiefly to Religion, and the Reformation of It,* I, i, 224.

4. L&P, VI, 491–92, 500.

5. Ibid.,VI, ii, 511, 556.

6. Ibid., VI, ii, 617.

7. Ibid.

8. Loades, *Mary Tudor,* 37.

9. L&P, VII, i, 69.

10. Ibid., VI, ii, 629.

11. Ibid., VII, i, 31.

12. Ibid., VI, ii, 465.

13. Ibid., VII, i, 31–32.

14. Ibid., VII, i, 68.

15. Ibid., VII, i, 31, 323.

16. Ibid., VII, i, 84.

17. CSPS, 1534–35, V, i, 72.

18. L&P, VII, i, 214.

19. Ibid., VII, i, 204.

20. Ibid.

21. Ives, *The Life and Death of Anne Boleyn,* 248.

22. L&P, IX, 189, 568.

23. Ibid., VII, i, 84, 142; IX, 197; Porter, *Mary Tudor,* 96–97.

24. L&P, IX, 424.

25. Ibid., IX, 463.

26. CSPS, 1534–35, V, i, 573.

27. L&P, VII, ii, 495; VIII, 58.

28. Ibid., VIII, 193.

29. Hibbert, *Elizabeth I*, 15.

30. L&P, X, 374.

31. Ibid.

32. Starkey, *Six Wives*, 584.

33. Weir, *The Six Wives of Henry VIII*, 345; L&P, X, 450.

34. L&P, XI, 17.

35. Weir, *Six Wives of Henry VIII*, 345.

36. L&P, VII, ii, 485.

37. Ibid., X, 70.

38. Weir, *Six Wives of Henry VIII*, 293.

39. L&P, X, 51.

40. Ibid.

41. Ibid., X, 102.

42. Ibid.

43. Ibid.

44. Ibid., X, 134.

45. CSPS, 1536–38, V, ii, 84.

46. CSPF, Elizabeth 1558–59, 527–28.

47. L&P, X, 361–62.

48. Ibid., X, 333.

49. Hamilton, *A Chronicle of England During the Reigns of the Tudors*, 137–38.

50. L&P, X, 330.

51. Ibid., X, 381.

52. Ibid., X, 453.

53. Mary had scornfully rejected Anne's conciliatory overtures, still full of hatred toward the woman who had caused her such misery over the years.

CHAPTER 3: The Royal Nursery

1. Wiesener, *The Youth of Queen Elizabeth*, I, vi.

2. L&P, XI, 130.

3. HMC, *Rutland*, I, 310.

4. Weir, *The Lady in the Tower*.

5. CSPF, Elizabeth 1558–59, 528.

6. Jenkins, *Elizabeth the Great*, 95–96.

7. Taylor-Smither, "Elizabeth I: A Psychological Profile," 53.

8. Strype, *Ecclesiastical Memorials*, I, i, 436.

9. L&P, X, 403.

10. Ibid., X, 467.

11. Ibid., XI, 7.

12. Ibid., X, 494.
13. Ibid., XI, 55.
14. Ibid., XI, 96.
15. Ibid., XI, 132.
16. Ibid., XI, 90.
17. Ibid., XI, 190.
18. Ibid.
19. Strickland, *Life of Queen Elizabeth*, 10.
20. Plowden, *The Young Elizabeth*, 55.
21. L&P, X, 374.
22. Ibid., X, 377.
23. Ibid., X, 452.
24. Ibid., X, 504.
25. Ibid., XI, 24.
26. Ibid., X, 374.
27. Ibid., XI, 346.
28. Ibid., XII, ii, 339.
29. Wood, *Letters of Royal and Illustrious Ladies of Great Britain*, iii, 112.
30. Ibid.
31. Collins, *Letters and Memorials of State*, II, 200–3.
32. Madden, *Privy Purse Expenses of the Princess Mary*, 85, 88.
33. Lady Anne Shelton was replaced by Lady Mary Kingston, wife of Sir William, later lieutenant of the Tower. When he was promoted to the king's household in 1539, he and his wife were replaced by Sir Edward and Lady Baynton.
34. Graves, *A Brief Memoir of the Lady Elizabeth Fitzgerald*, 6.
35. Adams and Rodríguez-Salgado, "The Count of Feria's Dispatch to Philip II of 14 November 1558," 330.
36. Erickson, *The First Elizabeth*, 41.
37. The hitherto unknown influence of Lady Troy in Elizabeth's household has recently been brought to light by Ruth Richardson's excellent biography of Blanche Parry. This biography has also made an invaluable contribution to our knowledge of Blanche's life and career.
38. Powel, *Historie of Cambria* (1584), preface.

CHAPTER 4: Stepmothers

1. L&P, XI, 253.
2. The household accounts for Princess Elizabeth in 1551–52 indicate that Kat was then being paid £7.15s. per year. This was rather less than Blanche Parry, who earned £10 a year from a more junior post.

3. Camden, *The Elizabethan Woman*, 41.

4. CSPD, Edward VI, 82.

5. Weir, *Six Wives of Henry VIII*, 406–7.

6. Warnicke, *The Marrying of Anne of Cleves*, 427.

7. Ibid., 408.

8. Strickland, *Lives of the Queens of England*, III, 59.

9. Wood, *Letters*, iii, 161.

10. Weir, *Six Wives of Henry VIII*, 427.

11. Plowden, *Tudor Women*, 95.

12. Weir, *Six Wives of Henry VIII*, 413.

13. L&P, XVI, i, 5.

14. Strickland, *Life of Queen Elizabeth*, 13.

15. L&P, XVI, i, 391.

16. Ibid., XVI, ii, 636.

17. Plowden, *Tudor Women*, 101.

18. Denny, *Katherine Howard*, 237.

19. Weir, *Children of England*, 9; Denny, *Katherine Howard*, 253.

20. In so doing, she would have used parchment and a "hornbook." The hornbook originated in the mid-fifteenth century and consisted of a sheet containing the letters of the alphabet (in both small and capital letters), as well as the Lord's Prayer and a cross. It was mounted on wood, bone, or leather and protected by a thin sheet of transparent horn or mica. The wooden frame often had a handle, and it was usually hung at the child's girdle so that he or she could refer to it at any time. Not all daughters of royal blood received even this rudimentary education. Margaret Douglas, Countess of Lennox, had such poor handwriting that it was barely legible. The same was true of her mother, Margaret Tudor, sister of Henry VIII.

21. Starkey, *Elizabeth*, 26.

22. Strickland, *Life of Queen Elizabeth*, 21; L&P, XI, 55.

23. Haynes, *Collection of State Papers Relating to Affairs in the Reigns of King Henry VIII, King Edward VI, Queen Mary and Queen Elizabeth*, 95.

24. Mary was then twenty-six and Katherine, thirty.

25. James, *Kateryn Parr*, 127.

26. Weir, *Six Wives of Henry VIII*, 493.

27. L&P, XI, 55.

28. Cerovski, *Sir Robert Naunton, Fragmented Regalia*, 40.

29. Perry, *The Word of a Prince*, 30–31. The original letter is held at the British Library: BM Cotton MS C X, f.235.

30. James, *Kateryn Parr*, 172.

31. Strickland, *Life of Queen Elizabeth*, 23.

32. Perry, *Word of a Prince*, 23.

33. Strickland, *Lives of the Queens of England*, III, 328.

34. CSPV, VI, ii, 1059.

35. Ibid., VI, iii, 1538; ibid., Elizabeth 1558–80, VII, 601; ibid., VI, ii, 1058.

36. The impact of Elizabeth's stay at court during Katherine Parr's regency is set out in Starkey, *Elizabeth*, 38–41.

37. BM Add. MS 39288, f.5.

38. Levin and Watson, *Ambiguous Realities*, 145.

39. Perry, *Word of a Prince*, 32–34 (BM Cotton MS Nero C X, f.13).

40. Astley is often spelled "Ashley" or "Ashlay." Kat herself tended to spell her married name "Aschely."

41. Starkey, *Elizabeth*, 81.

42. L&P, XVIII, ii, 283.

43. Loades, *The Tudor Court*, 73.

44. Marcus, Mueller, and Rose, *Elizabeth I*, 10.

45. Ballard, *Memoirs of Several Ladies of Great Britain*, 115–16.

46. Strickland, *Life of Queen Elizabeth*, 14.

CHAPTER 5: Governess

1. CSPD, Edward VI, 92.

2. Perry, *Word of a Prince*, 45.

3. Haynes, *Collection of State Papers*, 61.

4. BM Lansdowne MS 1236, f.26.

5. Weir, *Six Wives of Henry VIII*, 544.

6. Martienssen, *Queen Katherine Parr*, 233.

7. CSPS, Edward VI 1547–49, IX, 48–49, 52.

8. A ward was someone who was placed under the protection of a legal guardian. Often, he was the heir to an estate whose father (the natural legal guardian) had died. In persuading the Duke of Suffolk to sell his rights over his daughter Jane, Seymour was speculating on the prospect that he would be able to secure a prestigious marriage for her and thus make a substantial profit for himself because guardians always received a fee as part of the marriage settlement.

9. Wiesener, *Youth of Queen Elizabeth*, 94–95.

10. BM Lansdowne MS 1236, f.39.

11. Haynes, *Collection of State Papers*, 99.

12. Ibid.

13. Marcus, Mueller, and Rose, *Elizabeth I*, 30.

14. CSPD, Edward VI, 92.

15. Haynes, *Collection of State Papers*, 96.

16. Ibid., 99.

17. Ibid., 96.

18. TNA, SP 10/2, no. 25.

19. BM Cotton MS Otho C X f.236v.
20. Ibid.
21. CSPD, Edward VI, 49.
22. Haynes, *Collection of State Papers*, 103–4.
23. CSPD, Edward VI, 91.
24. Perry, *Word of a Prince*, 58.
25. CSPD, Edward VI, 91.
26. Haynes, *Collection of State Papers*, 100.
27. Ibid., 98.
28. Ibid., 70; Marcus, Mueller, and Rose, *Elizabeth I*, 23–24.
29. Haynes, *Collection of State Papers*, 96–97.
30. Ibid., 96, 100.
31. CSPD, Edward VI, 92.
32. Ibid., 82.
33. Haynes, *Collection of State Papers*, 70.
34. Ibid., 102.
35. CSPD, Edward VI, 92.
36. Haynes, *Collection of State Papers*, 99–101.
37. Ibid., 102.
38. Ibid., 107.
39. Ibid., 108–9.
40. Ibid.
41. BM Lansdowne MS 1236, f.35.
42. The surviving evidence does not tell us where they had been during that time. They may have lived with John's relatives in Norfolk, or with Kat's in Devon, or perhaps they stayed close to Hatfield in anticipation of their return to favor.
43. Although it was rumored that he, like his sister, was the illegitimate offspring of Henry VIII, he was born too long after the end of Mary's affair with the king for this to be likely.
44. CSPS, Mary I 1554–58, XIII, 387.
45. Somerset, *Elizabeth I*, 32–33.
46. L&P, XXI, i, 400.
47. CSPS, Edward VI 1550–52, X, 210–12.
48. Ibid., X, 101.
49. Ibid., X, 212.
50. Ibid.; CSPS, Edward VI 1547–49, IX, 489.

CHAPTER 6: Sister

1. CSPS, Mary I 1553–54, XI, 115.
2. Somerset, *Elizabeth I*, 71.

3. Ibid.

4. CSPV, 1556–57, VI, ii, 1058.

5. Ibid., VI, ii, 1054.

6. CSPS, Mary I 1554–58, XIII, 6.

7. Ibid., Mary I 1553–54, XI, 50.

8. Arber, *John Knoxe, First Blast of the Trumpet Against the Monstrous Regiment of Women,* 9–10.

9. CSPS, Mary I 1554–58, XIII, 61.

10. CSPD, Mary I, 10; CSPS, Mary I 1553–54, XI, 259; Nichols, *The Chronicle of Queen Jane and of Two Years of Queen Mary,* 28; Richards, "Mary Tudor as 'Sole Quene'? Gendering Tudor Monarchy," 902.

11. The Empress Matilda had not been crowned during her brief reign in 1141.

12. Richards, "Mary Tudor as 'Sole Quene'?" 908–9.

13. CSPV, VII, 329.

14. Ibid., VI, ii, 1056.

15. CSPS, Mary I 1553–54, XI, 252.

16. Plowden, *Tudor Women,* 138.

17. CSPV, VII, 601.

18. Redworth, "Matters Impertinent to Women," 598.

19. Richards, "Mary Tudor as 'Sole Quene'?" 908–9.

20. BM Cotton MS Vespasian F iii, f.23.

21. CSPS, Mary I 1554–58, XIII, 3–4, 61.

22. Perry, *Word of a Prince,* 88–89; CSPS, Mary I Jan.–Jun. 1554, XII, 50.

23. CSPS, Mary I 1553–54, XI, 169.

24. Ibid., XI, 410.

25. Plowden, *Tudor Women,* 137.

26. Perry, *Word of a Prince,* 85.

27. CSPS, Mary I 1553–54, XI, 274.

28. CSPF, Elizabeth 1562, 16.

29. Marshall, *Queen Mary's Women,* 108.

30. L&P, VII, 7.

31. Perry, *Word of a Prince,* 23.

32. H. Boethius, *The Historie and Cronicles of Scotland,* II, 16–18.

33. Marshall, *Queen Mary's Women,* 110.

34. CSPV, VI, ii, 1058.

35. Ibid., VI, ii, 1059.

36. Plowden, *Tudor Women,* 149.

37. CSPS, Mary I 1553–54, XI, 196.

38. Ibid., XI, 220.

39. Ibid., XI, 220–21.

40. Ibid., XI, 221.

41. Ibid., XI, 252–53.

42. Ibid., XI, 188.

43. Ibid., XI, 393.

44. Ibid., XI, 411.

45. CSPS, Mary I 1553–54, XI, 418; Perry, *Word of a Prince*, 87.

46. CSPS, Mary I 1553–54, XI, 436, 446.

47. Loades, *Mary Tudor*, 204.

48. Although Wyatt always maintained that he was loyal to Queen Mary and that the rebellion was against her councillors, there is also evidence to suggest that many of the rebels wished to place Elizabeth on the throne.

49. Foxe, *The Acts and Monuments of John Foxe*, VI, 414–15.

50. CSPS, Mary I 1554, XII, 42.

51. Strype, *Ecclesiastical Memorials*, I, i, 126.

52. CSPD, Mary I, 53–54.

53. Nichols, *Chronicle*, 70–71.

54. CSPS, Mary I 1554, XII, 140.

55. Ibid.

56. Strickland, *Lives of the Queens of England*, III, 501–2.

57. CSPS, Mary I 1554, XII, 218.

58. Marcus, Mueller, and Rose, *Elizabeth I*, 66.

59. Dasent, *Acts of the Privy Council of England*, 1554–56, 129.

60. Perry, *Word of a Prince*, 99.

61. BM Lansdowne MS 1236, f.37.

62. CSPV, VI, ii, 1060.

63. Ibid., VI, i, 475, 479–80.

64. CSPD, Mary I, 208.

65. CSPV, VI, i, 475.

66. Ibid., VI, i, 480.

67. Ibid., VI, i, 484.

68. Ibid., VI, i, 718–19.

69. It is not clear what Kat Astley did after she had been banished from Elizabeth's household; she and her husband disappear from the records. Perhaps John, always more sensible than his wife, persuaded her to take comfort from the fact that of the two alternatives—retirement and imprisonment—the former was preferable.

70. Weir, *Children of England*, 331.

71. Ibid.

72. CSPS, Mary I 1554–58, XIII, 28.

73. Ibid., XIII, 2–3, 6.

74. Ibid., Elizabeth 1554–58, XIII, 2–3, 13.

75. Nichols, *The Diary of Henry Machyn*, 76; CSPS, Mary I 1554–58, XIII, 124.

76. CSPV, VI, ii, 1055. The treatments for symptoms such as those displayed by Mary could be brutal. They included inserting a tube into the vagina

and sending up steam from a boiling liquid in order to "fumigate" the uterus. Worse still, horse leeches could be inserted into the neck of the womb: Erickson, *Bloody Mary*, 127–28. A number of theories have been put forward by modern-day medical historians about the cause of Mary's condition. She may have suffered a phantom pregnancy, although the ill health she endured throughout her life suggests that there was something else wrong. Another theory is that she had an ovarian tumor, which would have explained her lack of periods, swollen abdomen, and frequent abdominal pains. Alternatively, she may have had prolactinoma, which is a benign tumor of the pituitary gland. This condition often causes infertility and changes in menstruation, with some women losing their periods altogether. Women who are not pregnant or nursing may begin producing breast milk. An excellent summary of the various theories to explain Mary's symptoms is provided by Medvei. See also Brewer.

77. CSPS, Mary I 1554–58, XIII, 51, 226.
78. Foxe, *Acts and Monuments*, III, 619–21.
79. CSPF, Mary I, 165–66.
80. Ibid., 172.
81. CSPS, Mary I 1554–58, XIII, 224.
82. CSPF, Mary I 1554–58, 174.
83. HMC, *Rutland*, I, 310–11.
84. CSPS, Mary I 1554–58, XIII, 250
85. CSPV, VI, ii, 1059.
86. CSPS, Elizabeth 1558–67, I, 9.
87. CSPV, VI, i, 558.
88. Ibid., VI, i, 887.
89. Ibid.
90. Camden, *The Historie of the Most Renowned and Victorious Princess Elizabeth*, 9.
91. Perry, *Word of a Prince*, 86. The original letter is in the British Library, BM Lansdowne MS 94, f.21.
92. CSPV, VI, ii, 1060.
93. Loades, *Mary Tudor*, 189.
94. CSPV, VI, ii, 1060.
95. Camden, *Historie*, 8.
96. Loades, *Mary Tudor*, 143, quoting John Foxe.
97. CSPV, VI, iii, 1549.
98. Marcus, Mueller, and Rose, *Elizabeth I*, 66.
99. Adams and Rodríguez-Salgado, "Count of Feria's Dispatch," 329–30.
100. Merton, "The Women Who Served Queen Mary and Queen Elizabeth," 160. William Paget was Lord Privy Seal and a member of Mary's Privy Council.
101. Adams and Rodríguez-Salgado, "Count of Feria's Dispatch," 331, 334.

102. Ibid., 335.
103. Camden, *Historie*, 10; CSPV, VI, iii, 1538; Tytler, *England Under the Reigns of Edward VI and Mary*, II, 497.
104. CSPV, VI, ii, 1058.
105. Ibid., VI, i, 201.
106. CSPS, Mary I 1554–58, XIII, 416.
107. CSPS, Elizabeth 1558–67, I, 34.
108. Erickson, *Bloody Mary*, 481.
109. CSPV, VI, iii, 1538.
110. CSPS, Mary I 1554–58, XIII, 438.
111. Adams and Rodríguez-Salgado, "Count of Feria's Dispatch," 328.
112. Harington, *Nugae Antiquae*, II, 312.
113. Strickland, *Life of Queen Elizabeth*, 122.
114. Starkey, *Elizabeth*, 311.
115. BM Cotton MS Vespasian D XVIII, f.104.
116. Redworth, "Matters Impertinent to Women," 599.

CHAPTER 7: The Queen's Hive

1. Only when James I came to the throne was this rectified. Elizabeth's own coffin was placed in the same vault as her half sister's, and James ordered a magnificent monument to be erected above them both. This bore the inscription: "Partners both in throne and grave, here rest we, two sisters, Elizabeth and Mary, in the hope of resurrection."
2. CSPV, VI, iii, 1559.
3. CSPS, Elizabeth 1558–67, I, 7.
4. Somerset, *Elizabeth I*, 7.
5. See for example Harrison, *The Letters of Queen Elizabeth*, 83.
6. Adams and Rodríguez-Salgado, "Count of Feria's Dispatch," 331.
7. CSPS, Elizabeth 1558–67, I, 7.
8. CSPV, VII, 659.
9. *Holinshead's Chronicle*, quoted in Plowden, *Young Elizabeth*, 209.
10. Strype, *Ecclesiastical Memorials*, III, ii, 166.
11. CSPS, Mary I 1553–54, XI, 393.
12. Cerovski, *Sir Robert Naunton*, 38.
13. Levin, *The Heart and Stomach of a King*, 10.
14. Hibbert, *Elizabeth I*, 67.
15. Warnicke, *Women of the English Renaissance and Reformation*, 62–63.
16. An excellent analysis of the link between Anne Boleyn's coronation and the symbolism adopted by Elizabeth is provided by Ives, *Life and Death of Anne Boleyn*, 222–30.

17. CSPV, VII, 659.
18. Somerset, *Elizabeth I*, 65; Nichols, *Diary of Henry Machyn*, 263.
19. Watkins, *In Public and Private*, 182.
20. Boyle, *Memoirs of the Life of Robert Carey*, 73n.
21. Watkins, *In Public and Private*, 58–59.
22. Rye, *England as Seen by Foreigners in the Days of Elizabeth and James the First*, II, 18.
23. Ibid.
24. Ibid., II, 17.
25. The palace was incomplete when Henry VIII died in 1547. In 1556 his daughter Mary had leased it to the 19th Earl of Arundel, who completed it. It returned to royal hands in 1592, when Elizabeth acquired it.
26. Loades, *Tudor Court*, 119.
27. Cerovski, *Sir Robert Naunton*, 41–42.
28. Perry, *Word of a Prince*, 113.
29. The whole incident is chronicled in CSPF, Elizabeth 1561–62, 244, 303–4, 309, 311, 329, 344, 356, 361.
30. Adams, "Eliza Enthroned?" 64.
31. Kat's husband was also appointed to a position of great prestige as master of the Jewel House.
32. TNA, LC2/4/3, f.53.
33. Equivalent to around £5,600 ($8,945) today.
34. Nichols, *The Progresses and Public Processions of Queen Elizabeth*, I, 38.
35. Bruce, *Annals*, 15.
36. Wilson, *Queen Elizabeth's Maids of Honour and Ladies of the Privy Chamber*, 3.
37. Birch, *Memoirs of the Reign of Queen Elizabeth from the Year 1581 till Her Death*, 120–21.
38. Frye, *Maids and Mistresses, Cousins and Queens*, 132.
39. Weir, *The Life of Elizabeth*, 51.
40. TNA, LC 2/4 (3), ff.104–5.
41. CSPV, Elizabeth 1558–80, VII, 91–92.
42. Merton, "The Women Who Served," 64.
43. CSPV, Elizabeth 1558–80, VII, 611.
44. Rye, *England as Seen by Foreigners*, III, 107.
45. Harrison and Jones, *Andre Hurault de Maisse*, 35.
46. Hibbert, *Elizabeth I*, 109.
47. HMC, *Salisbury*, II, 159.
48. BM Cotton MS Galba C IX, f.128.
49. Harrison and Jones, *Andres Hurault de Maisse*, 95.
50. Somerset, *Elizabeth I*, 64.
51. Thurley, *The Royal Palaces of Tudor England*, 173.
52. Merton, "The Women Who Served," 121.

53. Somerset, *Elizabeth I*, 3, 62.

54. Harington, *Nugae Antiquae*, 125.

55. Perry, *Word of a Prince*, 153; Harington, *Nugae Antiquae*, 95.

56. Merton, "The Women Who Served," 7.

57. CSPS, Elizabeth 1558–67, I, 21.

58. Merton, "The Women Who Served," 244.

CHAPTER 8: The Virgin Queen

1. Haigh, *Elizabeth I*, 20.

2. Pryor, *Elizabeth I: Her Life in Letters*, 31; HMC, *Salisbury*, I, 158.

3. Erickson, *First Elizabeth*, 71.

4. Somerset, *Elizabeth I*, 90; Plowden, *Tudor Women*, 154.

5. CSPS, Elizabeth 1558–67, I, 77. See also VII, 611.

6. von Klarwill, *Queen Elizabeth and Some Foreigners*, 114.

7. CSPS, Elizabeth 1558–67, I, 95–96.

8. Ibid., I, 98–101.

9. Ibid., I, 106–7, 112–13, 115.

10. CSP Rome, I, 105.

11. Wilson, *Elizabeth's Maids of Honour*, 25.

12. Robert Dudley was created Earl of Leicester in 1564.

13. CSPF, Elizabeth 1560–61, 10.

14. von Klarwill, *Elizabeth and Some Foreigners*, op. cit., 113–15.

15. BM Add. MS 48,023, f.353.

16. CSPF, Elizabeth 1562, 217–24.

17. Ibid., 173.

18. Jenkins, *Elizabeth the Great*, 76–77.

19. CSPS, Elizabeth 1558–67, I, 63.

20. Weir, *Life of Elizabeth*, 47; Jenkins, *Elizabeth the Great*, 77.

21. Camden, *Historie*, 9.

22. CSPV, Elizabeth 1558–80, VII, 105.

23. CSPS, Elizabeth 1580–86, III, 252; Somerset, *Elizabeth I*, 96.

24. Jenkins, *Elizabeth the Great*, 123.

25. CSPV, Elizabeth 1558–80, VII, 601.

26. Marcus, Mueller, and Rose, *Elizabeth I*, 157; Weir, *Life of Elizabeth*, 48; HMC, *Salisbury*, II, 245.

27. Murdin, *A Collection of State Papers Relating to Affairs in the Reign of Queen Elizabeth, 1571–96*, 558; CSP Scotland, 1584–85, VII, 5; Erickson, *First Elizabeth*, 262; Somerset, *Elizabeth I*, 101; Johnson, *Elizabeth I: A Study in Power and Intellect*, 115; Weir, *Life of Elizabeth*, 48–49; Laing, *Notes of Ben Jonson's Conversations with William Drummond of Hawthornden*, 23.

28. Arthur Dudley's account implicated Kat Astley and her husband, who were said to have spirited the child away from Hampton Court as soon as it was born, and had hushed up the affair ever since. There is no other evidence to corroborate the story, but it suited Philip's interests to make sure that it was repeated far and wide. For that reason, it was given more credence than it perhaps deserves.

29. Levin, *Heart and Stomach of a King,* 82–83; Laing, *Ben Jonson's Conversations,* 23.

30. Gristwood, *Elizabeth and Leicester,* 132–33.

31. Jenkins, *Elizabeth the Great,* 141–42.

32. Ibid.

33. Eccles, *Obstetrics and Gynaecology in Tudor and Stuart England,* 26–27.

34. Alison Weir provides an excellent assessment of Bloch's theory in *Elizabeth the Queen,* 49.

35. Francis Steuart, *Sir James Melville,* 94; CSPV, VII, 594; Levin, *Heart and Stomach of a King,* 172.

36. Weir, *Life of Elizabeth,* 46.

37. CSPS, III, 252; Somerset, *Elizabeth I,* 96.

38. Weir, *Life of Elizabeth,* 47.

39. Frye, *Elizabeth I,* 12.

40. Weir, *Life of Elizabeth,* 51.

41. Marcus, Mueller, and Rose, *Elizabeth I,* 168; Heisch, "Elizabeth I and the Persistency of Patriarchy," 50.

42. Harington, *Nugae Antiquae,* 124; Merton, "The Women Who Served," 144.

43. Harington, *Nugae Antiquae,* 124.

44. Erickson, *Mistress Anne,* 187.

45. CSPF, Elizabeth 1586–88, 86.

46. Murdin, *Collection of State Papers,* 558.

CHAPTER 9: Cousins

1. Lovell, *Bess of Hardwick,* 31.

2. CSPS, Mary I 1553–54, XI, 40.

3. Ibid., XI, 46.

4. Perhaps because the groom was said to be "very ill." Ibid., XI, 40.

5. Davey, *The Sisters of Lady Jane Grey and Their Wicked Grandfather,* 109.

6. CSPF, Elizabeth 1558–59, 443; CSPS, Elizabeth 1558–67, I, 45.

7. CSPS, Elizabeth 1558–67, I, 45.

8. CSPF, Elizabeth 1558–59, 443.

9. CSPS, Elizabeth 1558–67, I, 45.

10. Ibid., I, 116.
11. Ibid., I, 176. See also HMC, *Salisbury,* I, 197.
12. Fraser, *Mary, Queen of Scots,* 3.
13. Dunn, *Elizabeth and Mary,* 94–95.
14. Camden, *Historie,* 34.
15. CSPS, Elizabeth 1588–67, I, 122.
16. HMC, *Salisbury,* I, 158.
17. CSPS, Elizabeth 1558–67, I, 114.
18. CSPF, Elizabeth 1560–61, 291.
19. Their kinship derived from Lady Cecil's brother's marriage to a cousin of Katherine Grey.
20. BM Harley MS 6286, ff.37–37v. It was said that Frances died giving birth to her young husband's child, but this is doubtful, given that she knew she was already seriously ill when she wrote the letter.
21. Ibid., f.20.
22. Another account says that it was Eltham, not Greenwich, that the Queen had decided to visit.
23. Katherine's message must have been relayed to the Queen by Jane Seymour, because Katherine claimed to have "a swelling in her face," which would have been exposed as an obvious lie if she had presented her excuses in person.
24. "As circles five by art compact, shewe but one ring in sight / So trust uniteth faithfull mindes, with knott of secret might / Whose force to break but greedie death no wight possesseth power / As tyme and sequele well shall prove, my ring can saie noe more."
25. BM Harley MS 6286, f.28v.
26. Ibid., f.25.
27. Ibid., f.21v.
28. Wilson, *Elizabeth's Maids of Honour,* 78.
29. BM Add. MS 35830, f.104.
30. Nichols, *Diary of Henry Machyn,* 253–54.
31. Levine, *The Early Elizabethan Succession Question,* 142–43.
32. CSPF, Elizabeth 1561–62, 159.
33. BM Harleian MS 6286, f.22.
34. Ibid., ff.44v–45v.
35. Perry, *Word of a Prince,* 166; Haynes, *Collection of State Papers,* 369.
36. Wright, *Queen Elizabeth and Her Times,* I, 68–69.
37. CSPF, Elizabeth 1561–62, 322; CSPS, Elizabeth 1558–67, I, 213; CSPF, Elizabeth 1562, 46.
38. HMC, *Salisbury,* XIII, 62.
39. BM Harley MS 6286, ff.23v, 45v.
40. CSPF, Elizabeth 1561–62, 335.

41. CSPS, Elizabeth 1558–67, I, 216; CSPF, Elizabeth 1561–62, 277.
42. CSPS, Elizabeth 1558–67, I, 213; CSP Rome, I, 51.
43. CSP Rome, I, 51–52.
44. CSPF, Elizabeth 1562, 13–14.
45. Ibid., 13, 24.
46. Plowden, *Two Queens in One Isle*, 61.
47. Perry, *Word of a Prince*, 170.
48. Ibid.; CSPF, Elizabeth 1561–62, 251.
49. CSPF, Elizabeth 1561–62, 161.
50. CSP Scotland, 1547–62, I, 559.
51. Somerset, *Elizabeth I*, 150.
52. Dunn, *Elizabeth and Mary*, 177; Plowden, *Two Queens*, 74.
53. Plowden, *Two Queens*, 69.
54. Dunn, *Elizabeth and Mary*, 175; CSPF, Elizabeth 1561–62, 357.
55. Marcus, Mueller, and Rose, *Elizabeth I*, 65–66.
56. CSPF, Elizabeth 1561–62, 477; CSP Scotland, 1547–62, I, 587.
57. CSP Scotland, 1547–62, I, 639.
58. Ibid., I, 591, 594.
59. Ibid., I, 622–23, 639.
60. CSPF, Elizabeth 1562, 14.
61. Ibid., 14, 23.
62. CSPS, Elizabeth 1558–67, I, 220–1.
63. CSPF, Elizabeth 1561–62, 580.
64. Ibid., 641–42.
65. CSPF, Elizabeth 1562, 15–16.
66. Ibid., 172.
67. Ibid., 90, 172, 258, 397.
68. CSPF, Elizabeth 1561–62, 277.
69. Camden, *Historie*, 58.
70. Haynes, *Collection of State Papers*, I, 396.
71. Francis Steuart, *Sir James Melville*, 114.
72. CSP Rome, I, 52.
73. CSPS, Elizabeth 1558–67, I, 263.
74. Ibid., I, 273.
75. TNA SP 12/159, f.38v. Lady Mary was not the only victim. Elizabeth's childhood nurse, Sybil Penne, who had been in attendance at Hampton Court, also contracted the disease. Sadly, for her it proved fatal, and she died shortly afterward. Her unquiet spirit is said to still haunt the palace.
76. CSPS, Elizabeth 1558–67, I, 263.
77. Ibid., I, 196. Although she was still only twenty-nine, Elizabeth was "old" by Tudor standards, to be unmarried.
78. Nichols, *Diary of Henry Machyn*, 300.

79. When Seymour pleaded that he was unable to pay the fine, it was later commuted to £3,000.

80. His dismissal had apparently done nothing to diminish his sympathy for Lady Katherine, for he wrote to Burghley, bemoaning the state of the furnishings that had adorned her apartments, all of which were "torn and tattyred." Wright, *Queen Elizabeth and Her Times*, I, 140.

81. *Third Report of the Royal Commission on Historical Manuscripts*, appendix, 47.

82. Haynes, *Collection of State Papers*, I, 404–5.

83. Chapman, *Two Tudor Portraits*, 222.

84. Somerset, *Elizabeth I*, 152.

85. Bourdeille, *The Book of the Ladies*, 91.

86. Francis Steuart, *Sir James Melville*, 95–97.

87. Ibid., 94, 101.

88. Ibid., 78, 101.

89. CSPS, Elizabeth 1568–79, II, 36.

90. CSPF, Elizabeth 1563, VI, 617, 637.

91. Ibid., VI, 509.

92. Francis Steuart, *Sir James Melville*, 81.

93. Gristwood, *Elizabeth and Leicester*, 158.

94. Robert Dudley was Elizabeth's master of horse, an important official of the royal household.

95. Francis Steuart, *Sir James Melville*, 91.

96. CSP Scotland, 1563–69, II, 49.

97. CSPF, Elizabeth 1563, 463.

98. Ibid.

99. Francis Steuart, *Sir James Melville*, 99.

100. Robert Dudley had been created Earl of Leicester by Elizabeth in September 1564. Ibid., 101.

101. Schutte, *A Biography of Margaret Douglas*, 193.

102. CSPF, Elizabeth 1564–65, 387.

103. CSPS, Elizabeth 1558–67, I, 477; CSPF, Elizabeth 1564–65, 417.

104. CSP Scotland, 1563–69, II, 81.

105. CSP Rome, I, 173.

106. Ibid., 176. Although he was referred to as the Earl of Lennox, this title still belonged to his father. Darnley himself would never inherit it.

107. Francis Steuart, *Sir James Melville*, 81.

108. CSP Scotland, 1563–69, II, 225.

109. CSPS, Elizabeth 1558–67, I, 468.

110. Wright, *Queen Elizabeth and Her Times*, I, 207.

111. CSPF, Elizabeth 1561–62, 506.

112. CSPS, Elizabeth 1558–68, I, 468; Wright, *Queen Elizabeth and Her Times*, I, 207.

113. This later became the country residence of the British prime ministers.

114. CSPF, Elizabeth 1564–65, 410; CSPF, Elizabeth 1566–68, 72.

115. CSPF, Elizabeth 1564–65, 428; CSPF, Elizabeth 1566–68, 17; CSPS, Elizabeth 1568–79, II, 192.

116. Plowden, *Two Queens*, 96.

117. Francis Steuart, *Sir James Melville*, 107.

118. Ibid., 92.

119. Schutte, *Margaret Douglas*, 199.

120. CSPF, Elizabeth 1566–68, 33.

121. Weir, *Life of Elizabeth*, 174.

122. Francis Steuart, *Sir James Melville*, 131.

123. CSPS, Elizabeth 1558–67, I, 562.

124. CSPF, Elizabeth 1566–68, 45.

125. Francis Steuart, *Sir James Melville*, 132.

126. The earl himself apparently found the prospect distasteful and eventually persuaded the Countess of Argyll, Mary's bastard sister, to represent Elizabeth in his place, giving her a diamond "worth 500 crowns" as a bribe. CSP Rome, I, 226; CSPS, Elizabeth 1558–67, I, 562.

127. CSPS, Elizabeth 1558–67, I, 597.

128. Darnley was officially ill with smallpox, but it may have been syphilis.

129. Weir, *Mary, Queen of Scots*, 481.

130. Marcus, Mueller, and Rose, *Elizabeth I*, 116–17.

131. CSPS, Elizabeth 1558–67, I, 620n.

132. Perry, *Word of a Prince*, 208n.

133. BM Lansdowne MS 8, ff.67–68.

134. Wilson, *Elizabeth's Maids of Honour*, 69.

135. Francis Steuart, *Sir James Melville*, 299.

136. Plowden, *Two Queens*, 59.

137. Starkey, *Elizabeth*, 178.

138. Bassnett, *Elizabeth I: A Feminist Perspective*, 68.

139. Somerset, *Elizabeth I*, 195.

140. Marcus, Mueller, and Rose, *Elizabeth I*, 118.

141. Ibid., 119.

142. Wilson, *Elizabeth's Maids of Honour*, 32.

143. Two copies of this account survive in the British Library: BM Harleian MS 39, f.380; Cotton Titus MS 107, ff.124, and 131.

144. CSPS, Elizabeth 1558–67, I, 4.

145. In 1595 he secretly caused a record to be put into the Court of Arches to prove that his former marriage had been lawful and his children were therefore legitimate. This earned him another spell in the Tower when the Queen found out, and his elder son, Edward, was stripped of his title of Lord Beauchamp. However, the lack of documentary proof on both sides meant that the door was left open for supporters of the boys to

push their claim in the future. Finally, in 1606, Seymour succeeded in having his sons' legitimacy confirmed by James I, which allowed Edward to inherit his father's title when the latter died fifteen years later. In 1610 Edward's son William married Arbella Stuart, the granddaughter of Lady Margaret Douglas and Bess of Hardwick. This uniting of royal blood proved just as threatening to the crown as that of his grandparents a half century before, and they too found themselves in the Tower. Nevertheless, William continued to be seen as a contender, and even as late as 1688, his descendants were still considered to have a genuine claim to the English throne.

CHAPTER 10: Faithful Servants

1. CSPS, Elizabeth 1558–67, I, 45.
2. CSPF, Elizabeth 1566–68, 130.
3. Richardson, *Mistress Blanche,* 61.
4. Quoted from Thomas Newton's epitaph to Katherine.
5. W. Knollys, "Papers Relating to Mary, Queen of Scots," *Philobiblon Society Miscellanies* 14 (London, 1872–76), 14–69.
6. Wright, *Queen Elizabeth and Her Times,* I, 308.
7. Ibid., I, 308–9.
8. Richardson, *Mistress Blanche,* 4.
9. Ibid., 58.
10. CSPD, Elizabeth 1566–79, Addenda, 3.
11. HMC, *Rutland,* I, 134; HMC, *De L'Isle & Dudley,* II, 327.
12. CSPS, Elizabeth 1558–67, I, 475.
13. Bradford, *Helena, Marchioness of Northampton,* 48.
14. CSPS, Elizabeth 1558–67, I, 475.
15. Bradford, *Helena,* 48.
16. Camden, *Historie,* 31.
17. James, *Kateryn Parr,* 395.
18. Bradford, *Helena,* 52.
19. CSPS, Elizabeth 1558–67, I, 546.
20. Bradford, *Helena,* 178.
21. Ibid., 52–54.
22. Merton, "The Women Who Served," 186.
23. CSPD, Elizabeth 1547–80, 442.
24. HMC, *Bath,* V, 166.
25. BM Cotton MS Vespasian F XII, f.179.
26. CSPS, Elizabeth 1568–79, II, 682.
27. CSPF, Elizabeth 1566–68, 460–61.
28. Francis Steuart, *Sir James Melville,* 170.

29. Pryor, *Elizabeth I*, 51.
30. CSP Scotland, 1563–69, II, 430.
31. CSPF, Elizabeth 1566–68, 466.
32. CSPS, Elizabeth 1568–79, II, 36.
33. CSP Scotland, 1563–69, II, 449; HMC, *Salisbury*, XIII, 87.
34. CSPS, Elizabeth 1568–79, II, 57.
35. HMC, *Salisbury*, I, 549, 571.
36. Weir, *Life of Elizabeth*, 201.
37. CSP Rome, I, 289.
38. Ibid., I, 291.
39. CSP Scotland, 1563–69, II, 428.
40. Chamberlain, *The Sayings of Queen Elizabeth*, 233, 246.
41. Ibid., 235.
42. Somerset, *Elizabeth I*, 207.
43. CSPV, VII, 427.
44. Marcus, Mueller, and Rose, *Elizabeth I*, 121–23.
45. CSPS, Elizabeth 1568–69, II, 180.
46. CSPV, VII, 468–69.
47. CSP Rome, I, 401–2.
48. The cushion was used in evidence during Norfolk's trial. See Frye, *Maids and Mistresses*, 170.
49. CSP Rome, II, 3.
50. CSPS, Elizabeth 1568–79, II, 340.
51. Bassnett, *Elizabeth I*, 111.
52. CSPS, Elizabeth 1568–79, II, 342.
53. Marcus, Mueller, and Rose, *Elizabeth I*, 130.

CHAPTER 11: "That She-Wolf"

1. Schutte, *Margaret Douglas*, 230.
2. Ibid.
3. Ibid., 231.
4. Ibid.
5. Mary's total annual income amounted to about £17,400 ($27,800) in today's money.
6. HMC, *Salisbury*, II, 99.
7. BM Lansdowne MS 27, f.31.
8. Holles, *Memorials of the Holles Family, 1493–1656*, 70.
9. Ibid., 70–71.
10. Others included Margaret Douglas, Countess of Lennox, and the first Earl of Essex, husband of Lettice Knollys.
11. Gristwood, *Elizabeth and Leicester*, 226.

12. Nichols, *Progresses*, I, 328.

13. Read, "A Letter from Robert, Earl of Leicester, to a Lady," 24–25.

14. Gristwood, *Elizabeth and Leicester*, 228.

15. Ibid.

16. Ibid., 229.

17. See for example "An Agreement for Plastering Work to Be Done at the Dowager Lady Sheffield's House near Blackfriars Bridge, London, 1575," TNA E210/10340.

18. Holles, *Memorials*, 71.

19. Craik, *The Romance of the Peerage, or Curiousities of Family History*, III, 91.

20. Gristwood, *Elizabeth and Leicester*, 251.

21. CSPS, Elizabeth 1568–79, II, 511; Somerset, *Ladies-in-Waiting*, 85.

22. Gristwood, *Elizabeth and Leicester*, 253.

23. Somerset, *Ladies-in-Waiting*, 85.

24. See for example Wilson, *Sweet Robin: A Biography of Robert Dudley, Earl of Leicester, 1533–1588*, 226.

25. HMC, *Bath*, V, 205–6.

26. Camden, *Historie*, 81.

27. HMC, *Bath*, V, 206.

28. Dovey, *An Elizabethan Progress*, 148.

29. Another theory is that Leicester himself had admitted his betrayal to the Queen as early as April 1579. The Spanish ambassador reported that Elizabeth had cancelled an audience with him at short notice and gone with great haste to the Earl of Leicester's house, where she had stayed until ten o'clock at night. A few days later, Leicester had left court for Buxton, ostensibly to take the waters, and stayed away an unusually long time. Wilson, *Sweet Robin*, 229; Weir, *Life of Elizabeth*, 312–13.

30. Somerset, *Ladies-in-Waiting*, 318.

31. CSPD, Elizabeth 1580–1625, Addenda, 137.

32. Somerset, *Ladies-in-Waiting*, 318.

33. Yorke, *Miscellaneous State Papers*, I, 214–16. Douglas was eventually allowed back to court after her husband sent her home from France for her protection in 1588—the situation there was becoming volatile in the wake of the execution of Mary, Queen of Scots, and the onset of the armada. She apparently remained in the Queen's service for most of the 1590s, although little mention is made of her, and it is unlikely, given her history, that she became one of Elizabeth's intimates.

34. Robert Dudley would later try to prove that his parents had been lawfully married. His case was eventually rejected by the court of the Star Chamber, but in such a way that the question was left in doubt, and even as late as the nineteenth century, Dudley claimants were still arguing over it.

35. Gristwood, *Elizabeth and Leicester*, 314.
36. HMC, *Bath*, V, 44.
37. Weir, *Elizabeth the Queen*, 308.
38. CSP Scotland, 1574–81, V, 229.
39. CSPS, Elizabeth 1580–86, III, 426.
40. CSPD, Elizabeth 1580–1625, Addenda, 137.
41. Bruce, *Correspondence of Robert Dudley*, 112, 144.

CHAPTER 12: The "Bosom Serpent"

1. Weir, *Life of Elizabeth*, 351.
2. The results of their labors can still be seen at Hardwick Hall today.
3. Murdin, *Collection of State Papers*, 559; CSP Scotland, 1584–85, VII, 5.
4. CSP Scotland, 1584–85, VII, 5; Murdin, *Collection of State Papers*, 558; Erickson, *First Elizabeth*, 262; Somerset, *Elizabeth I*, 101; Johnson, *Elizabeth I*, 115.
5. Durant, *Arbella Stuart*, 44.
6. Williams, *Elizabeth, Queen of England*, 256; CSPF, Elizabeth 1581–82, 589; CSPS, Elizabeth 1580–86, III, 495.
7. BM Lansdowne MS 1236, f.32.
8. HMC, *Salisbury*, XIII, 254–55, 309.
9. Weir, *Life of Elizabeth*, 355.
10. CSPF, Elizabeth 1583–84, 596.
11. Ibid., Elizabeth 1571–74, 373; HMC, *Salisbury*, II, 428; CSPF, Elizabeth 1584–85, 166–67.
12. Labanoff, V, *Lettres, Instructions et Memoires de Marie Stuart, Reine d'Ecasse*, 436; Wilson, *Sweet Robin*, 244; Lovell, *Bess of Hardwick*, 307.
13. CSP Scotland, 1584–85, VII, 373; Strickland, *Life of Queen Elizabeth*, 478.
14. Labanoff, *Lettres*, VII, 168; Marshall, *Queen Mary's Women*, 179.
15. Gristwood, *Arbella*, 39.
16. HMC, *Bath*, V, 55.
17. CSPS, Elizabeth 1580–86, III, 426.
18. Ibid., III, 546.
19. HMC, *Salisbury*, III, 152, 166; HMC, *Bath*, V, 70.
20. HMC, *Salisbury*, III, 152.
21. CSP Scotland, 1585–86, VIII, 657.
22. Bassnett, *Elizabeth I*, 113.
23. CSPV, VIII, 206.
24. HMC, *Salisbury*, III, 199.
25. Perry, *Word of a Prince*, 273.
26. Strickland, *Life of Queen Elizabeth*, 476.

27. CSP Scotland, 1587–88, IX, 251–52.

28. Marcus, Mueller, and Rose, *Elizabeth I*, 186–88, 199–202; Chamberlain, *Sayings*, 240–43.

29. Marcus, Mueller, and Rose, *Elizabeth I*, 187.

30. Perry, *Word of a Prince*, 272–73; Johnson, *Elizabeth I*, 291.

31. Strickland, *Life of Queen Elizabeth*, 476.

32. Ibid., 477.

33. BM Cotton MS Titus C VII, f.49; Francis Steuart, *Sir James Melville*, 313–14.

34. BM Cotton MS Titus C VII, ff.48–53.

35. Ibid., f.50v.

36. Camden, *Historie*, 103; CSPS, Elizabeth 1587–1603, IV, 35.

37. Ballard, *Memoirs*, 175.

38. Longford, *The Oxford Book of Royal Anecdotes*, 244. The dog died soon afterward, apparently from pining for his dead mistress.

39. CSP Scotland, 1587–88, IX, 274–75; CSPF, Elizabeth 1586–87, 688.

40. Camden, *Historie*, 115.

41. BM Lansdowne MS 1236, f.32.

42. Camden, *Historie*, 115.

43. Even Shakespeare got in on the act, penning a speech in the words of King John to defend the Queen: "It is the curse of kings to be attended / By slaves, that take their humours for a warrant." *King John*, act 4, scene 2, lines 208–10.

44. Camden, *Historie*, 115; CSPV, VIII, 256.

45. Weir, *Life of Elizabeth*, 381.

46. CSPF, Elizabeth 1586–88, 227.

47. HMC, *Salisbury*, III, 230.

48. CSP Scotland, 1587–88, IX, 285.

49. HMC, *Salisbury*, III, 230; CSPF, Elizabeth 1586–88, 276; Francis Steuart, *Sir James Melville*, 315.

50. Perry, *Word of a Prince*, 272.

CHAPTER 13: Gloriana

1. CSPD, Elizabeth 1566–79, Addenda, 315.

2. Haigh, *Elizabeth I*, 173.

3. Weir, *Life of Elizabeth*, 221.

4. Read and Plummer, *Elizabeth of England*, 70.

5. Marcus, Mueller, and Rose, *Elizabeth I*, 70.

6. Ibid., 97; Haigh, "Elizabeth I", 21.

7. Haigh, *Elizabeth I*, 23; Heisch, "Elizabeth I," 53.

8. Marcus, Mueller, and Rose, *Elizabeth I*, 326.

9. Haigh, *Elizabeth I*, 21–22.
10. Marcus, Mueller, and Rose, *Elizabeth I*, 97; Weir, *Life of Elizabeth*, 222.
11. Haigh, *Elizabeth I*, 22.
12. Wilson, *Elizabeth's Maids of Honour*, 5.
13. Taylor-Smither, "Elizabeth I," 71–72.
14. Harington, *Nugae Antiquae*, 123.
15. CSPS, Elizabeth 1558–67, I, 63.
16. Harrison, *Letters*, 268; Pasmore, *The Life and Times of Queen Elizabeth I at Richmond Palace*, 56.
17. Marcus, Mueller, and Rose, *Elizabeth I*, 361–62.
18. Frye, *Elizabeth I*, 13.
19. This is preserved in the National Archive, Kew.
20. HMC, *Bath*, V, 221.
21. CSPD, Elizabeth 1591–94, 386; ibid., Elizabeth 1601–3, 22–23; HMC, *Bath*, V, 221–23.
22. Craik, *Romance of the Peerage*, I, 207–8.
23. BM Lansdowne MS 34, f.1.
24. Ibid., f.53; CSP Scotland, 1581–83, VI, 119.
25. CSP Scotland, 1581–83, VI, 505.
26. BM Lansdowne MS 34, f.53; CSP Scotland, 1581–83, VI, 119.
27. CSPV, IX, 541.
28. Somerset, *Elizabeth I*, 561.
29. See for example Gristwood, *Arbella*, 375.
30. CSPV, IX, 541.
31. Durant, *Arbella Stuart*, 45.
32. Ibid., 46; Edwards, *The Life of Sir Walter Ralegh*, I, 298.
33. CSPV, VII, 541.
34. Durant, *Arbella Stuart*, 52.
35. Gristwood, *Arbella*, 130.
36. CSP Scotland, 1581–83, VI, 413; ibid., 1587–88, IX, 661. The Duke of Parma was himself put forward as a potential husband in July 1590. Ibid., 1589–93, X, 360.
37. HMC, *Salisbury*, III, 268; CSP Scotland, 1589–93, X, 17, 687.
38. CSP Scotland, 1589–93, X, 605.

CHAPTER 14: "Witches"

1. Ashdown, *Ladies-in-Waiting*, 60.
2. Weir, *Life of Elizabeth*, 260.
3. The notable exceptions are Merton, Frye, and Richardson, all of whom offer excellent insights into the role of women at Elizabeth's court.
4. Haigh, *Elizabeth I*, 99.

5. Wright, "A Change in Direction," 165.

6. See for example Richardson, *Mistress Blanche,* 78.

7. Merton, "The Women Who Served," 244.

8. Bradford, *Blanche Parry, Queen Elizabeth's Gentlewoman,* 26–27.

9. BM Add. MS 70093.

10. Blanche Parry's magnificent monument in Bacton Church can still be visited today. It is remarkably well preserved, as is the tapestry that adorns the church wall, which was said to have been worked by Blanche ·herself. This exquisite piece, which includes embroideries of frogs, deer, foliage, and flowers on a fine silver cloth, may have formed part of a dress for the Queen.

11. Somerset, *Ladies-in-Waiting,* 71.

12. HMC, *Salisbury,* I, 345.

13. See for example BM Add. MS 12506, ff.47, 72; ibid., 12507, f.131.

14. Merton, "The Women Who Served," 54–55.

15. Williamson, *Lady Anne Clifford, Countess of Dorset, Pembroke & Montgomery, 1590–1676,* 37.

16. Merton, "The Women Who Served," 168, 171, 180, 194, 197; TNA SP 46/125, f.236. For other examples of Lady Warwick's influence at court, see: BM Add. MS 27401, f.21; ibid., 12406, ff.41, 80; BM Lansdowne MS 128, f.12; HMC, *Salisbury,* IV, 199; ibid., V, 53, 444, 481; ibid., VI, 402; ibid., IX, 21; ibid., X, 86, 319; ibid., XIV, 16–17.

17. Merton, "The Women Who Served," 165, 168.

18. HMC, *Salisbury,* V, 484.

19. HMC, *De L'Isle & Dudley,* II, 163.

20. Ibid., II, 201–2.

21. Ibid., II, 203–5.

22. Ibid., II, 179, 186, 202–4.

23. BM Add. MS 15552, f.5.

24. HMC, *De L'Isle & Dudley,* II, 244.

25. Clifford, *The Diaries of Lady Anne Clifford,* 22.

26. HMC, *De L'Isle & Dudley,* II, 313.

27. Ibid., II, 391.

28. Ibid., II, 444, 474.

29. Ibid., II, 391, 425, 440, 474, 487, 488.

30. HMC, *Salisbury,* IX, 21; ibid., X, 86.

31. CSPD, Elizabeth 1598–1601, 252; HMC, *De L'Isle & Dudley,* II, 471.

32. HMC, *De L'Isle & Dudley,* II, 274.

33. Ibid., II, 314, 317. See also ibid., II, 422.

34. Ibid., II, 465, 472.

35. Ibid., II, 472.

36. Merton, "The Women Who Served," 100. NB: Aglets were pieces of

metal fixed to the end of ribbon to tie pieces of clothing together, and were often highly ornamental. Meanwhile, bezoar stones were believed to have the power of a universal antidote against any poison.

37. Wilson, *Elizabeth's Maids of Honour,* 133.
38. BM Cotton MS Titus B II, f.346.

CHAPTER 15: "Flouting Wenches"

1. Thoms, *Anecdotes and Traditions, Illustrative of Early English History and Literature,* 70–71.
2. Haigh, *Elizabeth I,* 98.
3. Merton, "The Women Who Served," 140.
4. Edwards, *Life of Sir Walter Ralegh,* I, 137.
5. T. Fuller, *Anglorum speculum, or The Worthies of England in Church and State* (London, 1663), 262.
6. von Klarwill, *Elizabeth and Some Foreigners,* 336.
7. Somerset, *Ladies-in-Waiting,* 88.
8. Beer, *Bess,* 61.
9. Murdin, *Collection of State Papers,* 657.
10. Birch, *Memoirs,* I, 79.
11. HMC, *Salisbury,* XII, 84.
12. In mythology, a dangerous queen of the underworld.
13. Beer, *Bess,* 133.
14. In fact, the new reign would bring nothing but heartache. Her husband was arrested on suspicion of treason just four months after Elizabeth's death and thrown in the Tower. He was released many years later, only to be arrested again in 1618. This time there would be no pardon. Ralegh was executed on October 29. Legend has it that his grief-stricken widow had his head embalmed and kept it with her for the rest of her life.
15. Merton, "The Women Who Served," 145.
16. This theory was first put forward at a meeting of the New Shakespeare Society in 1884, when Thomas Tyler argued that Mary Fitton was the "dark lady" of Shakespeare's sonnets. In support of his argument, he claimed that "Mr WH," referred to in the sonnets, was almost certainly William Herbert. His theory subsequently gained wide acceptance. However, it was discounted by the *Dictionary of National Biography* in 1908, which pointed out that in the surviving portraits of Mary, she bears little resemblance to the "dark lady," with her brown hair and light eyes.
17. Hannay, *Philip's Phoenix,* 169.
18. Somerset, *Elizabeth I,* 552–53.
19. HMC, *De L'Isle & Dudley,* II, 265.

20. Harington, *Nugae Antiquae*, 90.

21. Camden, *Historie*, 172.

22. The Queen's wigs were made by her silk woman, Dorothy Spekarde. In 1602 it was recorded that she paid for "six heads of hair, twelve yards of hair curl and one hundred devices made of hair." Jenkins, *Elizabeth the Great*, 296.

23. Camden, *Elizabethan Woman*, 178.

24. Her concern over aging had apparently begun many years before. It was rumored that from her late thirties, she had secretly employed a Dutch alchemist to find the elixir of youth.

25. Merton, "The Women Who Served," 106.

26. Harington, *Nugae Antiquae*, 90; Perry, *Word of a Prince*, 316; CSPV, IX, 531–32.

27. Harrison and Jones, *Andre Hurault de Maisse*, 25–26, 36–39, 55. See also Rye, *England as Seen by Foreigners*, 104–5; Pasmore, *Life and Times*, 9; CSPV, VII, 628.

28. Clifford, *Diaries*, 27.

29. HMC, *Salisbury*, IV, 153.

30. Merton, "The Women Who Served," 128.

31. HMC, *De L'Isle & Dudley*, II, 311.

32. Ibid., II, 317–18.

33. Chamberlain, *Letters*, 44.

34. CSPD, Elizabeth 1598–1601, 97.

35. Merton, "The Women Who Served," 145.

36. Wilson, *Elizabeth's Maids of Honour*, 235–36.

37. HMC, *Salisbury*, VIII, 355, 357.

38. Wilson, *Elizabeth's Maids of Honour*, 237–38.

39. HMC, *Salisbury*, IX, 197; Camden, *Historie*, 142.

40. Craik, *Romance of the Peerage*, I, 30, 148–52.

41. Birch, *Memoirs*, II, 362.

42. HMC, *De L'Isle & Dudley*, II, 325, 327.

43. Ibid., II, 328.

44. Ibid., II, 329–30.

45. Somerset, *Ladies-in-Waiting*, 88; Perry, *Word of a Prince*, 304.

46. HMC, *De L'Isle & Dudley*, II, 444.

47. Ibid., II, 442–43.

48. Gristwood, *Elizabeth and Leicester*, 258; BM Add. MS 32092, f.48.

49. TNA PROB 11/167, sig.1.

50. Wilson, *Society Women of Shakespeare's Time*, 120.

51. Ibid., 121.

52. Harington, *Nugae Antiquae*, 125.

53. Ibid.

54. Hibbert, *Elizabeth I,* 253.

55. Harington, *Nugae Antiquae,* 90.

56. Weir, *Life of Elizabeth,* 470.

57. Camden, *Historie,* 222.

58. CSPD, Elizabeth 1580–1625, Addenda, 407.

59. HMC, *Salisbury,* IV, 335; CSPD, Elizabeth 1601–3, 37.

60. Durant, *Arbella Stuart,* 68.

61. Ibid., 56.

62. CSPV, VII, 564.

63. HMC, *Salisbury,* XV, 65.

64. Ibid., XIV, 253.

65. Durant, *Arbella Stuart,* 95–96.

66. CSPV, VII, 564.

67. HMC, *Salisbury,* XII, 593–96.

68. Ibid., XII, 594–95.

69. Ibid., XII, 593–624.

70. Ibid., XII, 624.

71. Lovell, *Bess of Hardwick,* xii.

72. HMC, *Salisbury,* XII, 626–27.

73. Ibid., XII, 682–83.

74. Ibid., XII, 682–83, 691, 693.

75. Ibid., XII, 626.

76. Pryor, *Elizabeth I,* 131; HMC, *Salisbury,* XII, 681.

77. CSPD, Elizabeth 1601–3, 299.

78. Gristwood, *Arbella,* 156.

79. HMC, *Salisbury,* XII, 685–86.

80. Ibid., XII, 690, 692, 693.

CHAPTER 16: "The Sun Now Ready to Set"

1. Somerset, *Elizabeth I,* 553; CSPS, Elizabeth 1587–1603, IV, 650; HMC, *De L'Isle & Dudley,* II, 475; CSPV, IX, 529.

2. Nichols, *Progresses,* III, 612; Merton, "The Women Who Served," 90.

3. Weir, *Life of Elizabeth,* 480; Haigh, *Elizabeth I,* 166; Harington, *Nugae Antiquae,* 96.

4. Harington, *Nugae Antiquae,* 96; Boyle, *Memoirs of Robert Carey,* 137–38.

5. Bassnett, *Elizabeth I,* 258.

6. Ibid., 149.

7. Merton, "The Women Who Served," 90; Birch, *Memoirs,* II, 506–7; Boyle, *Memoirs of Robert Carey,* 140; Pasmore, *Life and Times,* 65.

8. Nichols, *Progresses,* III, 613.

9. Boyle, *Memoirs of Robert Carey*, 136–38.

10. CSPD, Elizabeth 1601–3, 301; Gristwood, *Arbella*, 188; CSPV, VII, 554.

11. CSPV, VII, 564; CSPD, Elizabeth 1601–3, 302.

12. HMC, *Salisbury*, XII, 693.

13. CSPV, VII, 562; Edwards, *Life of Sir Walter Ralegh*, I, 296.

14. Boyle, *Memoirs of Robert Carey*, 137.

15. Levine, *Early Elizabethan Succession*, 29.

16. CSPD, Elizabeth 1601–3, 298, 301; CSPV, IX, 554. See also HMC, *Salisbury*, XII, 670.

17. The story is seductively dramatic, but it is almost certainly false. It is not referred to in any of the sources at the time of Essex's imprisonment and execution, and it only appeared some twenty years later when it was included in J. Webster's *The Devil's Law Case*. It was then re-created in a work of fiction toward the end of the seventeenth century. Elizabeth's first biographer, William Camden, knew of the story and declared it to be false. More compelling is the fact that the Queen was devastated by Katherine's death and never got over her grief. She would hardly have felt this way if she had just discovered an act of such treachery. The tale is related in a number of more recent publications, notably: Strickland, *Life of Queen Elizabeth*, 673–74; Wilson, *Elizabeth's Maids of Honour*, 274–76; Ashdown, *Ladies-in-Waiting*, 71–72.

18. CSPD, Elizabeth 1601–3, 303.

19. Ibid., 302.

20. Read and Plummer, *Elizabeth of England*, 99; Kenny, *Elizabeth's Admiral*, 257.

21. J. Bruce (ed.), *The Diary of John Manningham* (London, 1868), entry for 23 March 1603.

22. Clifford, *Diaries*, 21.

23. Somerset, *Elizabeth I*, 569; Bruce, op. cit., 159.

24. Johnson, *Elizabeth I*, 438.

25. Clifford, *Diaries*, 21; Nichols, *Progresses*, III, 613.

26. Nichols, *Progresses*, III, 613; Strickland, *Life of Queen Elizabeth*, 704–5.

27. Read and Plummer, *Elizabeth of England*, 112.

Bibliography

MANUSCRIPT SOURCES

British Museum Manuscripts: Add. 12506, 12507, 15552, 24783, 27401, 32092, 39288, 63543, 70093, 73965; Cotton Otho C X, Titus B II, and C VII; Vespasian F XII; Harley 283, 6286; Lansdowne 8, 34, 82, 158, 1236; Royal Appendix 68.
The National Archives: C115/01, C270/33, and 34; E/210/10340 and 10738, E211/421, E214/989; KB 8/9; SP 46/10, 34, and 125, SP 88/1.

PRINTED PRIMARY SOURCES

Adams, S., and M. J. Rodríguez-Salgado. "The Count of Feria's Dispatch to Philip II of 14 November 1558." *Camden Miscellany* 28 (London, 1984).
Arber, E., ed. *John Knoxe, First Blast of the Trumpet Against the Monstrous Regiment of Women*. London: 1878.
Bain, J., J. D. Mackie, et al., eds. *Calendar of the State Papers Relating to Scotland and Mary, Queen of Scots, 1547–1603*. Vols. 1–13, part 2. Edinburgh: 1898–1969.

Bell, J. *Queen Elizabeth and a Swedish Princess. Being an Account of the Visit of Princess Cecilia of Sweden to England in 1565.* London: 1926.

Birch, T. *Memoirs of the Reign of Queen Elizabeth from the Year 1581 till Her Death.* 2 vols. London: 1754.

Bourdeille, P. de, Seigneur de Brantôme. *The Book of the Ladies . . . with Elucidations on Some of Those Ladies.* London: 1899.

———. *The Lives of Gallant Ladies.* London: 1965.

Boyle, J., ed. *Memoirs of the Life of Robert Carey . . . Written by Himself.* London: 1759.

Brewer, J. S., and W. Bullen, eds. *Calendar of the Carew Manuscripts, preserved in the Archiepiscopal Library at Lambeth, 1515–1603.* 4 vols. London: 1867–70.

Brown, R., et al., eds. *Calendar of State Papers and Manuscripts, Relating to English Affairs, Existing in the Archives and Collections of Venice.* Vols. 4–9. London: 1871–97.

Bruce, J., ed. *Annals of the First Four Years of the Reign of Queen Elizabeth,* by J. Hayward. Camden Society, Old Series, VII. London: 1840.

———. *The Correspondence of Robert Dudley, Earl of Leycester, During His Government of the Low Countries, in the Years 1585 and 1586.* Camden Society, XXVII. London: 1844.

Camden, W. *The Historie of the Most Renowned and Victorious Princesse Elizabeth, Late Queene of England.* London: 1630.

Cerovski, J. S., ed. *Sir Robert Naunton, Fragmentia Regalia, or Observations on Queen Elizabeth, Her Times and Favourites.* London and Toronto: 1985.

Clifford, D. J. H., ed. *The Diaries of Lady Anne Clifford.* Stroud: 1992.

Clifford, H. *The Life of Jane Dormer, Duchess of Feria.* London: 1887.

Collins, A., ed. *Letters and Memorials of State, in the Reigns of Queen Mary, Queen Elizabeth, Etc . . . Written and Collected by Sir Henry Sidney, Etc.* 2 vols. London: 1746.

Collins, A. J., ed. *Jewels and Plate of Queen Elizabeth I. The Inventory of 1574.* London: 1955.

Craik, G. L. *The Romance of the Peerage, or Curiosities of Family History.* Vols. 1–4. London: 1849.

Dasent, J. R. *Acts of the Privy Council of England.* New Series. London: 1890.

Edwards, E. *The Life of Sir Walter Ralegh. Based on Contemporary Documents . . . Together with His Letters.* 2 vols. London: 1868.

Ellis, H., ed. *Original Letters Illustrative of English History, Including Numerous Royal Letters.* 3rd Series, Vols. 2–4. London: 1846.

Foxe, J. *The Acts and Monuments of John Foxe.* 3 vols. London: 1853–55.

Francis Steuart, A., ed. *Sir James Melville: Memoirs of His Own Life, 1549–93.* London: 1929.

Halliwell, J. O., ed. *The Private Diary of John Dee.* Camden Society, XIX. London: 1842.

Hamilton, W. D., ed. *A Chronicle of England During the Reigns of the Tudors, from AD 1485 to 1559,* by C. Wriothesley. 2 vols. Camden Society, New Series, II. London: 1875.

Harington, Sir J. *Nugae Antiquae: Being a Miscellaneous Collection of Original Papers in Prose and Verse: Written in the Reigns of Henry VIII, Queen Mary, Elizabeth, King James, Etc.* London: 1779.

Harrison, G. B., *The Letters of Queen Elizabeth.* London: 1935.

Harrison, G. B., and R. A. Jones. *Andre Hurault de Maisse: A Journal of All That Was Accomplished by Monsieur de Maisse, Ambassador in England from King Henri IV to Queen Elizabeth, 1597.* London: 1931.

Haynes, A. *Collection of State Papers Relating to Affairs in the Reigns of King Henry VIII, King Edward VI, Queen Mary and Queen Elizabeth, from the Year 1542 to 1570 . . . Left by William Cecil, Lord Burghley . . . at Hatfield House.* London: 1740.

Hearne, T. *Syllogue Epistolarum.* London: 1716.

Heath, J. B. "An Account of Materials Furnished for the Use of Queen Anne Boleyn, and the Princess Elizabeth, by William Loke, the King's Mercer, Between the 20th January 1535 and the 27th April, 1536." *Miscellanies of the Philobiblon Society.* Vol. 7. London: 1862–3.

Hentzner, P. *Travels in England During the Reign of Queen Elizabeth.* London: 1889.

Historical Manuscripts Commission, *Calendar of the Manuscripts of the Marquis of Salisbury, Preserved at Hatfield House, Herts.* Vols. 1–15. London, 1883–1930.

Historical Manuscripts Commission, *Calendar of the Manuscripts of the Most Honourable the Marquess of Bath, Preserved at Longleat, Wiltshire, 1533–1659.* Vol. V. London: 1980.

Historical Manuscripts Commission, *The Manuscripts of His Grace the Duke of Rutland, Preserved at Belvoir Castle.* Vol. 1. London: 1888.

Historical Manuscripts Commission, *Report on the Manuscripts of Lord De L'Isle & Dudley, Preserved at Penshurst Place.* Vols. 1 and 2. London: 1925.

Historical Manuscripts Commission, *Third Report of the Royal Commission on Historical Manuscript.* London: 1842. Appendix.

Hoby, Sir T. *The Book of the Courtier, from the Italian of Count Baldessare Castiglione, 1561.* London: 1900.

Hume, M. A. S., ed. *Calendar of Letters and State Papers Relating to English Affairs, Preserved Principally in the Archives of Simancas, Elizabeth I.* 4 vols. London: 1892–99.

Hume, M. A. S., R. Tyller, et al., eds. *Calendar of Letters, Despatches, and State Papers, Relating to the Negotiations Between England and Spain, Preserved in the Archives at Simancas and Elsewhere, 1547–1558.* London: 1912–54.

James, H., ed. *Facsimiles of National Manuscripts from William the Conqueror to Queen Anne.* 2 vols. Southampton: 1865.

Jordan, W. K., ed. *The Chronicle and Political Papers of King Edward VI.* 2 vols. London, 1966.

Labanoff, A., ed. *Lettres, Instructions et Memoires de Marie Stuart, Reine d'Ecosse.* 7 vols. London: 1844.

Laing, D., ed. *Notes of Ben Jonson's Conversations with William Drummond of Hawthornden.* Vol. 1. London: 1842.

Lerer, S. *Courtly Letters in the Age of Henry VIII: Literary Culture and the Arts of Deceit.* Cambridge University Press, 1997.

Loades, D. M., ed. *Elizabeth I: The Golden Reign of Gloriana. English Monarchs: Treasures from the Archives.* Richmond: 2003.

McClure, N. E. *The Letters and Epigrams of Sir John Harington.* London: 1930.

Manning, C. R. "State Papers Relating to the Custody of the Princess Elizabeth at Woodstock." *Norfolk Archaeology* 4 (1855).

Marcus, L. S., J. Mueller, and M. B. Rose. *Elizabeth I: Collected Works.* Chicago and London: 2002.

Murdin, W. *A Collection of State Papers Relating to Affairs in the Reign of Queen Elizabeth, 1571–96 . . . Left by William Cecil Lord Burghley . . . at Hatfield House.* London: 1759.

Nichols, J. *The Progresses and Public Processions of Queen Elizabeth.* 3 vols. London: 1823.

Nichols, J. G., ed. *The Chronicle of Queen Jane and of Two Years of Queen Mary,* Camden Society, 48. London: 1850.

———. *The Diary of Henry Machyn: Citizen and Merchant-Taylor of London, from AD 1550 to AD 1563.* London: 1848.

Noialles, A. de. *Ambassades de Monsieur de Noailles en Angleterre.* Leyden: 1763.

Perry, M. *The Word of a Prince.* London: 1990.

Prescott, A. L., ed. *The Early Modern Englishwoman: A Facsimile Library of Essential Works, Series I, Printed Writings, 1500–1640, Part 2 Volume 5, Elizabeth and Mary Tudor.* Aldershot: 2001.

Pryor, F. *Elizabeth I: Her Life in Letters.* California: 2003.

Read, C., and E. Plummer, eds. *Elizabeth of England. Certain Observations Concerning the Life and Reign of Queen Elizabeth, by John Chapman.* Philadelphia: 1951.

Rigg, J. M., ed. *Calendar of State Papers, Relating to English Affairs, Preserved Principally at Rome, in the Vatican Archives and Library, 1558–71 and 1572–78.* 2 vols. London: 1916 and 1926.

Rye, W. B., ed. *England as Seen by Foreigners in the Days of Elizabeth and James the First.* London: 1865.

Sawyer, E., ed. *Memorials of Affairs of State in the Reigns of Queen Elizabeth and King James I, Collected (Chiefly) from the Original Papers of the Right Honourable Sir Ralph Winwood.* Vol. 1. London: 1725.

Seaton, E. *Queen Elizabeth and a Swedish Princess, Being an Account of the Visit of*

Princess Cecilia of Sweden to England in 1565 from the Original Manuscript of James Bell. London: 1926.

Stevenson, J., ed. *The Life of Jane Dormer, Duchess of Feria, by Henry Clifford: Transcribed from the Ancient Manuscript in the Possession of the Lord Dormer.* London: 1887.

Stevenson, J., et al., eds. *Calendar of State Papers, Foreign Series, of the Reign of Elizabeth I, 1558–1591.* London: 1863–1969.

Strangford, Viscount. "Household Expenses of the Princess Elizabeth During her Residence at Hatfield, October 1, 1551 to September 30, 1552." Camden Miscellany, II (London: 1853).

Strype, J. *Ecclesiastical Memorials, Relating Chiefly to Religion, and the Reformation of It . . . Under King Henry VIII, King Edward VI and Queen Mary I.* 3 vols. Oxford: 1822.

Thoms, W. J., ed. *Anecdotes and Traditions, Illustrative of Early English History and Literature.* Camden Society. London: 1850.

Traherne, J. M., ed. *Stradling Correspondence: A Series of Letters Written in the Reign of Queen Elizabeth.* London: 1840.

Tytler, P. F. *England Under the Reigns of Edward VI and Mary, Illustrated in a Series of Original Letters.* 2 vols. London: 1839.

Wernham, R. B. *List and Analysis of State Papers Foreign Series, Elizabeth I, Preserved in the Public Record Office, June 1591–December 1596.* London: 1980–2000.

Wood, M. A. E. *Letters of Royal and Illustrious Ladies of Great Britain.* 3 vols. London: 1846.

Wright, T. *Queen Elizabeth and Her Times, A Series of Original Letters, Selected from the Inedited Private Correspondence of the Lord Treasurer Burghley, the Earl of Leicester, the Secretaries Walsingham and Smith, Sir Christopher Hatton, Etc.* 2 vols. London: 1838.

Wyatt, G. "Extracts from the Life of the Virtuous, Christian and Renowned Queen Anne Boleyn," in Singer, S. W., ed., and G. Cavendish, *The Life of Cardinal Wolsey.* London: 1827.

Yorke, P., ed. *Miscellaneous State Papers: From 1501 to 1726.* Vol. 1. London: 1778.

SECONDARY SOURCES

Adams, S. "Eliza Enthroned? The Court and Its Politics," in Haigh, C., ed., *The Reign of Elizabeth I.* London: 1984.

Ashdown, D. M. *Ladies-in-Waiting.* London: 1976.

———. *Tudor Cousins: Rivals for the Throne.* Sutton: 2000.

Baldwin Smith, L. *A Tudor Tragedy: The Life and Times of Catherine Howard.* London: 1961.

Ballard, G. *Memoirs of Several Ladies of Great Britain Who Have Been Celebrated for Their Writings or Skill in the Learned Languages, Arts, and Sciences.* Detroit: 1985.

Bassnett, S. *Elizabeth I: A Feminist Perspective.* Oxford and New York: 1988.

Beer, A. *Bess: The Life of Lady Ralegh, Wife to Sir Walter.* London: 2005.

Bradford, C. A. *Blanche Parry, Queen Elizabeth's Gentlewoman.* London: 1935.

———. *Helena, Marchioness of Northampton.* London: 1936.

Bradford, G. *Elizabethan Women.* New York: 1969.

Brewer, C. *The Death of Kings: A Medical History of the Kings and Queens of England.* London: 2004.

Bruce, M. L. *Anne Boleyn.* London: 1972.

Burton, E. *The Early Tudors at Home.* London: 1976.

———. *The Elizabethans at Home.* London: 1970.

Camden, C. *The Elizabethan Woman.* New York: 1975.

Carleton Williams, C. *Bess of Hardwick.* Bath: 1959.

Chamberlain, F. *The Private Character of Queen Elizabeth.* London: 1921.

———. *The Sayings of Queen Elizabeth.* London and New York: 1923.

Chapman, H. W. *Anne Boleyn.* London: 1974.

———. *Two Tudor Portraits: Henry Howard, Earl of Surrey, and Lady Katherine Grey.* London: 1960.

Cockayne, G. E., ed. *The Complete Peerage of England, Scotland, Ireland, Great Britain, and the United Kingdom.* 12 vols. London: 1910–59.

Davey, R. *The Sisters of Lady Jane Grey and Their Wicked Grandfather.* London: 1911.

Denny, J. *Katherine Howard: A Tudor Conspiracy.* London: 2005.

Doran, S. *Elizabeth: The Exhibition at the National Maritime Museum.* London: 2003.

Dovey, Z. *An Elizabethan Progress: The Queen's Journey into East Anglia, 1578.* Sutton: 1999.

Dunlop, I. *Palaces and Progresses of Elizabeth I.* London: 1962.

Dunn, J. *Elizabeth and Mary: Cousins, Rivals, Queens.* New York: 2004.

Durant, D. N. *Arbella Stuart: A Rival to the Queen.* London: 1978.

———. *Bess of Hardwick: Portrait of an Elizabethan Dynast.* London, 1999.

Dutton, R. *English Court Life: From Henry VII to George II.* London: 1963.

Eccles, A. *Obstetrics and Gynaecology in Tudor and Stuart England.* London: 1982.

Erickson, C. *Bloody Mary.* London: 2001.

———. *The First Elizabeth.* London: 1999.

———. *Mistress Anne.* New York: 1984.

Fraser, A. *Mary, Queen of Scots.* London: 1994.

———. *The Six Wives of Henry VIII.* London: 1996.

Friedmann, P. *Anne Boleyn: A Chapter of English History.* 2 vols. London: 1884.

Frye, S. *Elizabeth I: The Competition for Representation.* New York and Oxford: 1993.

———. *Maids and Mistresses, Cousins and Queens: Women's Alliances in Early Modern England.* New York and Oxford: 1999.

Furnivall, F. J. "Shakespeare and Mary Fitton." *The Theatre* (December 1897).

Gairdner, J. "Mary and Anne Boleyn." *English Historical Review* 8 (1893).

Gilson, J. P. *Lives of Lady Anne Clifford, Countess of Dorset, Pembroke and Montgomery, 1590–1676, and of Her Parents.* London: 1916.

Goff, C. *A Woman of the Tudor Age.* London: 1930.

Graves, J. *A Brief Memoir of the Lady Elizabeth Fitzgerald, Known as the Fair Geraldine.* Dublin: 1874.

Gristwood, S. *Arbella.* London: 2003.

———. *Elizabeth and Leicester.* London: 2007.

Gross, P. M. *Jane the Quene, Third Consort of King Henry VIII.* Lewiston, Queenston, Lampeter: 1999.

Guy, J. *"My Heart Is My Own": The Life of Mary, Queen of Scots.* London: 2004.

Haigh, C., ed. *Elizabeth I.* London and New York: 1988.

Handover, P. M. *Arbella Stuart: Royal Lady of Hardwick and Cousin to King James.* London: 1957.

Hannay, M. P. *Philip's Phoenix: Mary Sidney, Countess of Pembroke.* Oxford: 1990.

Hannay, M. P., N. J. Kinnamon, and M. G. Brennan. *The Collected Works of Mary Sidney Herbert, Countess of Pembroke.* 2 vols. Oxford: 1998.

Heisch, A. "Elizabeth I and the Persistence of Patriarchy." *Feminist Review* 4 (1980).

Hibbert, C. *Elizabeth I: A Personal History of the Virgin Queen.* London: 1992.

Holles, G. *Memorials of the Holles Family, 1493–1656.* Camden Society, Third Series, Vol. 4. London: 1937.

Hopkins, L. *Queen Elizabeth I and Her Court.* London and New York: 1990.

Howe, B. *A Galaxy of Governesses.* London: 1954.

Hume, M. *The Courtships of Queen Elizabeth: A History of the Various Negotiations for Her Marriage.* London: 1904.

———. *Two English Queens and Philip.* London: 1908.

Hurstfield, J. *The Queen's Wards: Wardship and Marriage Under Elizabeth I.* London: 1958.

Ives, E. *Anne Boleyn.* Oxford: 1986.

———. *The Life and Death of Anne Boleyn: "The Most Happy."* Oxford: 2004.

James, S. E. "A Tudor Divorce: The Marital History of William Parr, Marquess of Northampton." *Cumberland and Westmorland Antiquarian and Archaeological Society, New Series* 90 (1990).

———. *Catherine Parr: Henry VIII's Last Love.* Stroud: 2008.

———. *Kateryn Parr: The Making of a Queen.* Aldershot: 1999.

Jenkins, E. *Elizabeth the Great.* London: 1965.

Johnson, P. *Elizabeth I: A Study in Power and Intellect.* London: 1974.

Jones, N. *The Birth of the Elizabethan Age: England in the 1560s.* Oxford: 1993.

Kenny, R. W. *Elizabeth's Admiral: The Political Career of Charles Howard, Earl of Nottingham, 1536–1624.* Baltimore and London: 1990.

Lever, T. *The Herberts of Wilton.* London: 1967.

Levin, C. *The Heart and Stomach of a King: Elizabeth I and the Politics of Sex and Power.* Philadelphia: 1994.

Levin, C., and J. Watson. *Ambiguous Realities: Women in the Middle Ages and Renaissance.* Detroit: 1987.

Levine, M. *The Early Elizabethan Succession Question.* California: 1966.

Loades, D. *Henry VIII and His Queens.* Sutton: 2000.

————. *Mary Tudor: The Tragical History of the First Queen of England.* Richmond: 2006.

————. *The Tudor Court.* Oxford: 2003.

Longford, E., ed. *The Oxford Book of Royal Anecdotes.* Oxford: 1989.

Lovell, M. S. *Bess of Hardwick: First Lady of Chatsworth, 1527–1608.* London: 2005.

McCaffrey, W. T. *Elizabeth I.* London: 1993.

Mackie, J. D. *The Later Tudors.* London: 1952.

Madden, F. *Privy Purse Expenses of the Princess Mary.* London: 1831.

Marshall, R. K. *Queen Mary's Women: Female Relatives, Servants, Friends, and Enemies of Mary, Queen of Scots.* Edinburgh: 2006.

Martienssen, A. K., *Queen Katherine Parr.* London: 1973.

Medvei, V. C. "The Illness and Death of Mary Tudor." *Journal of the Royal Society of Medicine* 80, no. 12 (December 1987).

Merton, C. "The Women Who Served Queen Mary and Queen Elizabeth: Ladies, Gentlewomen and Maids of the Privy Chamber, 1553–1603." Cambridge PhD thesis: 1992.

Montagu, W. *Court and Society from Elizabeth to Anne.* 2 vols. London: 1864.

Murphy, J. "The Illusion of Decline: The Privy Chamber, 1547–1558," in Starkey, D., ed., *The English Court: From the Wars of the Roses to the Civil War.* Longman, 1987.

Neale, J. E. "The Accession of Elizabeth I." *History Today* 3 (May 1953).

————. *Queen Elizabeth I.* London: 1998.

————. "The Sayings of Queen Elizabeth." *History* 10 (October 1925).

Newdigate-Newdigate, A. *Gossip from a Muniment Room: Being Passages in the Lives of Anne and Mary Fitton, 1574 to 1618.* London: 1898.

Notestein, W. "The English Woman, 1580–1650," in Plumb, J. H., ed., *Studies in Social History: A Tribute to G. M. Trevelyan.* London: 1958.

Pasmore, S. *The Life and Times of Queen Elizabeth I at Richmond Palace.* Richmond Local History Society: 2003.

Percival, R. and A. *The Court of Elizabeth the First.* London: 1976.

Plowden, A. *Elizabethan England: Life in an Age of Adventure.* London: 1982.

————. *Marriage with My Kingdom: The Courtships of Elizabeth I.* London: 1977.

————. *Tudor Women: Queens and Commoners.* Sutton: 2002.

————. *Two Queens in One Isle: The Deadly Relationship Between Elizabeth I and Mary, Queen of Scots.* Sutton: 1999.

————. *The Young Elizabeth: The First Twenty-Five Years of Elizabeth I.* Sutton: 1999.

Porter, L. *Mary Tudor: The First Queen.* London: 2007.

Prescott, H. F. M. *Mary Tudor.* London: 1952.

Read, C. "A Letter from Robert, Earl of Leicester, to a Lady." *Huntingdon Library Quarterly* (April 1936).

Redworth, G. "Matters Impertinent to Women: Male and Female Monarchy Under Philip and Mary." *English Historical Review* 40, no. 4 (December 1997).

Richards, J. M. "Love and a Female Monarch: The Case of Elizabeth Tudor." *Journal of British Studies* 38 (April 1999).

————. "Mary Tudor as 'Sole Quene'? Gendering Tudor Monarchy." *Historical Journal* 40, no. 4 (December 1997).

————. "'To Promote a Woman to Beare Rule': Talking of Queens in Mid-Tudor England." *Sixteenth Century Journal* 28, no. 1 (1997).

Richardson, A. *Famous Ladies of the English Court.* London: 1899.

Richardson, R. E. *Mistress Blanche: Queen Elizabeth I's Confidante.* Herefordshire: 2007.

Ridley, J. *Elizabeth I: The Shrewdness of Virtue.* New York: 1987.

Rowse, A. L. "The Coronation of Queen Elizabeth I." *History Today* 3 (May 1953).

————. *The England of Elizabeth.* MacMillan: 1953.

Schutte, K. *A Biography of Margaret Douglas, Countess of Lennox, 1515–1578, Niece of Henry VIII and Mother-in-Law of Mary, Queen of Scots.* Lampeter: 2000.

Schutte, W. M. "Thomas Churchyard's 'Dolefull Discourse' and the Death of Lady Katherine Grey." *Sixteenth Century Journal* 15, no. 4 (1984).

Seymour, W. *Ordeal by Ambition: An English Family in the Shadow of the Tudors.* London: 1972.

Sitwell, E. *The Queens and the Hive.* London: 1991.

Skidmore, C. *Edward VI: The Lost King of England.* London: 2007.

Somerset, A. *Elizabeth I.* London: 1991.

————. *Ladies-in-Waiting: From the Tudors to the Present Day.* London: 1984.

Starkey, D. *Elizabeth: Apprenticeship.* London: 2001.

————. *Six Wives: The Queens of Henry VIII.* London: 2003.

————, ed. *The English Court: From the Wars of the Roses to the Civil War.* Longman: 1987.

Stone, L. *The Family, Sex, and Marriage in England, 1500–1800.* London: 1977.

Stopes, C. C. *Henry, 3rd Earl of Southampton.* London: 1922.

Strickland, A. *The Life of Queen Elizabeth*. London: 1910.

———. *Lives of the Queens of England*. Vols. II and III. London: 1851.

Strong, R. *The Cult of Elizabeth: Elizabethan Portraiture and Pageantry*. London: 1977.

———. *Gloriana: The Portraits of Queen Elizabeth I*. London: 1987.

Taylor-Smither, L. J. "Elizabeth I: A Psychological Profile." *Sixteenth Century Journal* 15 (1984).

Thurley, S. *Hampton Court: A Social and Architectural History*. New Haven and London: 2003.

———. *The Royal Palaces of Tudor England: Architecture and Court Life, 1460–1547*. New Haven and London: 1993.

Trill, S. "Sixteenth-Century Women's Writing: Mary Sidney's Psalmes and the 'Femininity of Translation,'" in Zunder, W., and S. Trill, *Writing and the English Renaissance*. London and New York: 1996.

von Klarwill, V. *Queen Elizabeth and Some Foreigners*. London: 1928.

Waller, G. F. *Mary Sidney, Countess of Pembroke: A Critical Study of her Writings and Literary Milieu*. Austria: 1979.

Ward, B. M. *The Seventeenth Earl of Oxford, 1550–1604*. London: 1928.

Warnicke, R. M. *The Marrying of Anne of Cleves: Royal Protocol in Early Modern England*. Cambridge: 2000.

———. *The Rise and Fall of Anne Boleyn*. Cambridge: 1991.

———. *Women of the English Renaissance and Reformation*. Connecticut: 1983.

Watkins, S. *In Public and Private: Elizabeth I and Her World*. London: circa 1998.

Weir, A. *Children of England: The Heirs of King Henry VIII, 1547–1558*. London: 1996.

———. *Elizabeth the Queen*. London: 1999.

———. *Henry VIII: King and Court*. London: 2001.

———. *The Lady in the Tower: The Fall of Anne Boleyn*. London: 2009.

———. *The Life of Elizabeth*. New York: 1998.

———. *Mary, Queen of Scots and the Murder of Lord Darnley*. London: 2003.

———. *The Six Wives of Henry VIII*. London: 1991.

Wheeler, E. D. *Ten Remarkable Women of the Tudor Courts and Their Influence in Founding the New World, 1530–1630*. Lewiston, Queenston, Lampeter: 2000.

Wiesener, L. *The Youth of Queen Elizabeth, 1533–1558*. 2 vols. London: 1879.

Williams, N. *Elizabeth, Queen of England*. London: 1984.

Williams, P. *The Later Tudors: England 1547–1603*. Oxford: 1995.

Williamson, G. C. *Lady Anne Clifford, Countess of Dorset, Pembroke & Montgomery, 1590–1676: Her Life, Letters, and Work*. Wakefield: 1967.

Wilson, D. *Sir Francis Walsingham: A Courtier in an Age of Terror*. London: 2007.

———. *Sweet Robin: A Biography of Robert Dudley, Earl of Leicester, 1533–1588*. London: 1981.

Wilson, V. A. *Queen Elizabeth's Maids of Honour and Ladies of the Privy Chamber.* London: 1922.

———. *Society Women of Shakespeare's Time.* London: 1924.

Wright, P. "A Change in Direction: The Ramifications of a Female Household, 1558–1603," in Starkey, D., ed., *The English Court: From the Wars of the Roses to the Civil War.* Longman: 1987.

Index

ABOUT THE AUTHOR

TRACY BORMAN studied and taught history at the University of Hull in England and was awarded a Ph.D. in 1997. She has worked for various historic properties and national heritage organizations, including Historic Royal Palaces, the National Archives, and English Heritage. She is now chief executive of the Heritage Education Trust and is a regular contributor to history magazines, such as *BBC History Magazine,* and a frequent guest on television and radio.